MW01599224

ADVANCES IN SPATIAL ANALYSIS AND DECISION MAKING

International Society for Photogrammetry and Remote Sensing (ISPRS) Book Series

Book Series Editor

Maxim Shoshany
Faculty of Civil and Environmental Engineering
Technion, Israel Institute of Technology
Haifa, Israel

A SELECTION OF PEER-REVIEWED PAPERS PRESENTED AT THE
ISPRS WORKSHOP ON SPATIAL ANALYSIS AND DECISION MAKING,
3–5 DECEMBER 2003, HONG KONG, CHINA

Advances in Spatial Analysis and Decision Making

Edited by

Zhilin Li
*Department of Land Surveying and Geo-Informatics,
Hong Kong Polytechnic University, Hong Kong, China*

Qiming Zhou
Department of Geography, Hong Kong Baptist University, Hong Kong, China

Wolfgang Kainz
Department of Geography and Regional Research, University of Vienna, Austria

A.A. BALKEMA PUBLISHERS   LISSE / ABINGDON / EXTON (PA) / TOKYO

Published by: A.A. Balkema, a member of Swets & Zeitlinger Publishers
www.balkema.nl and www.szp.swets.nl

ISBN 90 5809 652 1
ISSN 1572-3348

Printed in the Netherlands

Table of Contents

Spatial analysis and decision support systems

Spatial representation

Preface

It is my pleasure to write the preface for the first book in newly launched ISPRS Book Series. The ISPRS Book Series is intended to comprise high quality scientific contributions in the photogrammetry, remote sensing and spatial information sciences, with the aim of promoting the quality and range of the scientific output of ISPRS. The contributions will be rigorously peer-reviewed by experts in the field and are expected to be leading articles in the topic of each book. The Series may also include textbooks or translations of textbooks and books on advanced scientific topics that are not directly related to ISPRS events, and high quality tutorials if appropriate.

The activities of ISPRS cover a broad range of topics in the photogrammetry, remote sensing and spatial information sciences. This book is a contribution from ISPRS Technical Commission II – Systems for Spatial Data Processing, Analysis and Representation. Commission II covers topics which include design and development of systems for measurement, processing, analysis, representation and storage of image and geospatial data; system integration and modeling aspects for data and information processing; analysis of systems and their components for automated and semi-automatic digital processing systems; systems for production and update of geoinformation; and interoperability of spatial information systems.

This book is the result of the hard work of the editors Zhilin Li, Qiming Zhou and Wolfgang Kainz, Chairs and Co-Chairs of ISPRS Working Groups II/5 and II/6 in Commission II, in attracting a review panel and compiling the final papers. They deserve recognition for their hard work in producing this book. The book was prepared for the workshop on Joint Workshop of ISPRS Working Groups II/5 and II/6 in Commission II on Spatial Analysis and Decision Making Spatial Analysis Methodology, held in Hong Kong from 2–5 December 2003. As expressed in the introduction by the editors, while the traditional aspects of GIS have been growing rapidly over recent years, new developments have focused on the geographic information service and delivery, which will realise the benefits of spatial information to the community. The analysis and application of spatial information for decision support systems is an important development in realising these benefits.

I hope that readers will find this book of benefit in understanding the developments in the emerging areas of the analysis and application of spatial information for decision support.

John Trinder
President, ISPRS

Advances in Spatial Analysis and Decision Making, Li, Zhou & Kainz (eds)
© 2004 Swets & Zeitlinger, Lisse, ISBN 90 5809 652 1

Advances in spatial analysis and decision making: an introduction

Zhilin Li
Department of Land Surveying and Geo-Informatics, Hong Kong Polytechnic University, Hong Kong, China

Qiming Zhou
Department of Geography, Hong Kong Baptist University, Hong Kong, China

Wolfgang Kainz
Department of Geography and Regional Research, University of Vienna, Austria

ABSTRACT: This paper serves a brief introduction to this book, which consists of 30 peer-reviewed papers selected from the ISPRS Workshop on Spatial Analysis and Decision Making, held in Hong Kong during 3–5 December 2003. The background of this workshop is presented, the developments in spatial analysis and decision making is briefly reviewed and the content of this book is introduced.

1 BACKGROUND

Geographical Information System (GIS) has been one of the most rapidly developed field in science and technology in recent years. With the rapid diffusion of the technology, the fast growing GIS industry has now focused on geographical information service and delivery of the benefit of the technology adoption, in addition to somewhat 'traditional' actions such as spatial database development and mapping.

The 19th Congress of the International Society of Photogrammetry and Remote Sensing (ISPRS) recognized 'the need for efficient processing and presentation of such data in a value added form', and recommended 'the development and validation of end-to-end processing systems for specific applications, making use of a range of imaging systems, a range of components from the spatial information sciences and paying particular attention to techniques for the delivery and presentation of information'. Designing a spatial image-based decision support system for solving user's specific problems is one of the tendencies now. The other direction is to develop spatial analysis systems for interpreting and mining the raw and historical data and visualization systems for exploration of spatial data, such as the complex remotely sensed datasets.

Observing the above trend of development, the ISPRS Technical Commission II 'Systems for Spatial Data Processing, Analysis and Representation' in 2000 has established two corresponding working groups (WG), namely:

- WG II/5 'Design and Operation of Spatial Decision Support Systems' and
- WG II/6 'Spatial Analysis and Visualization Systems'.

In the past 3 years, these two working groups have co-organized and participated in a number of activities, e.g.

- The Third International Workshop on Dynamic and Multi-dimensional GIS (DMGIS) in 2001 at Bangkok, Thailand,

- International Workshop on Mobile and Internet GIS in 2002 at Wuhan, China, and
- ISPRS Commission II Symposium in 2002, Xi'an, China.

Due to the increasing demands for more spatial analysis functionality in GIS and more intelligent spatial decision support systems, the colleagues of these two working groups believe it is pertinent time to organize a more focused event on spatial analysis and decision making, leading to the International Workshop on Spatial Analysis and Decision Making, and thus this volume.

2 DEVELOPMENT OF SPATIAL DECISION SUPPORT SYSTEMS (SDSS)

Spatial decision support systems have been in use for a long time. In recent years we have seen significant developments related to the evolution of general purpose geographic information systems. These are mainly related to miniaturization, networking, interoperability, telematics services, and societal penetration (or ubiquitous GIS).

In general, computers shrink in size with increasing processor power, and at the same time, the requirements for storage increase. Networks, in particular the Internet, link millions of computers, and its use has become common for average households. With so many different software- and hardware-systems on the net, these systems must be able to exchange and share data and programs. Therefore, interoperability, or the ability of heterogeneous systems to work together, is a must in current computer systems. Widespread use of telecommunication services enhanced with added functionality is as common as simple voice communication over telephone lines. All these developments have lead to a situation where geo-information services have penetrated our everyday lives to such an extent that often we do not even realize when we use or are exposed to them (car navigation systems, flight tracking systems, route planning, etc.).

In the domain of decision making based on spatial data we have also seen quite interesting developments. We can summarize these developments under three major headings:

- Formalism for representing exact and vague spatial knowledge,
- Mechanism for spatial reasoning (inference and common-sense reasoning), and
- Facilities to (automatically) acquire knowledge.

The representation and use of vague or uncertain knowledge in spatial decision support systems has received broad attention, because it is much closer to the real world situation than normally applied in exact and crisp representations. Fuzzy logic has proven to be an excellent basis for the treatment of uncertainty and vagueness in spatial features and reasoning.

The reasoning part of SDSS has greatly benefited by the use of novel techniques to derive and refine a rule base. Neural networks have proven to be excellent tools in a spatial reasoning process often based on vague and uncertain data.

One of the most promising developments, however, is the advent of component based systems. This means that no complete system resides on a single computer, but a rather thin client requests services (functions and operations) from the net. These services are provided by component servers (Figure 1).

The major difference to common client/server architectures is that not only data and application programs are served but rather GIS functional components. None of these components constitutes a complete GIS. A client user puts together only those functions and services that he or she needs for a particular task. No expensive, full-fledged systems are needed that often possess functions that are not needed. A component broker will assist the users in locating and serving the required services (Figure 2).

To make such a system work we need standards and metadata. These are data about data. In our sense we do not only mean metadata concerning spatial data but especially data about components and defined procedures, protocols and interfaces to share and use components. Metadata could be extremely useful in determining what types of services are needed and useful for particular applications. If, for instance, a user requests the overlay of two data sets with incompatible data quality

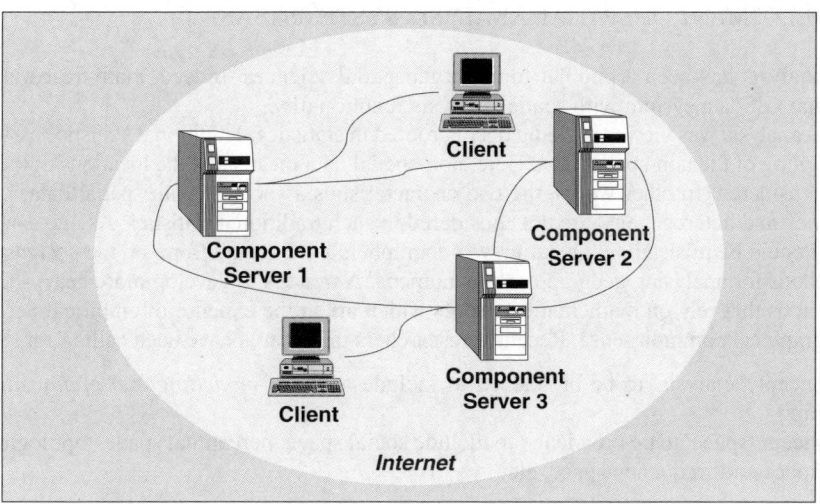

Figure 1.   Component servers.

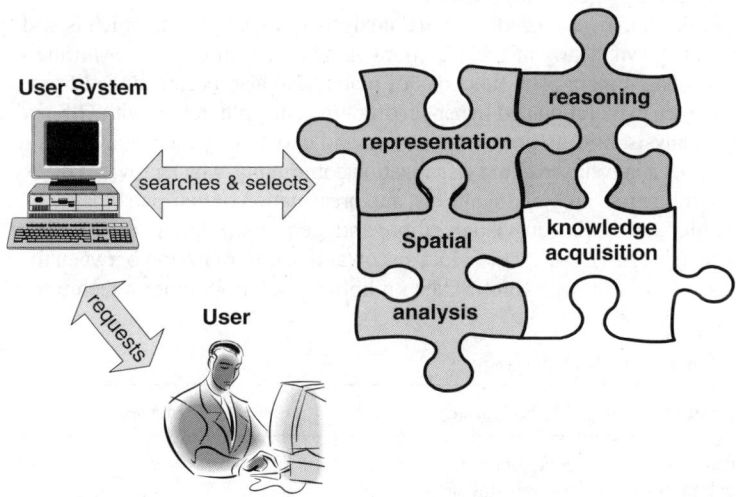

Figure 2.   Component-based system.

(such as positional accuracy) the system could refuse such operation simply based on an analysis of the metadata even before any overlay service has been located or applied.

New professional services will emerge in the context of component based systems. Geo-information service providers (GISPs) will emerge and providing services related to

- offering components and data,
- clearinghouses for component metadata,
- value added services, and
- validation and metadata processing.

Service brokers will become enablers and facilitators of new and novel spatial decision support services.

3

3 DEVELOPMENT OF SPATIAL ANALYSIS SYSTEMS (SAS)

Spatial analysis has been a popular topic in geospatial sciences. Indeed, many researchers have argued that GIS is a system with spatial analysis functionality.

Spatial analysis was viewed as being deeply rooted in statistics. Traditional statistics assumes random sampling of the non-spatial data. Here, non-spatial data means that the locations of the samples are not considered. In other words, the two characteristics associated with spatial data, i.e. spatial dependency and heterogeneity, are not considered in such traditional statistics. As one can imagine, the result could be misleading. Quantitative geographers have spent efforts on the designs of statistical methods for analysing geographical phenomena. A trend is to develop more heavy-duty statistical methods that rely on mathematical proofs which are at the expense of empirical performance and geographical common sense. Recently, researchers in this area have been calling for

- the concept 'analysis' to be broadened to include some of *modelling* and *optimization* functions, and
- the concept 'space' to be broadened to include social space, perceptual space, topological space, scale space and frequency space, etc.

The development of spatial analysis is summarized Table 1.

The terms 'other spaces' refers to those mentioned above and 'spatial operations' refers to some of the optimization methods, modelling functions and other computational methods, leading to a popular new terminology geo-computation.

Another rapid development related to spatial analysis is visualization, which is widely regarded as visual analysis. As shown clearly in Table 2, more variables are in use while shifting from traditional paper map to web-based maps. The functions of maps have also been changed from traditional data storage and display to exploration and hyper-linking. With the integration with GIS and other technology (e.g. image analysis, information visualization, and exploratory data analysis), a new term 'Geo-visualization' has been introduced in the visualization community, which provides theory, methods, and tools for visual exploration, analysis, synthesis, and presentation of geospatial data.

It is noticeable that both geo-visualization and geo-computation are moving together into exploratory data analysis (EDA), which focuses on active collaboration between the visual and the numeric analysis. Data mining, which relies on both visual and numeric techniques in search for

Table 1.   Development of spatial analysis.

Stage of development	Methods used	Spaces
Statistical analysis	Statistics	None
Spatial statistical analysis	Spatial statistics	Euclidean space
Spatial analysis	Spatial statistics	Euclidean space
Generalized spatial analysis	Spatial statistics	Other spaces
	Spatial operations	Euclidean space
		Other spaces

Table 2.   Variables in use at different development stage of visualization.

Map form	Variables in use				
Paper maps	Visual variables				
Digital maps	Visual variables	Screen variables			
Cartographic visualization	Visual variables	Screen variables	Dynamic variables	Exploration acts	
Web-based visualization	Visual variables	Screen variables	Dynamic variables	Exploration acts	Web variables

anomalies, outliers and patterns, is regarded as a typical example. Current trend also shows that web-based exploratory data analysis (Web-EDA) has been fashionable.

4 ABOUT THE ISPRS WORKSHOP ON SPATIAL ANALYSIS AND DECISION MAKING

Observing the above trend of development, we believe that it is now necessary to review the progress, exchange the research experience, and discuss the future direction among scholars and researchers in fields of spatial analysis and spatial decision support systems. The ISPRS Workshop on Spatial Analysis and Decision Making is then organized and held in Hong Kong on 3–5 December 2003.

The workshop covers wide a range of topics regarding methodology and application development in spatial analysis and spatial decision support, including:

- Spatial Decision Support System (SDSS) design and implementation,
- Image-based SDSS, spatial analysis and visualization,
- Knowledge-based spatial decision processes and systems,
- Artificial neural networks and fuzzy set theory for spatial analysis and decision making,
- Spatial data integration and modelling for SDSS,
- Multi-scale spatial analysis,
- Spatial relations and spatial reasoning,
- Web-based systems for value-added data analysis and visualization,
- Mobile-based systems for visualization and value-added data analysis,
- Systems for on-demand visualization and value-added data analysis, and
- Integration of 3-dimensional, temporal and dynamic aspects into spatial analysis and visualization systems.

To ensure quality of this workshop, the number of participants was confined and all submitted articles were peer reviewed. An international programme committee consisting of 26 experts was formed to take the responsibility to review the submitted manuscripts.

5 ABOUT THIS VOLUME

After peer-reviewing and revision processes, a collection of 30 articles has been included in this volume. This volume is structured into four parts, namely,

- Spatial analysis methodology,
- Integrated techniques and application-oriented models,
- Spatial analysis and decision support systems, and
- Spatial representation.

Such a division is rather arbitrary and there must be many alternatives. The main consideration was to put papers into balanced sections. As a result, the first part consists of nine papers, the second part of eight, the third part of seven and the fourth part of six.

The first part consists of nine articles and discuss general issues in spatial analysis, such as point data analysis, time series analysis, knowledge framework, spatial relation analysis, TIN analysis, web-based service for spatial analysis and decision making, fuzzy object formation, techniques for improving image classification and pattern extraction from grid DEM.

Eight articles are included in the second part that discusses methodology, with a focus on integration and application. The topics include integration of spatial statistics into multi-resolution classification framework, integration of photogrammetry with GIS, cybercity models for urban analysis, wavelet-based noise model for visualization of spatial activity anomalies of earthquakes, Theil coefficient for regional economic disparity analysis, cellular automata for urban planning, true-orthophoto for urban planning and management, and analysis of spatio-temporal pattern of arid environment.

The third part includes seven articles focusing on application systems. The systems are for network traffic transport analysis and management, oil and pipeline route planning, ice forecasting, environment analysis and management, temporal analysis, risk analysis for children, and land use monitoring.

Six articles are in the fourth part concerning spatial representation, in warehouses, in rich media, on sphere, on the web, and in multi-scale.

The articles in each of these four parts are ordered in the alphabetic order of the first author's surname.

ACKNOWLEDGEMENTS

The editors would like to express thanks to the Croucher Foundation, Hong Kong Baptist University and Hong Kong Polytechnic University for providing sponsorship and financial support. Thanks also go to members of the programme committee for reviewing the papers, to Dr. Josephine Khu for English editing, to Dr. T. Ai, Dr. W.B. Zou and Mr. K. Khoshelham for formatting the articles, to the ISPRS Council, the President of ISPRS Technical Commission II, Prof. J. Chen, and the ISPRS Book Series Editor, Prof. Maxim Shoshany, for their encouragement and support, to the publisher, Dr. Janjaap Blom, for the fast production of this volume. All credits and responsibilities of each included article belong to the corresponding contributor(s).

Spatial analysis methodology

Advances in Spatial Analysis and Decision Making, Li, Zhou & Kainz (eds)
© 2004 Swets & Zeitlinger, Lisse, ISBN 90 5809 652 1

Analysis and simplification of point cluster based on Delaunay triangulation model

Tinghua Ai

Department of Cartography and Geo-sciences, School of Resource and Environment Sciences,
Wuhan University, China
Department of Land Surveying and Geo-Informatics, The Hong Kong Polytechnic University,
Hong Kong, China

Yaolin Liu

School of Resource and Environment Sciences, Wuhan University, China

ABSTRACT: Point cluster object contains much structured information in spatial distribution, which is interesting for the research of spatial analysis and map generalization. This paper divides the spatial distribution information of point cluster into three categories: existing, metrical structure and topological structure, and focuses the discussion on metrical structure. Based on the Delaunay triangulation and Voronoi diagram model, the paper defines four characteristic parameters for describing metrical structure: distribution range, distribution density, distribution centre and distribution axis. Considering Gestalt principles in visual adjacency cognition, the presented method finds the distribution range polygon by progressively stripping the outside triangles. The distribution density is represented by Voronoi cell size and visualized as a grey image. Applying an image processing method, the distribution centre can be extracted from the grey image. A method of point cluster simplification is provided in the paper on the basis of Voronoi diagram establishment in a dynamic way. The relative distribution properties above are preserved in the simplification method.

1   INTRODUCTION

A point cluster object plays an important role in the field of spatial analysis and map generalization. The analysis aiming at this kind of object focuses on the cluster structure information rather than individual information such as object size, geometrical shape. The goal of point cluster analysis is to extract spatial distribution principles and then to support decision-making. In the history of GIS, we have examples of the sources of disease being uncovered by case study based on point cluster analysis. In urban planning, the planner decides the location of point type facility through an analysis of functions of the facility and its interrelation within a certain region, which can also be regarded as a type of point cluster analysis. To describe the spatial information structure contained in point cluster, related studies define the distribution parameters of point cluster and presents corresponding operations. Guo (1997) presented four parameters based on spatial statistics: distribution density, distribution centre, distribution axis and discrete degree. From the perspective of visual recognition, Yukio (1997) gave the definitions of three distribution concepts: proximity relation, distribution centre and density change. For the point cluster oriented distribution analysis, the existing methods mainly apply the method of spatial statistics on the basis of distance factor (Burrough, 1986).

In map representation there are many instances of point cluster objects, such as residential area (point), island cluster, discrete distributed facility and others. The point cluster oriented generalization

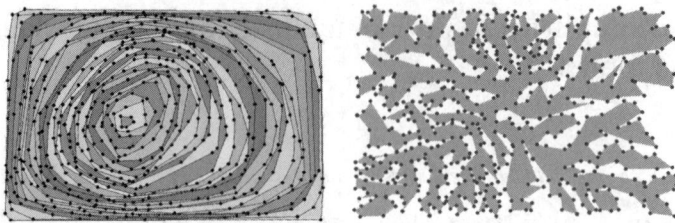

Figure 1.   Multiple enveloping convex hull of a point cluster, and the progressive projection result.

and spatial analysis has been an interesting issue over years. The well-known Töpfer's law answers the question of the selection ratio for a point cluster in the sense of statistics (Töpfer and Pillewizer, 1966). But it can not determine which one will be selected and which one will not. In the simplification of residential points, some researchers have developed methods such as gravity modelling, distribution-coefficient-control, circle-growth and others, in which the basic ideas are using a circle to represent the impact range of the point entity and then decide which points to be remained or to be removed through the interrelation judgment between neighbor circles (van Kreveld, 1995). Wu (1995) developed a method in which the convex-hull model is introduced into the point cluster generalization. The method builds a multiple enveloping convex hull of point cluster and transfers the simplification of point cluster to the simplification of points in different loops (see Figure 1). This method actually converts the generalization question of 2-dimension to that of 1-dimension. In an actual application the simpler method of simplification is to randomly resample point entities based on a grid structure. To evaluate the quality of the simplification result, we need to investigate the degree to which the spatial distribution properties have been preserved. After generalization, the number of point entities decreases but the spatial distribution properties should not be distorted. The simplified point cluster should cover the same spatial range as before, neither increasing nor decreasing. The relative change of distribution density should be preserved among different areas. As the main constraint of point cluster generalization, the maintenance of spatial distribution properties becomes an interesting problem in this field. Traditionally, the spatial analysis of a point cluster is based on spatial statistics and uses geometric distance-related parameters to describe spatial distribution. From the point of view of computation geometry, point cluster distribution belongs to the question of spatial proximity. Delaunay triangulation and its dual Voronoi diagram have been acted as the important support models (Ai and van Oosterom, 2001). A lot of generalization algorithms associated with proximity analysis have been achieved in the community of map generalization (Ai et al., 2000; Bader et al., 1997; Jones et al., 1995). This study attempts to apply the above two models to define the distribution parameters of a point cluster including distribution range, distribution density, distribution centre and distribution axis. The computation methods of these parameters are also respectively presented. Finally, the study gives an algorithm to generalize the point cluster based on Voronoi diagram dynamically organization, which well preserves the above spatial distribution characteristics.

2   SPATIAL DISTRIBUTION PARAMETERS OF POINT CLUSTER

2.1   *Information contained in point cluster*

The theory of information regards information as a decrease in uncertainty and uses the parameter entropy to represent the quantity of information. When the map user reads a map, his or her ignorance of spatial cognition is decreased, and then we can say he or she obtains the 'spatial information'. The greater the decrease of ignorance of spatial cognition, the more information he or she has obtained. The theory of information is produced from the transmission and receipt of electronic signals, emphasizing the 'process'. This means that information can be obtained through a comparison

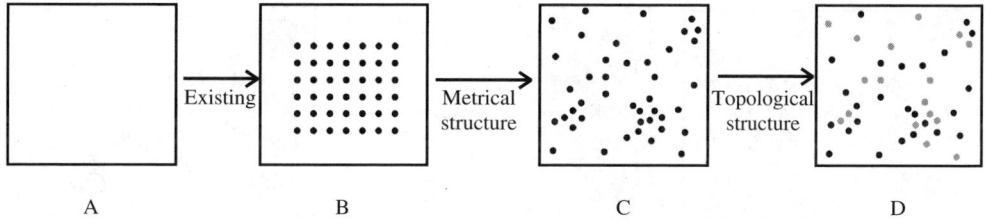

Figure 2.   Three kinds of structured information contained in a point cluster distribution.

between before and after transmission. In spatial analysis and map generalization the theory of information can also be applied but the information is transferred in different ways. The transfer of spatial information involves three worlds from the viewpoint of spatial cognition and spatial information transactions. The first transfer is from the real world to the conceptual world, which appears as spatial database or map. The second is from conceptual world one to conceptual world two (from the former with high resolution to the latter with low resolution). And the third is from the conceptual world to the user world. Within every transfer process an uncertainty change occurs, and so the information quantity changes with either the increase or decrease in entropy. Combined with the theory of linguistics, spatial information can be divided as three kinds of information: semantic, syntactic and pragmatic. Based on information theory Bjørke (1996) separated spatial syntactic information into three parts: existing entropy, metrical entropy and topological entropy. Based on this idea, this study divides the information contained in a point cluster into three sorts: existing information, metrical information and topological information. In Figure 2, from A to B one sees that the region has n points without considering the concrete distribution of points. In this process, the obtained information behaves as the existing ones. From B to C one sees the further spatial structure associated with metrical distance, behaving as the metrical information. From C to D one gets the attribute difference between the neighbor point entities, which has nothing to do with distance or direction, and this kind of information acts as topological information. In the simplification of a point cluster from a large scale to a small scale, the existing information must decrease in quantity (and the uncertainty entropy increase) but the metrical information and topological information have to be maintained as much as possible. This means using fewer points to represent the distribution density, distribution range, distribution centre and other structured information as that of original state. The Töpfer radical law resolves the simplification of existing information on point clusters. But the question of which point is retained and which point is removed belongs to the generalization of metrical information. This study tries to resolve the latter question. The generalization of the topological information of point clusters is related to the difference in point attributes, and will not be considered in this study.

2.2   Parameters of the metrical structure of point cluster

The metrical structure within a point cluster acts as the spatial proximity. In computation geometry, the Delaunay triangulation and Voronoi diagram are powerful support tools in proximity analysis. The edge link in Delaunay triangulation represents the adjacency relationship between neighbor points. The polygon cell in a Voronoi diagram describes the impact range of one point based on the principle of equally partitioning the competition space. On the basis of these two models the following text attempts to define four distribution parameters for a point cluster: distribution range, distribution density, distribution centre and distribution axis.

2.2.1   Distribution range
The determination of the distribution range of a point cluster is an uncertainty question just as it is with the determination of urban range or vegetable boundary. To select the proper boundary

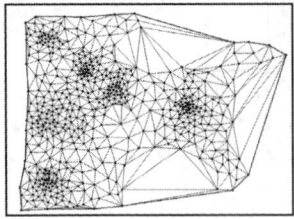 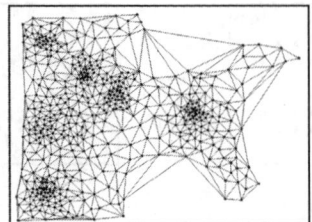 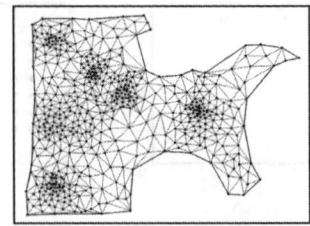

Figure 3.   Stripping the outside triangles in the triangulation and getting different ranges of a point cluster.

polygon enveloping all points as the distribution range has to respect the spatial visual recognition principles, such as the Gestalt principle. In a simple way, we traditionally use the convex hull of a point cluster to represent the distribution range of a point cluster. This method is not accurate, as it is possible to include the large concave area which is actually not covered by the points as illustrated in the left of Figure 3. The distribution range of a point cluster is composed of the connection of proximity points, which visually respects the Gestalt principle of the adjacency link. This means that only when the distance between two neighbor points is shorter than the visual distance can our eyes regard the link as the boundary. Based on this idea, we define an operation stripping in Delaunay triangulation. The stripping of triangle is to progressively remove the outside triangle whose outside edge length is longer than a tolerance. The removal procedure continues until the edge links of all outside triangles are within the tolerance distance in the sense of Gestalt visual recognition. The sequent link of all remaining outside triangles then makes the distribution range polygon of a point cluster. Figure 3 illustrates the different distribution ranges of the same point cluster under different visual recognition tolerances. This representation of a point cluster makes use of the property that the triangle link of Delaunay triangulation is as close as possible among the links between neighbor points (Preparata & Shamos, 1985). Here the visual tolerance distance can be determined by psychological experiments or by pre-experience. The greater the tolerance distance, the fewer the number of times progressive stripping will occur, and the closer the obtained range polygon is to the convex hull. The smaller the tolerance , the more often the stripping will occur, and the more the concave area the range polygon will have. In particular, determining a proper tolerance distance will enable different clusters to be separated and the point groups in a spatial distribution to be obtained.

2.2.2   *Distribution density*

Density is usually defined as the number of points within the unit area. In another way, it can also be defined as the space area covered by determined number of points. Here we consider that each point competes to get the growth space and the result of the competition is to partition the space equally between neighbor points (without the consideration of partitioning ability difference between points). This idea is exactly consistent with the partitioning of Voronoi diagram model. The cell polygon of Voronoi diagram represents the growth space of the corresponding inside point. As the Voronoi diagram is the dual of the Delaunay triangulation, we can link the circumcenter of the triangles that remain after the above stripping process and then get the Voronoi cell polygons in the valid distribution range of the point cluster. As the circumcenter of the outside triangles may locate out of the distribution range, the linked partitioning result needs to be clipped with the distribution range polygon. We can now compute the density for each cell polygon as follows. In a Voronoi diagram we define the density as the inverse of the area of the cell polygon of the corresponding inside point, which is represented as $1/a_i$. The more loosely the point is distributed, the larger the area covered by the cell polygon covers and the smaller the value of $1/a_i$. To get obvious visualization, we can establish the lineal relation between the density $1/a_i$ and the degree of greyness. Figure 4 is the image representation of the point cluster in Figure 3 based on this definition of density. From the grey image, we find that the smaller the Voronoi cell is, the darker the grey region is.

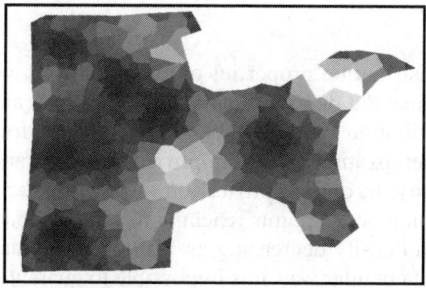

Figure 4. Representation of the distribution density of a point cluster using Voronoi cell size, and a visualization of corresponding grey image.

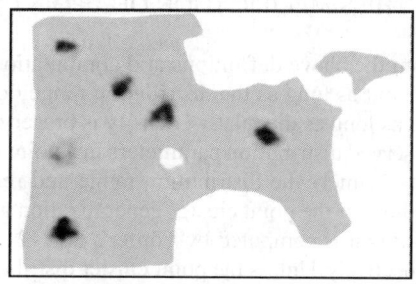

Figure 5. Applying image process method to extract distribution centre of point cluster.

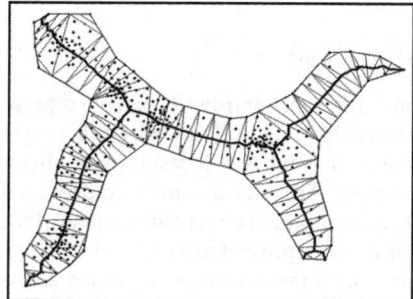

Figure 6. Constructing the skeleton line of Delaunay triangulation to obtain the distribution axis of a point cluster.

2.2.3 *Distribution centre*

The distribution centre is computed based on the degree of the density of the point cluster distribution and the context situation. Obviously, it is not the geometrical centre of the distribution range polygon. Within one point cluster the distribution centre may have several locations. Based on the density image, we use the image process method to obtain the locations of local maximum greyness and let them as the distribution centers. The result is represented in Figure 5.

2.2.4 *Distribution axis*

The distribution range of a point cluster acts as a two-dimensional polygon. The distribution axis is the extraction of a one-dimensional characteristic from the two-dimensional polygon, which can be regarded as the question of inverse buffer in GIS operation. This implies that the distribution medial line is retrieved from the range polygon and that generally the buffer width in different locations is different. The extraction of distribution axis is described as follows. Within the distribution range polygon, we construct the constrained Delaunay triangulation in which we just let the polygon boundary points participate the construction. Here, the constrained conditions mean that the edge of the triangle does not intersect with the edge of the boundary. As illustrated in Figure 6, by extracting the skeleton in this triangulation the distribution axis of point cluster will be obtained. Since the distribution axis describes the main distribution trend of a point cluster in a lineal structure, those hair lines in the skeleton extraction should be removed through the way of comparison of tolerance distance. Usually, the skeleton of one polygon acts as the tree structure. But when the point cluster distributes in an obvious strap structure, the distribution axis behaves as a strong lineal structure.

3 THE SIMPLIFICATION OF A POINT CLUSTER

From the above definitions and computations of the distribution properties of a point cluster, we find out as long as the distribution range does not change the distribution axis will be fixed, and that as long as the relative density is preserved the distribution centre will be fixed. Thus, the four preserved distribution parameters in a point cluster generalization can be simplified as two parameters, namely the distribution range and relative density. In our simplification method the constraints for the point cluster generalization are: (1) the number of points reaching to a determinate point that is computed by Töpfer's law, (2) every point density decreasing as 1/p of the original, respectively. Unless the point cluster distributes in a quite regular way, it is impossible to guarantee that every point density will decrease as 1/p exactly after generalization. This means that the second condition is only respected in statistics. We divide the point cluster simplification into two steps: the simplification of boundary points and the simplification of inside points. The former requires that the distribution range remain unchanged, and that the latter retain the relative density through the Voronoi diagram organization in a dynamic way.

3.1 *Simplification of boundary points*

The points on the boundary have different contributions in the representation of distribution range. As illustrated in Figure 7, the removal of point A or C has little impact on the distribution range of the point cluster. But the removal of point B will result in an obvious change in the distribution range. Here, we call points like point B that have a strong impact on the distribution range, the key points. To detect the key point, we compute the vertical distance of the current point to the link line between the pre-point and post-point, represented as d_i, and then compare it with the pre-defined tolerance λ. If d_i is larger than λ, then the current point serves as the key point. For the selection of boundary points, all key points are selected unconditionally. A proper number of points from the remaining non-key points is then selected to satisfy the number decrease condition. If there are n points located on the boundary and the number of key points is n_0, to select $n/p - n_0$ points from $n - n_0$ non-points yields the selection probability $(n - pn_0)/(pn - pn_0)$. Suppose the selection event is a binomial distribution. We use a Monte Carlo simulation method to determine the points to be selected or removed. For every non-key point, a random number R is generated that distributes between 0 and 1.0. If $R < (n - pn_0)/(pn - pn_0)$, then the current point is selected; otherwise, it is removed. The above generalization strategy guarantees that the number of boundary points will decrease to a proper number and simultaneously ensures that the distribution range will not change.

3.2 *Simplification of inside points*

The basic algorithm idea is described as follows. With the scale decrease, the points located in the densest area have the priority in removal, and the space area covered by the removed point will be

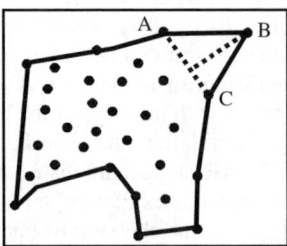

Figure 7.   Point B, a key point located on the distribution boundary of a point cluster.

14

partitioned by its neighbors under the condition of Voronoi diagram partitioning principle. To guarantee that the adjacency points are not continuously removed to distort the change in density, once a point is removed its neighbor points will be solidified. The next point that has the maximum density $1/a_i$ among the remaining points and that is not solidified should then be found. Continue the procedure of finding the points to be removed until all of the remaining points have been solidified. The simplification of the point cluster is then finished once. For the remaining points, update the density computation according to the new Voronoi diagram partitioning and sort the density again. Apply this iterative sample method to remove the points twice, three times, four times or more, until the retained points satisfy the condition that the number of points is n/p. Here, n refers to the number of original points and p to the selection probability in the whole simplification process. In an actual application, this condition is simply converted to the number of points as close to n/p as possible. We define a function $f(t)$ as representing the number of points remaining after the removal of points t times. Obviously, the function $f(t)$ is a monotone degressive one. In the first steps of removal, the smaller t is, the larger the number of points that are removed, and the faster the decrease in the function value. This means that the function curve has a concave shape. When $f(t_i) < n/p$, compare the value $|f(t_{i-1}) - n/p|$ and $|f(t_i) - n/p|$ and take the step in which the value is smaller as the final end condition. The points remaining after this step act as the final result of the point cluster simplification.

The core of the algorithm above is the dynamic organization of Voronoi diagram. Once a point is removed, the Voronoi diagram needs to be maintained in real time. In this study, the construction of Voronoi diagram is based on the circumcenter connection of triangles in the Delaunay triangulation. Therefore, the maintenance procedure is carried out through an adjustment of the Delaunay triangulation after the removal of points.

First, we investigate the impacted points in the Voronoi diagram after point P has been removed. In a Delaunay triangulation, we regard those points having a direct link to point P as first-order neighbor points represented as N_i $(i = 1,2,...)$. Those points whose link to point P will intersect the triangle edge once are regarded as second-order neighbors of point P and are represented as S_i $(i = 1,2,...)$. From Figure 8, we find that S_i faces point P through link $N_k N_{k+1}$.

In a Delaunay triangulation, one point P is removed, the impacted points will be first-order neighbors N_i and second-order neighbours S_i. In the local region, including neighbors N_i and S_i, update the construction of the triangulation according to Delaunay triangulation principles and correspondingly update the construction of the Voronoi diagram. Figure 8 illustrates the updating process after one point P has been removed. From the illustration, we find that the original cell polygon of inside point P is mainly separated by first-order neighbor points.

The simplification of point cluster reflects as the procedure that the size of the Voronoi diagram cell increases and the number of cell polygons decreases. The strategies that apply the iterative

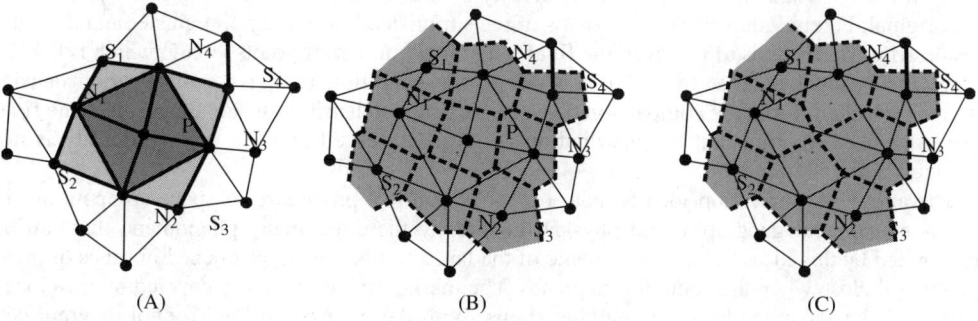

(A)   (B)   (C)

Figure 8. The illustration of Voronoi diagram re-construction after some points are removed. (A) An example of the impact region of point P on its neighbor area in a Delaunay triangulation, first-order neighbor N_1, N_2, N_3, N_4, and second-order neighbor S_1, S_2, S_3, S_4. (B) The Voronoi cell of point P and its neighbor points. (C) Point P cancelled, and its Voronoi cell being separated by neighbor points N_1, N_2, N_3, N_4, S_1, S_2, S_3, S_4.

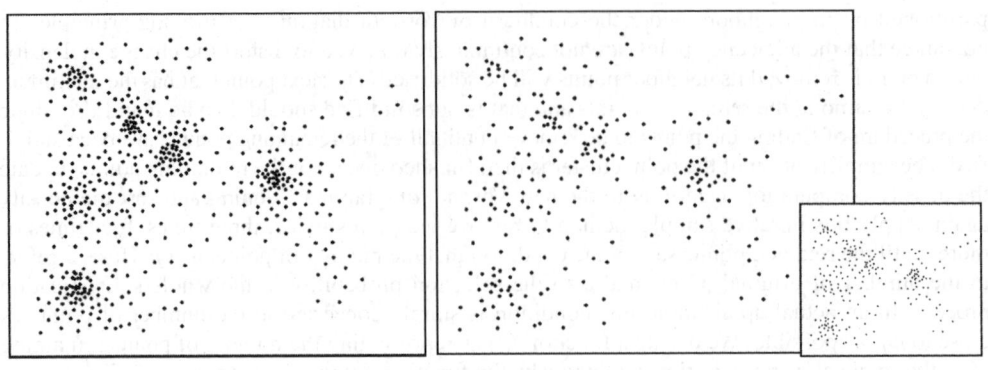

Original point cluster distribution (538 points)

The result after simplification (265 points)

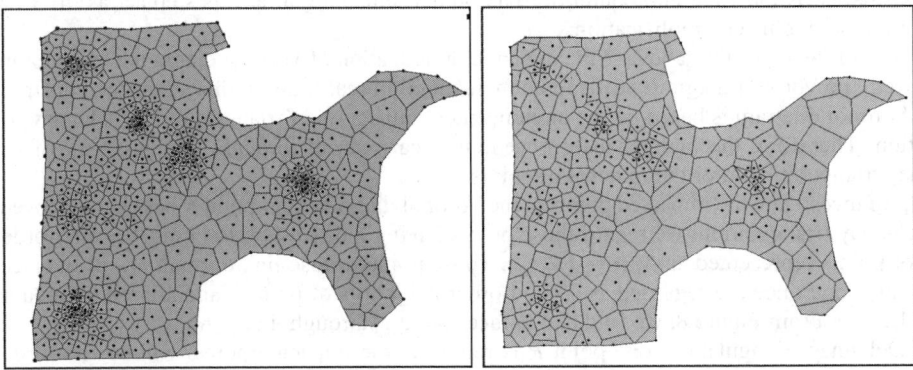

The VD of an original point cluster

The updated VD after having been re-sampled 5 rounds

Figure 9. The experiment of point cluster simplification.

removal and solidify some adjacency points in each step guarantee the removal in a relatively uniform way. This means that in statistics the density of each point will be decreased to $1/p$ to that of the original. Figure 9 describes some snaps of the simplification of the point cluster and the final result. After five rounds of removal, the final remaining number of points is 265, which is 0.4926 times the original number of 538 points. Here, the condition is that the distribution density decreases by half. From the comparison between the original distribution and the result of the final generalization, we find that the distribution range is preserved and the relative density is not changed in visual judgment.

The above simplification idea is that point selection is a procedure of space competition. In social–economical geography and physical geography there are many phenomena that can be interpreted by this idea. For example, some of the large number of shops opened in times of prosperity will close when the economy stagnates. The market space originally covered by the closed shops will be separated by the remaining shops. With the increase in the Voronoi diagram cell polygon (the market attraction range), to a certain degree the remaining shops are able to grow and a new balance is established. This is the same idea as the principle of the survival of the fittest in biological evolution. Therefore, to some degree the simplification idea in this study is consistent with natural principles in geo-sciences.

4   CONCLUSION

The metrical structure information contained in point cluster behaves as the distribution density in 3-dimension, the distribution range in 2-dimension, the distribution axis in 1-dimension and the distribution center in 0-dimension. These distribution parameters are interrelated to each other. In the polygon of distribution range, extracting the skeleton yields the distribution axis. Retrieving the local maximum amount of greyness from the image of the distribution density through an image processing method yields the distribution centre. The spatial characteristics of a point cluster are essentially determined by spatial proximity. In spatial proximity analysis the Delaunay triangulation and Voronoi diagram are two important geometrical models that can make great contributions. Thus, this study uses two geometrical models defining four distribution parameters and present corresponding computation methods. The distribution range considers the proximity connection in the sense of Gestalt principles. By progressively stripping away outside triangles in a Delaunay triangulation, the real range of a point cluster distribution is obtained. The distribution range obtained using this method is an obvious improvement over the representation of the convex hull. The distribution density is defined on the basis of the Voronoi diagram model and represented as the inverse of the area of the VD cell polygon. The distribution centre is detected through an image processing method.

To preserve the four distribution parameters above, this paper presents a method to simplify the point cluster by way of dynamically organizing the Voronoi diagram. This method has considered the statistical and geometrical characteristics of point cluster. The following aspects of the simplification method needs to be improved in the future: (1) smoothing the distribution density while avoiding a sharp change across the boundary of the VD cell polygon; (2) considering the attribute difference of each point and introducing the weighted VD model. The method focuses on considerations of spatial geometrics and statistics. In a real application of point cluster generalization, the semantic properties need to be combined with the spatial cluster properties. The geographic meaning and context situation of point entity play important roles in map generalization, including the simplification of point clusters. For example, the simplification of discrete residential points needs to take into account their significant grade and relation to road networks, river distribution and other geographic features with the exception of the cluster properties.

REFERENCES

Ai, T. & van Oosterom, P. 2001. A Map Generalization Model Based on Algebra Mapping Transformation. In Aref W. G. (ed.), *Proceedings of the 9th ACM-GIS*: 21–27. Atlanta, USA.

Ai, T. & Guo, R. 2000. A Binary Tree Representation of Bend Hierarchical Structure Based on Gestalt Principles. In He, J., Forer, P. & Yeh, A. (eds), *Proceedings of the 9th International Symposium on Spatial Data Handling*: 2a30–2a43. Beijing.

Ai, T. & Wu, H. 2000. Consistency Correction of Shared Boundary Between Adjacent Polygons. *Journal of Wuhan Technical University of Surveying and Mapping* 25(5): 426–431 (in Chinese).

Bader, M. & Weibel, R. 1997. Detecting and Resolving Size and Proximity Conflicts in the Generalization of Polygonal Maps. *Proceedings of the 18th ICC*, Vol. 3: 1525–1532. Stockholm, Sweden.

Burrough, P. A. 1986. *Principles of Geographical Information Systems for Land Resources Assessment*. England: Oxford, Clarendon Press.

Bjorke, J. T. 1996. Framework for Entroy-based Map Evaluation. *Cartography and Geographic Information Systems* 23(2): 78–95.

Guo, R. 1997. *Spatial Analysis* (Chapter 4, Spatial Distribution). Wuhan: Wuhan Technical University of Surveying and Mapping Press (in Chinese).

Jones, C. B., Bundy, G. L. & Ware, J. M. 1995. Map Generalization with a Triangulated Data Structure. *Cartography and Geographic Information System* 22(4): 317–331.

Preparata, F. P. & Shamos, M. I. 1985. *Computational Geometry an Introduction*. New York: Springer-Verlag.

Töpfer, F. & Pillewizer, W. 1966. The Principles of Selection. *The Cartographic Journal* 3(1):10–16.

van Kreveld, M. 1995. Efficient Settlement Selection for Interactive Display. *Proceedings of AutoCarto* 12: 287–296. Bethesda, Md.

Ware, J. M. & Jones, C. B. 1995. A Triangulated Spatial Model for Cartographic Generalization of Areal Objects. *Proceedings COSIT*: 173–192. Austria: Semmering.

Wu, H. 1995. The Structured Selection of Graphic Patch Group. *Journal of Wuhan Technical University of Surveying and Mapping* 20 (Supplement): 88–91 (in Chinese).

Yukio, S. 1997. Cluster Perception in the Distribution of Point Objects. *Cartographica* 34(1): 49–61.

Empirical Mode Decomposition (EMD) transform for spatial analysis

Zhilin Li, Kourosh Khoshelham & Xiaoli Ding
Dept. of Land Surveying and Geo-Informatics, Hong Kong Polytechnic University, Hong Kong, China

Dawei Zheng
Centre for Astro-Geodynamics Research, Shanghai Astronomical Observatory, Chinese Academy of Sciences, Shanghai, China

ABSTRACT: This paper describes a new method for spatial analysis – the Empirical Mode Decomposition (EMD) transform, which is the latest development in methods for signal decomposition. EMD is a data-driven method. In this transform, the basis of decomposition is derived from the data, instead of by using an *a priori* basis (sine, cosine and Wavelet) as is currently the case for all existing methods of data analysis; e.g., Fourier and Wavelet transforms. This adaptiveness has made EMD superior to some other methods in many ways. Examples of the use of EMD for spatial analysis such as signal extraction and data filtering are demonstrated. They show that EMD can be a powerful tool for spatial analysis.

1   INTRODUCTION

"Broadly speaking, spatial analysis can be defined as the formal quantitative study of phenomena that manifest themselves in space. This implies an attention to location, area, distance and interaction" (Anselin et al., 1993).

Spatial analysis has been a popular topic in geospatial sciences. Spatial analysis was viewed as being deeply rooted in spatial statistics, and quantitative geographers have devised statistical methods for analysing geographical phenomena. However, researchers have recently been calling for the concept to be broadened. Marble (2000) emphasized that some of the *modelling* functions should also be considered as spatial analyses of the concept. As an example, he mentioned transportation optimization models. Anselin (2000) also suggested that spatial analysis needs to go beyond dealing with Euclidean space and physical geographical locations to include location in 'social' and 'perceptual' space. The development of spatial analysis is summarized in Table 1.

Table 1.   Development of spatial analysis.

Stage of development	Methods used	Spaces
Statistical analysis	Statistics	None
Spatial statistical analysis	Spatial statistics	Euclidean Space
Spatial analysis	Spatial statistics	Euclidean Space Other spaces
Generalized spatial analysis	Spatial statistics Spatial operations	Euclidean Space Other spaces

Traditional statistics assumes random sampling of the non-spatial data. Here, non-spatial data means that the locations of the samples are not considered. In other words, the two characteristics associated with spatial data, i.e. spatial dependency and heterogeneity, are not considered in such traditional statistics. The term 'other spaces' refers to 'social' and 'perceptual' spaces, as suggested by Anselin (2000), as well as to topological space, scale space, frequency space, and so on. 'Spatial operations' refers to some of the optimization methods, modelling functions and computational methods, such as networking models.

This paper aims to demonstrate a new method for spatial analysis in spatio-temporal space, called the *Empirical Mode Decomposition* (EMD) transform. The method was developed by Huang et al. (1998) and has been widely used in engineering to replace the Fourier and Wavelet transforms, as the EMD transform has been shown to have advantages over the Fourier and Wavelet transforms. Examples of the application of this new method for spatial analysis are also given.

This introduction is followed by a section outlining the principle of the EMD transform; Section 3 describes the applications of EMD in spatial analysis. Section 4 briefly discusses the advantages of EMD over the Fourier and Wavelet transforms. Some conclusions are drawn in Section 5.

2   THE PRINCIPLE OF EMPIRICAL MODE DECOMPOSITION (EMD)

EMD is a new method of data analysis, designed for short span data, which are non-linear and non-stationary.

2.1   *The basic principle of EMD*

EMD is a data-driven method. That is, the basis of decomposition is derived from the data, instead of by using an *a priori* basis (sine, cosine and wavelet) as has been the case for all existing methods of data analysis; e.g. Fourier and Wavelet transforms (Huang et al., 1998). This adaptiveness has been found to be the most significant characteristic of EMD, allowing it to handle any kind of data.

The key idea of the EMD is to decompose the data into a finite number of intrinsic mode functions (IMFs). An IMF is defined as any function that has the same number of zero crossings and extrema and that also has symmetric envelopes defined by the local maxima and minima, respectively (Figure 1). In other words, an IMF can be considered to be a well-behaved signal with no riding and asymmetric waveforms.

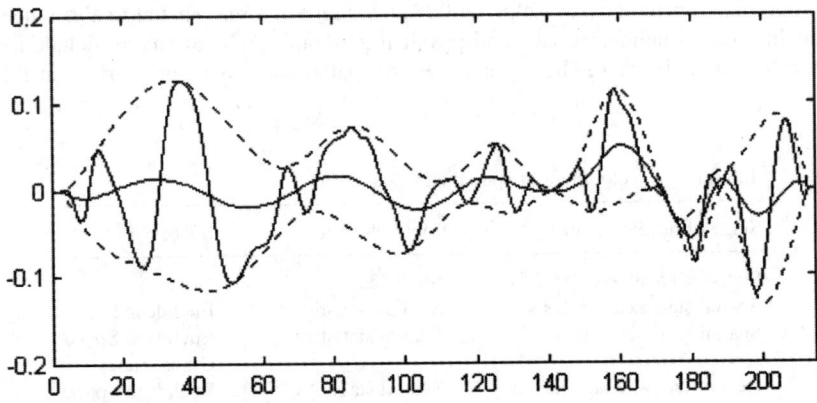

Figure 1.   The envelope concept in EMD.

2.2 The decomposition process in EMD: sifting

To decompose a set of data into IMFs, one should first find the local extrema (i.e., maxima and minima) from the data. This can be done simply by finding slope changes at points on the line. In other words, any point with a change in the sign slope (first derivative) is considered as a local extrema. Having located these points, the next step is to fit spline curves onto them; i.e., one curve for the maxima and the other for the minina. The first curve (passing through the maxima) defines the upper envelope and the second curve (passing through the minima) defines the lower envelope. Using the upper and lower envelopes, a set of mean values that define the mean envelope can be computed. Figure 1 shows such envelopes.

By subtracting the mean envelope from the signal, the first IMF is then generated. The mathematical formula could be written as

$$h_1 = x(t) - m_1 \tag{1}$$

where $x(t)$ is the original data function, m_1 is the mean envelope, and h_1 is the first component.

The resulting IMF is the first component of the data; i.e., $c_1 = h_1$. Ideally, h_1 should be an IMF, but depending on the degree of the non-stationarity of the data, new extrema might be generated in h_1. The process, therefore, can be iterated until the final result is completely an IMF. This process is called *sifting*, as it indeed sifts the data to extract its sharp oscillations.

The standard deviation (σ) between the two successive h values can be used as a criterion for terminating the iteration. That is, the iteration must stop if σ becomes smaller than a predefined threshold, for which a value of between 0.2 and 0.3 has been recommended (Huang et al., 1998). The threshold then becomes:

$$\text{Threshold:} \quad \sigma = \sum_{t=o}^{T} \left[\frac{\left| (h_{1(k-1)}(t) - h_{1k}(t)) \right|^2}{h_{1(k-1)}^2(t)} \right] \le 0.2$$

In order to make sure that the resulting component is a complete IMF, a test to check the process is required. The test checks whether the number of zero crossings and extrema points are equal to, or differ by, one at most.

After this first component is subtracted from the data, the remainder r_1 can be treated as a new set of input data to the sifting process. That is,

$$r_1 = x(t) - c_1 \tag{2}$$

The next component can then be extracted from the remainder r_1. Repeating this procedure several times, one can extract all of the components from the data as follows:

$$\begin{aligned} r_2 &= r_1 - c_2, \\ r_3 &= r_2 - c_3, \\ &\cdots, \\ r_n &= r_{n-1} - c_n \end{aligned} \tag{3}$$

The process for extracting components is continued until the remainder becomes a monotonic function, from which no more IMFs can be extracted.

Figure 2 shows an example of such a sifting process. The curve as shown in Figure 3a is presented in parametric form; i.e., by two functions $X(t)$, $Y(t)$. The sifting process is applied to these two functions and the results are shown in Figures 2a and 2b.

2.3 The reconstruction

The decomposition can be summarized by Equation (4) as follows:

$$x(t) = \sum_{i=1}^{n} c_i + r_n \tag{4}$$

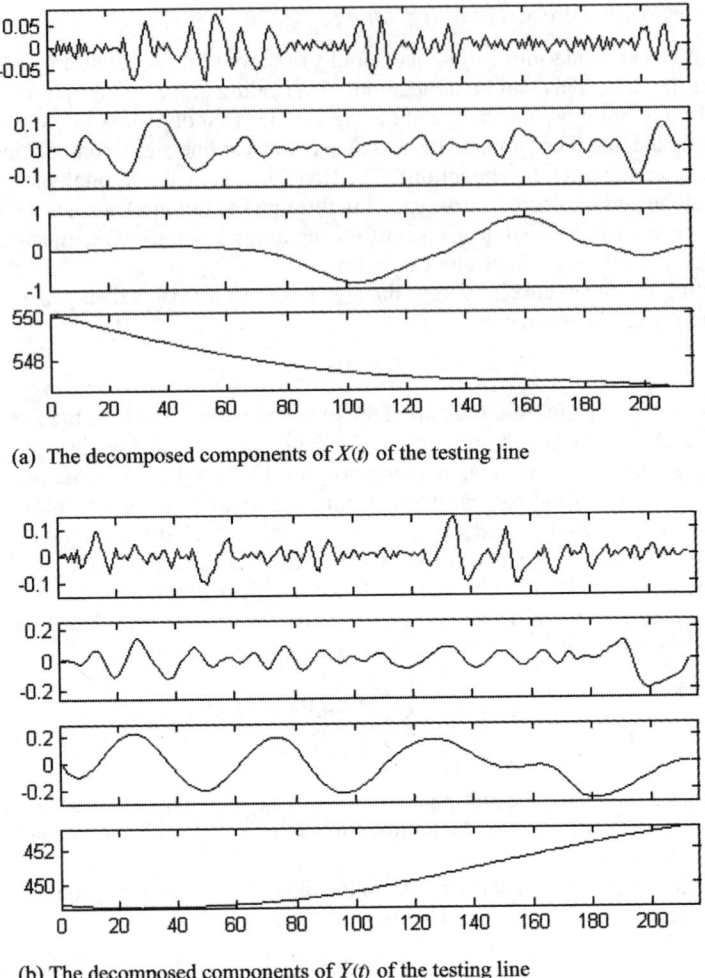

(a) The decomposed components of $X(t)$ of the testing line

(b) The decomposed components of $Y(t)$ of the testing line

Figure 2.   The decomposition of the X and Y components of the line shown in Figure 3a.

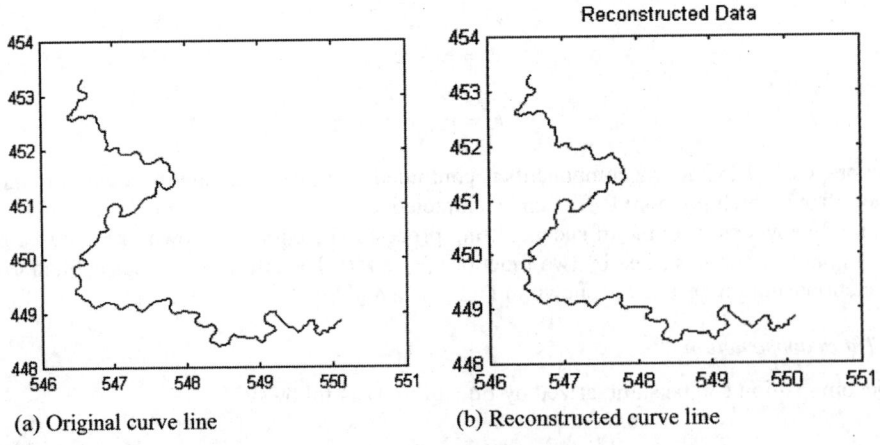

(a) Original curve line   (b) Reconstructed curve line

Figure 3.   A line for testing and the reconstruction of a line by EMD.

where $x(t)$ is the original data, c_i are the components and r_n is the last component, which is in fact a monotonic function showing the trend of the data. This equation indicates the completeness of the method because the original data can be reconstructed precisely as the components are summed (Figure 3). The two lines shown in Figure 3 should be identical. A possible difference between the original and reconstructed data would only be due to an error in rounding off the computations.

3 POTENTIAL OF EMD FOR SPATIAL ANALYSIS

Like the Fourier transform, EMD should also have great potential in spatial data analysis. Here, only two examples are illustrated: one on the filtering of data and the other on the extracting of components.

3.1 *EMD as a low-pass filter for data smoothing*

In this section, a linear feature shown in Figure 3a is used as an example of smoothing. This idea of using EMD for curve smoothing comes from the fact that what has to be done for curve smoothing is in fact a low-pass filtering, which is traditionally carried out in the domain of frequency. Both non-linear and non-stationary data can be used to generate harmonic waves of all ranges. Therefore, any type of filtering in frequency space will eliminate some of the harmonics, which will cause deformations in the filtered data. Using IMF, however, one can devise a time/space domain filtering.

A precise inspection of Figure 3a reveals that different components carry different frequency contents of data. One can also say the components are of different scales, as compared to wavelet analysis. Such components are shown in the top of both Figures 2a and 2b. Therefore, the removal of such components from the data is equivalent to the omission of oscillations at a specific frequency. With an analogy to frequency analysis, a smoother version of the line will be generated by removing the first component of the data, which is of the highest frequency (and the smallest scale). By the same logic, if a much smoother line is desired, the next components might be removed from the data. Figure 4 shows the different levels of smoothness obtained by removing different components from the original data, compared with the original line shown in Figure 3a.

The results in Figure 4 shows that only very small variations are removed with the first component in both $X(t)$ and $Y(t)$. The curve becomes greatly simplified and is thus smoothed when the first two components in both $X(t)$ and $Y(t)$ are removed. Finally, the curve is most simplified when only the last component in both $X(t)$ and $Y(t)$ are used for reconstruction. It can be seen clearly that the trend does follow the principle of data smoothing in a traditional sense.

3.2 *EMD for extraction of components of interest*

In many cases of spatial analysis, one may be interested in a particular component of the data. For example, an El-Nino event may be related to variations in the level of the sea. But one may be interested in a particular cycle; e.g., an interannual cycle. Then, one extracts data on the variations in sea level in an interannual cycle and correlates it to the El-Nino (or ENSO) cycle to see how high the correlation is. The EMD can be used to extract variations in sea level during any period such as an interannual cycle.

A time series of the monthly mean sea level in Hong Kong from 1954 to 2001 is shown at the top of Figure 5. The wavelet spectral results are given at the bottom of Figure 1. The frequency (periodic) and time distributions of the estimated spectral signals are shown on the ordinate and on the abscissa, respectively.

It can be seen from the figure that in addition to the rising trend term and more stable seasonal variations of annual and semi-annual signals, there are time variable higher frequency signals, quasi-biennial oscillations (QBO), interannual and decade fluctuations in the sea level time series.

Three methods have used to decompose the signals; i.e., wavelets, EMD and MSF (Zheng and Dong, 1986; Zheng and Luo, 1992). The results are shown in Figures 6–8. If picked up, the ENSO

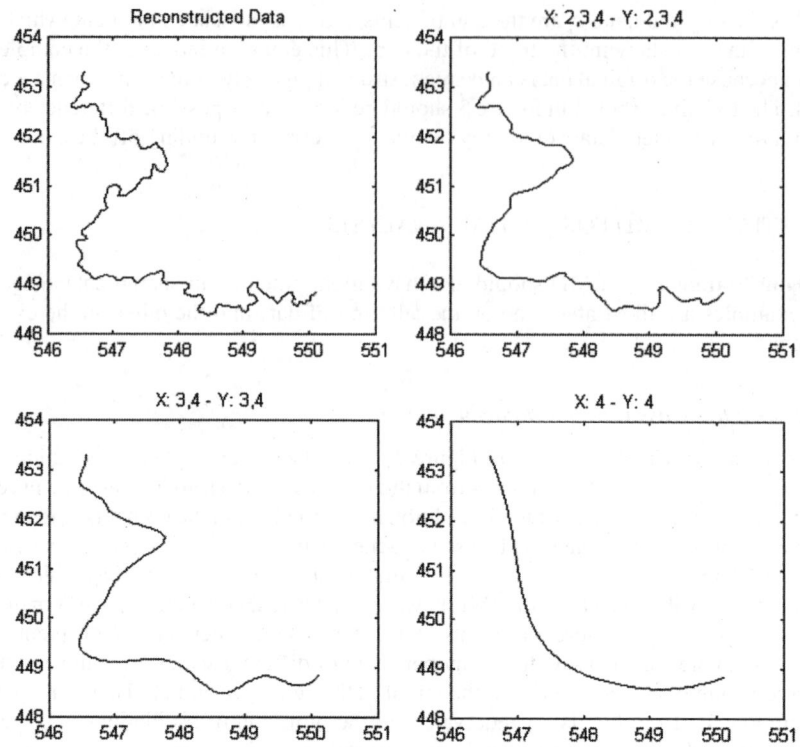

Figure 4. Smoothed lines which are reconstructed from various combinations of IMF components.

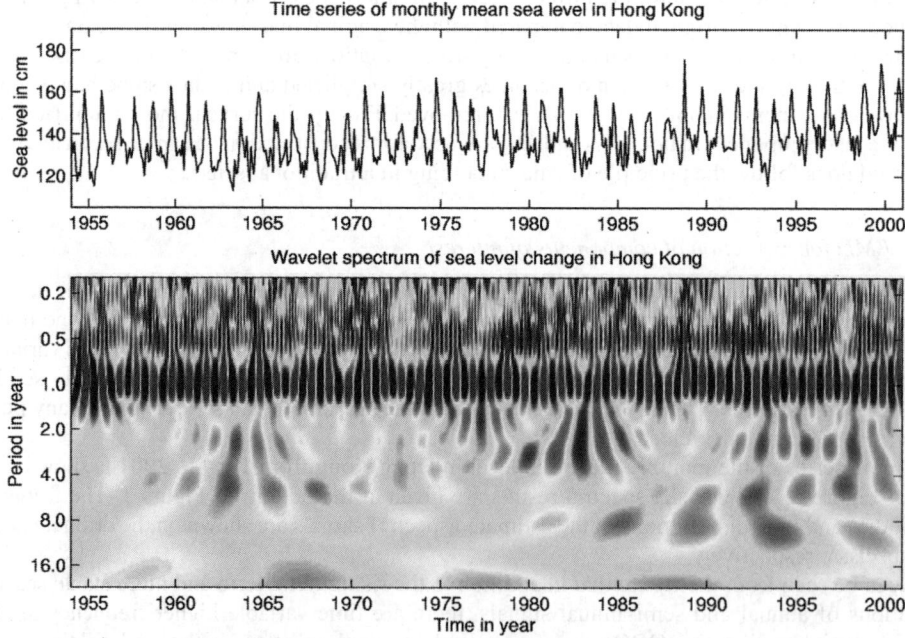

Figure 5. Time series of monthly mean sea level in Hong Kong of 1954–2001 is shown in the top of the figure and its wavelet spectral estimation is given in the bottom.

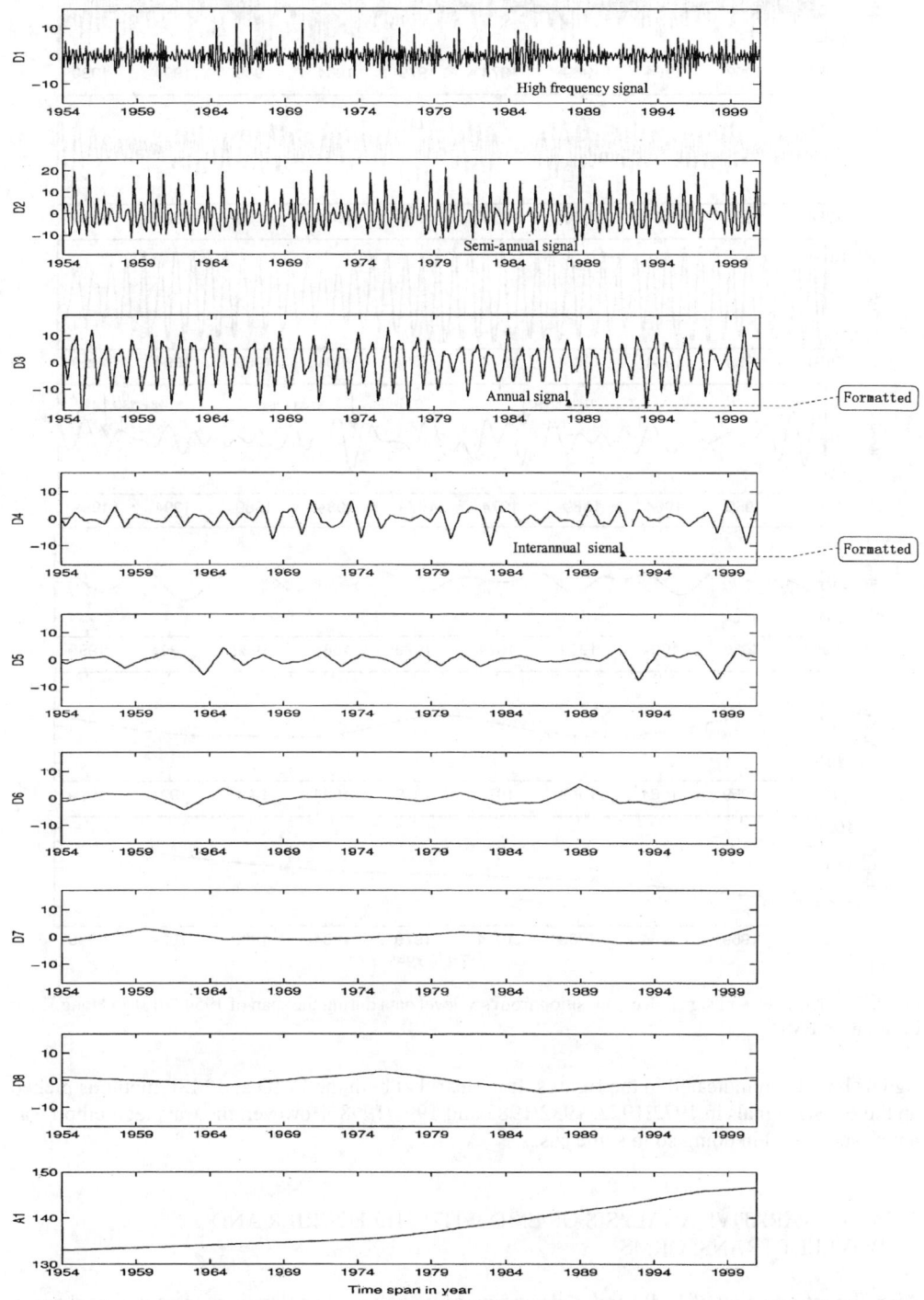

Figure 6. The results of signal decomposition from sea level data during the span of 1954–2001 in Hong Kong by means of wavelet transform.

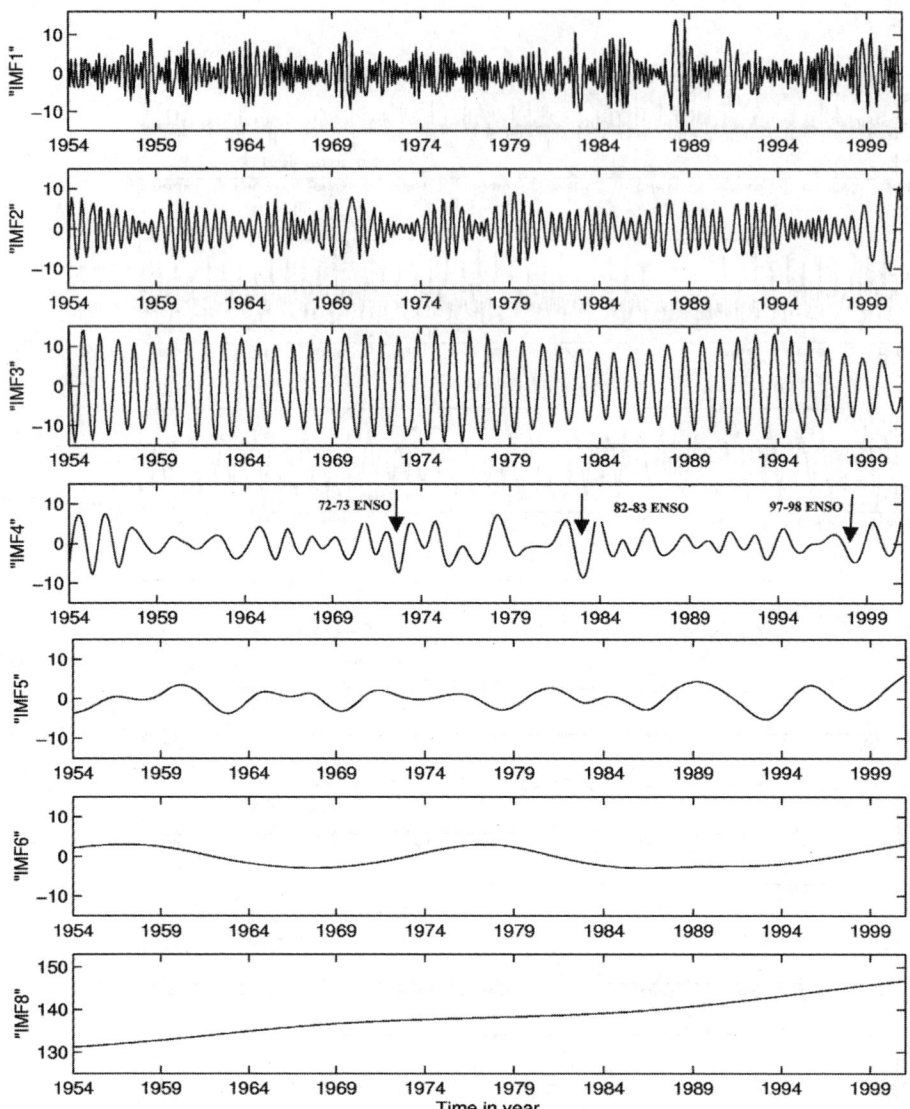

Figure 7. The results of signal decomposition from sea level data during the span of 1954–2001 in Hong Kong by means of EMD.

signals have been indicated in the figures. It is clear that both the EMD and MSF methods picked up the ENSO signals in 1972/1972, 1982/1983 and 1997/1998. However, the wavelet method was not as successful in doing so in some cases.

4   A COMPARATIVE ANALYSIS OF EMD WITH THE FOURIER AND WAVELET TRANSFORMS

After the introduction of EMD and a discussion of its potential applications in spatial analysis, a few points associated with the EMD itself should be clarified, especially when compared with the Fourier and Wavelet transform.

26

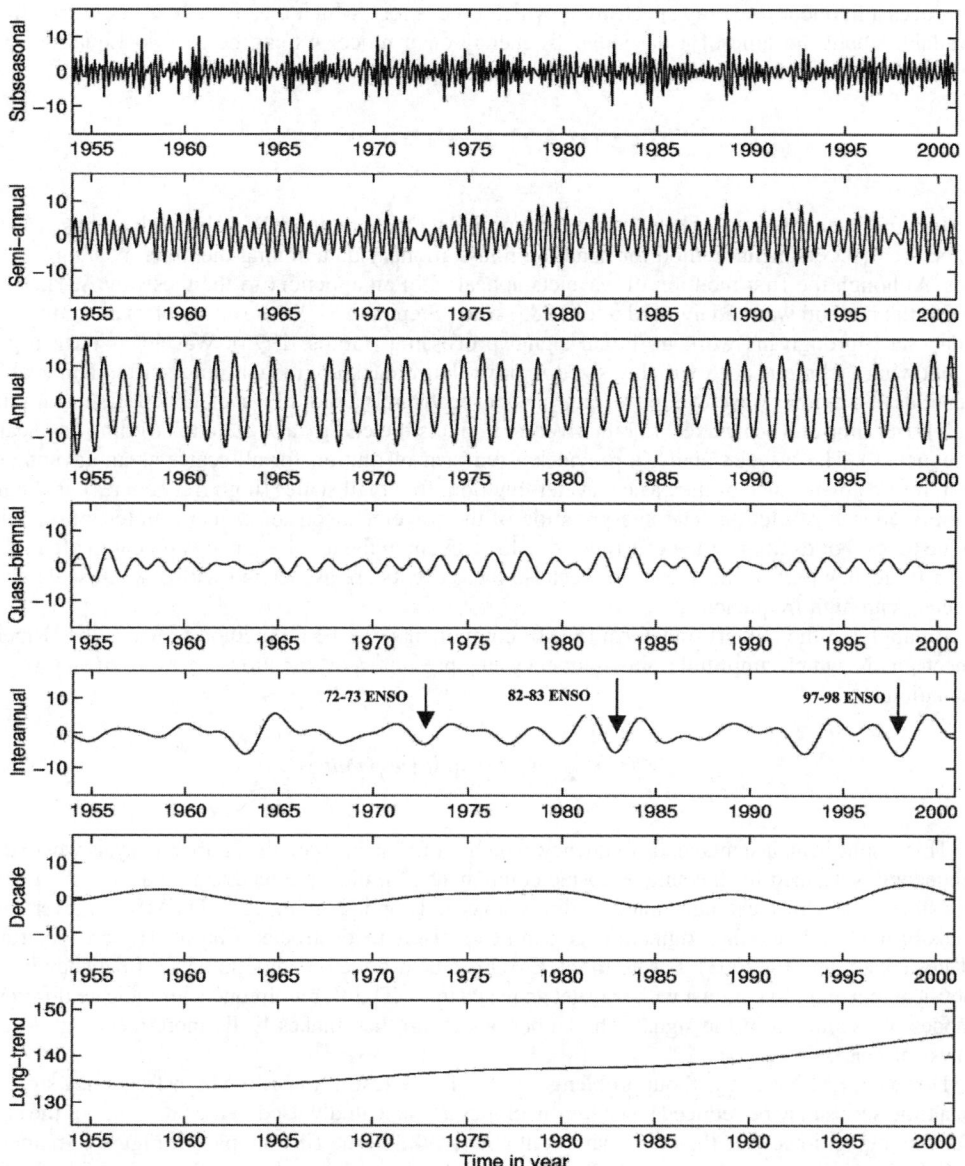

Figure 8. The results of signal decomposition from sea level data during the span of 1954–2001 in Hong Kong by means of MSF.

The Fourier transform is based on representing the data as a superposition of sine and cosine functions. Therefore, the linearity of the system and the stationarity of the data are essential conditions for the Fourier analysis to be valid. However the data generated from physical sources are often non-linear and non-stationary in nature. The Fourier transform produces a large number of harmonics to simulate such data. This leads to unwanted spurious energy and artificial frequencies. As a result, energy spreads over a wide range of frequencies. Constrained by the conservation of energy principle, these spurious harmonics and wide frequency spectrum cannot faithfully represent the true density of energy in the frequency and time domain (Huang, 1999). The other drawback with the Fourier analysis is the loss of the time component. The Fourier transform

27

produces a frequency–energy spectrum in which time is lost, which means there is no information available about the time of a particular frequency occurrence. As can be seen in Equation (5), amplitude and frequency are constants in the Fourier transform.

$$X(t) = \sum_{j=1}^{\infty} a_j \exp(i\omega_j t) \qquad (5)$$

This motivated the researchers to develop a more localized method. Wavelet analysis has proven to be the most efficient method for handling non-stationary data among the other existing methods. Although the first mention of wavelets appeared in an appendix to the thesis by A. Haar in 1909, the method was widely used until 1985, when Stephane Mallat gave wavelets an additional jump-start through his work in digital signal processing (Graps, 1995). Wavelet is basically a windowing technique with variably sized regions that breaks up a signal into shifted and scaled versions of a mother wavelet. Hence, the wavelet spectrum is a plot of wavelet coefficients (amplitude) over time and scale axes. In other words, it yields the energy as a function of time and scale (frequency). The wavelet analysis provides a uniform resolution for all scales as an advantage. But, limited by the size of the basic wavelet function, for small scales (high frequencies), it is uniformly poor in resolution. The interpretation of the wavelet spectrum is also counter-intuitive in some cases. For example, the occurrence of a local event in the low frequency range will affect the high frequency part of the spectrum, because local events are associated with a small scale and, hence, with high frequencies.

By applying the Hilbert transform to IMF components, the EMD method provides the Hilbert spectrum in which amplitude and frequency are presented as functions of time, as shown in Equation (6).

$$X(t) = \sum_{i=1}^{n} a_j(t) \exp\left(i \int \omega_j(t)dt\right) \qquad (6)$$

This results in an instantaneous frequency, which is impossible with the Fourier analysis where the frequency is defined for the sine or cosine components. Similar to Fourier and Wavelet, the EMD transform can also be extended into two dimensions and applied on images. 2D EMD is a reversible transform as well, and the original image can be precisely reconstructed from its Hilbert spectrum (Khoshelham and Li, 2003). Furthermore, EMD is a data-driven method in which frequency has a physical meaning. In contrast with wavelet analysis, in EMD different frequencies indicate different modes of oscillation of the signal. The Hilbert spectrum then makes EMD more meaningful in a physical sense.

However, EMD is not without problems. By EMD, the resolution of the high-frequency signals in a time series may be reduced since the envelopes are repeatedly used in the processes of the signal decompositions, (see the curve pattern at about 1988 in the first subplot of Figure 7; namely, IMF1). The influences of the edge effects on the decomposed signals are often induced by EMD (see the curve pattern at 2000 in the second to the fifth subplots of Figure 7; namely, IMF2, IMF3, IMF4 and IMF5). Furthermore, it is more difficult to determine the exact reasonable number of IMFs in a time series by means of EMD only. The confused phenomena for the near frequency signals in a time series may occur in the decomposed signals by EMD.

5 CONCLUSIONS

In this paper, the basic principle of EMD is first introduced. The application of EMD to spatial analysis is then explored; i.e., the filtering of spatial data and the extraction of the data component. It is clear that EMD will have great potential for different types of spatial analysis.

It was clarified that the EMD has some advantages over the Wavelet and Fourier transform.

ACKNOWLEDGEMENTS

The authors would like to thank Prof. Y. L. Xu and Dr J. Chen for producing Figure 7, and Dr W. B. Zhou for producing Figure 6 for us. The research funding from the RGC of the HKSAR is also acknowledged (Project No. PolyU 5073/01E).

REFERENCES

Anselin, L., 2000. GIS, spatial econometrics and social science research. *Journal of Geographical Systems*, 2: 11–15.

Anselin, L., Dodson, R. and Hudak, S., 1993. Linking GIS and Spatial Data Analysis in Practice. *Geographical Systems*, 1(1): 3–23.

Graps, Amara, 1995. An Introduction to Wavelets IEEE Computational Sciences and Engineering, 2(2): 50–61.

Huang, Norden E., 1999. *Computer implemented empirical mode decomposition method, Apparatus and article of manufacture*, US Patent 5, 983,162. Nov. 1999.

Huang, Norden E., Shen, Z., Long, S. R., Wu, M. C., Shih, H. H., Zheng, Q., Yen, N.- C., Tung, C. C. and Liu, H. H., 1998. The empirical mode decomposition and the Hilbert spectrum for nonlinear and non-stationary time series analysis. *Proceedings of the Royal Society of London*, A454, 903–995.

Khoshelham, K. and Li, Z. L., 2003. Image compression using empirical mode decomposition. *Proceedings of Picture Coding Symposium*, PCS 2003, St-Malo, France, April 2003.

Marble, D. F., 2000. Integration of spatial analysis and GIS. *Journal of Geographical Systems*, 2: 31–35.

Zheng, Dawei and Dong, Danan, 1986. Realization of narrow band filtering of the polar motion data with Multi-Stage Filter, *Acta Astronomica Sinica*, 27(4): 368–376 (in Chinese).

Zheng, Dawei and Luo, Shifang, 1992. Contribution of time series analysis to data processing of astronomical observations of Earth rotation in China, *Statistica Sinica*, 2(2): 605–618.

Advances in Spatial Analysis and Decision Making, Li, Zhou & Kainz (eds)
© 2004 Swets & Zeitlinger, Lisse, ISBN 90 5809 652 1

A knowledge framework for representing, manipulating and reasoning with geographic semantics

James O'Brien & Mark Gahegan
GeoVISTA Center, Department of Geography, The Pennsylvania State University, University Park, USA

ABSTRACT: This paper describes a programmatic framework for representing, manipulating and reasoning with geographic semantics. The framework enables visualizing knowledge discovery, automating tool selection for user defined geographic problem solving, and evaluating semantic change in knowledge discovery environments. Methods, data and human experts (our resources) are described using ontologies. An entity's ontology describes, where applicable: uses, inputs, outputs, and semantic changes. These ontological descriptions are manipulated by an expert system to select methods, data and human experts to solve a specific user-defined problem; that is, a semantic description of the problem is compared to the services that each entity can provide to construct a graph of potential solutions. A minimal spanning tree representing the optimal (least-cost) solution is extracted from this graph, and displayed in real-time. The semantic change(s) that result from the interaction of data, methods and people contained within the resulting tree are determined via expressions of transformation semantics represented within the JESS expert system shell. The resulting description represents the formation history of each new information product (such as a map or overlay) and can be stored, indexed and searched as required. Examples are presented to show (1) the construction and visualization of information products, (2) the reasoning capabilities of the system to find alternative ways to produce information products from a set of data methods and expertise, given certain constraints and (3) the representation of the ensuing semantic changes by which an information product is synthesized.

1 INTRODUCTION

The importance of semantics in geographic information is well documented (Bishr, 1998; Egenhofer, 2002; Fabrikant and Buttenfield, 2001; Kuhn, 2002). Semantics are a key component of interoperability between GIS; there are now robust technical solutions to interoperate geographic information in a syntactic and schematic sense (e.g., OGC, NSDI) but these fail to take account of any sense of meaning associated with the information. Visser et al., (2002) described how exchanging data between systems often fails due to confusion in the meaning of concepts. Such confusion, or semantic heterogeneity, significantly hinders collaboration if groups cannot agree on a common lexicon for core concepts. Semantic heterogeneity is also blamed for the inefficient exchange of geographic concepts and information between groups of people with differing ontologies (Kokla and Kavouras, 2002).

Semantic issues pervade the creation, use and re-purposing of geographic information. In an information economy we can identify the roles of information producer and information consumer and, in some cases, in national mapping agencies for example, datasets are often constructed incrementally by different groups of people (Gahegan, 1999) with an implicit (but not necessarily recorded) goal. The overall meaning of the resulting information products are not always obvious to those outside that group, existing for the most part in the creators' mental model. When solving a problem, a user may gather geo-spatial information from a variety of sources without ever encountering an explicit statement about what the data mean, or what they are (and are not) useful for. Without capturing the semantics of the data throughout the process of creation, the data may be misunderstood, used inappropriately, or not used at all. Geo-spatial semantics needs to explicitly cater for the particular way in which geo-spatial tasks are undertaken (Egenhofer, 2002). As a result, the underlying assumptions about methods used with data and the roles played by human expertise need to be represented in some fashion so that a meaningful association can be made between appropriate methods, people and data to solve a problem. It is not the role of this paper to present a definitive taxonomy of geographic operations or their semantics. To do so would trivialize the difficulties of defining geographic semantics.

2   BACKGROUND AND AIMS

This paper presents a programmatic framework for representing, manipulating and reasoning with geographic semantics. In general, semantics refers to the study of the relations between symbols and what they represent (Hakimpour and Timpf, 2002). In the framework outlined in this paper, semantics have two valuable and specific roles. First, to determine the most appropriate resources (method, data or human expert) to use in concert to solve a geographic problem; and second, to act as a measure of change in meaning when data are operated on by methods and human experts. Both of these roles are discussed in detail in Section 3. The framework draws on a number of different research fields, specifically: geographical semantics (Gahegan, 1999 and Kuhn, 2002), ontologies (Guarino, 1998) computational semantics (Sowa, 2000), constraint-based reasoning and expert systems (Honda and Mizoguchi, 1995) and visualization (MacEachren, in press) to represent aspects of these resources. The framework sets out to solve a multi-layered problem of visualizing knowledge discovery, automating tool selection for user-defined geographic problem solving and evaluating semantic change in knowledge discovery environments. The end goal of the framework is to associate with geo-spatial information products the details of their formation history and tools by which to browse, query and ultimately understand this formation history, thereby building a better understanding of meaning and the appropriate use of the information.

The problem of semantic heterogeneity arises due to the varying interpretations given to the terms used to describe facts and concepts. Semantic heterogeneity exists in two forms, cognitive and naming (Bishr, 1998). Cognitive semantic heterogeneity results from no common base of definitions between two (or more) groups. As an example, think of these as groups of scientists attempting to collaborate. If the two groups cannot agree on definitions for their core concepts, then collaboration between them will be problematic. Defining such points of agreement amounts to constructing a shared ontology or, at the very least, points of overlap (Pundt and Bishr, 2002).

Naming semantic heterogeneity occurs when the same name is used for different concepts or different names are used for the same concept. It is not possible to undertake any semantic analysis until problems of semantic heterogeneity are resolved. Ontologies, described below, are widely recommended as a means of rectifying semantic heterogeneity (Hakimpour and Timpf, 2002; Kokla and Kavouras, 2002; Kuhn, 2002; Pundt and Bishr, 2002; Visser et al., 2002). The framework presented in this paper utilizes that work and other ontological research (Brodaric and Gahegan, 2002; Chandrasekaran et al., 1997; Fonseca, 2001; Fonseca and Egenhofer, 1999; Fonseca et al., 2000; Guarino, 1997a; Guarino, 1997b; Mark et al., 2002) to solve the problem of semantic heterogeneity.

The use of an expert system for automated reasoning fits well with the logical semantics utilized within the framework. The Java Expert System Shell (JESS) is used to express diverse semantic aspects about methods, data and human experts. JESS performs string comparisons of resource attributes (parsed from ontologies) using backward chaining to determine interconnections between resources. Backward chaining is a goal-driven problem-solving methodology, starting from the set of possible solutions and attempting to derive the problem. If the conditions for the satisfaction of a rule are not found within that rule, the engine searches for other rules that have the unsatisfied rule as their conclusion, establishing dependencies between rules. JESS functions as a mediator system (Sotnykova, 2001), with a foundation layer where the methods, data and human experts are described (the domain ontology), a mediation layer with a view of the system (the task ontology) and a user interface layer (for receiving queries and displaying results).

2.1 Ontology

In philosophy, Ontology is the 'study of the kinds of things that exist' (Chandrasekaran *et al.*, 1997; Guarino, 1997b). In the artificial intelligence community, ontology has one of two meanings: as a *representation vocabulary*, typically specialised to some domain or subject matter; and as a *body of knowledge* describing some domain using such a representation vocabulary (Chandrasekaran *et al.*, 1997). The goal of sharing knowledge can be accomplished by encoding domain knowledge using a standard vocabulary based on an ontology (Chandrasekaran *et al.*, 1997; Kokla and Kavouras, 2002; Pundt and Bishr, 2002). The framework described here utilizes both definitions of ontology.

The representation vocabulary embodies the concepts that the terms in the vocabulary are intended to capture. Relationships described between conceptual elements in this ontology allow for the production of rules governing how these elements can be 'connected' or 'wired together' to solve a geographic problem. In our case, these elements are methods, data and human experts, each with their own ontology. In the case of datasets, a domain ontology describes salient properties such as location, scale, date, format, etc., as currently captured in meta-data descriptions (see Figure 1). In the case of methods, a domain ontology describes the services a method provides in terms of a transformation from one semantic state to another. In the case of human experts the simplest representation is again a domain ontology that shows the contribution that a human can provide in terms of steering or configuring methods and data. However, it should also be possible to represent a two-way flow of knowledge as humans learn from situations and thereby expand the number of services they can provide (we leave this issue for future work). The synthesis of a specific information product is specified via a task ontology that must fuse together elements of the domain and application ontologies to attain its goal. An innovation of this framework is the

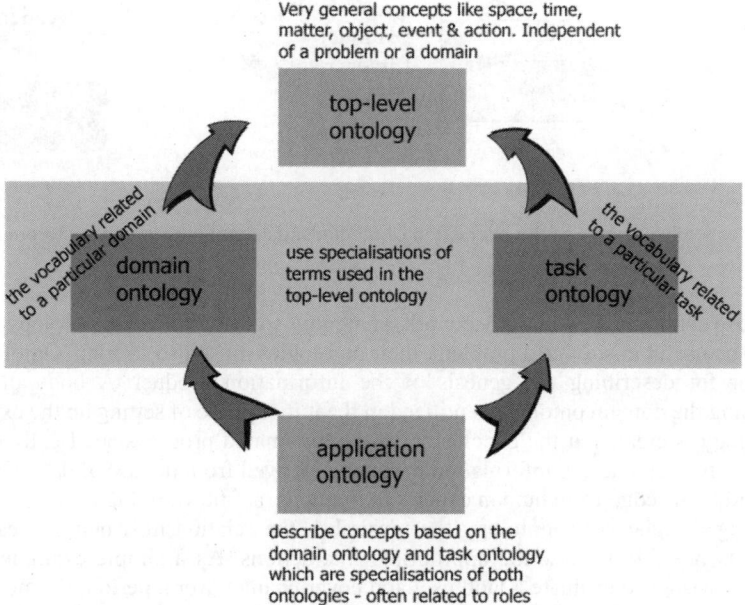

Figure 1.   Interrelationships between different types of ontology.

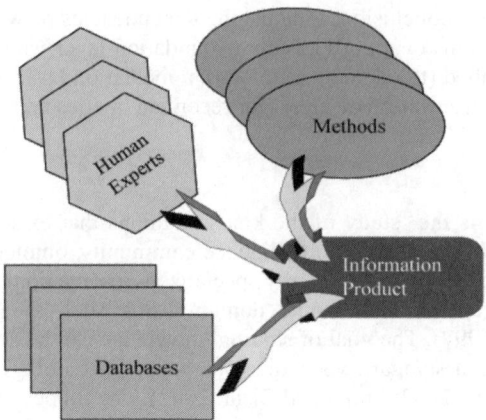

Figure 2.   Information products are derived from the interaction of entities.

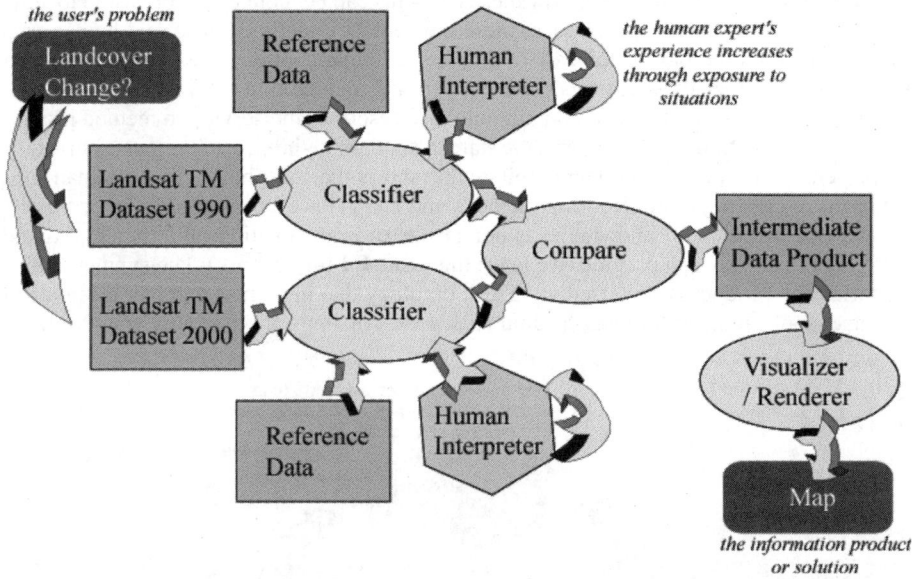

Figure 3.   A concrete example of the interaction of methods, data and human experts to produce an information product.

dynamic construction of the solution network, analogous to the application ontology. In order for resources to be useful in solving a problem, their ontologies must also overlap. Ontology is a useful metaphor for describing the genesis of the information product. A body of knowledge described using the domain ontology is utilized in the initial phase of setting up the expert system. A task ontology is created at the conclusion of the automated process specifically defining the concepts that are available. An information product is derived from the use of data extracted from databases and knowledge from human experts in methods, as shown in Figure 2.

By forming a higher-level ontology that describes the relationships between each of these resources it is possible to describe appropriate interactions. As a simple example (Figure 3), assume a user wishes to evaluate a land use/land cover change over a period of time. Classifying two LandsatTM images from 1990 and 2000 using reference data and expert knowledge, the user can compare the two resulting images to produce a map of the area(s) which have changed during

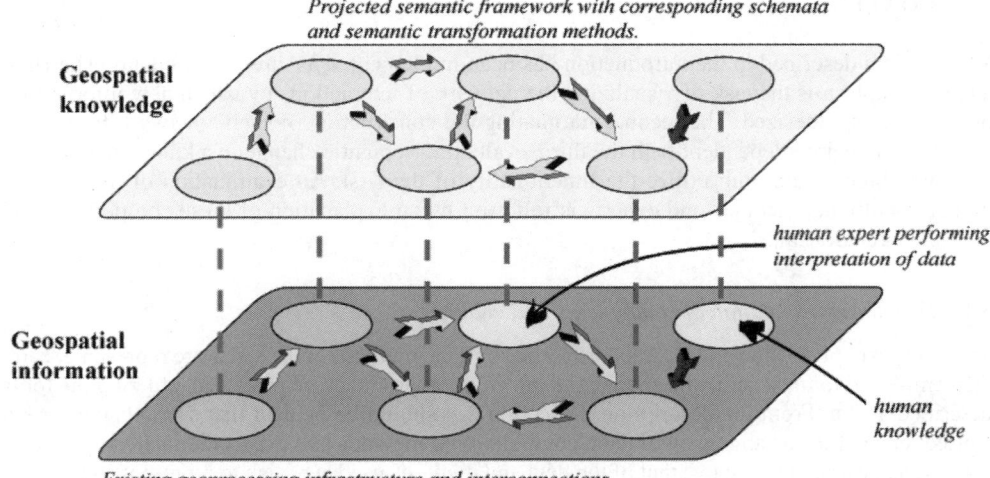

Projected semantic framework with corresponding schemata and semantic transformation methods.

Geospatial knowledge

human expert performing interpretation of data

Geospatial information

human knowledge

Existing geoprocessing infrastructure and interconnections.

Figure 4.   Interaction of the semantic layer and operational layer.

that time. One interesting feature demonstrated in this example is the ability of human experts to gain experience through repeated exposure to similar situations. Even in this simple example a basic semantic structure is being constructed and a lineage of the data can be determined. Arguably, an additional intermediate data product exists between the classifiers and the comparison; it has been removed for clarity.

2.2   *Semantics*

While the construction of the information product is important, a semantic layer sits above the operations and information (Figure 4). The geo-spatial knowledge obtained during the creation of the product is captured within this layer. The capture of this semantic information describes the transformations that the geo-spatial information undergoes, facilitating better understanding and providing a measure of repeatability of analysis, and improving communication in the hope of promoting best practices in bringing geo-spatial information to bear.

Egenhofer (2002) noted that the challenge remains of how best to make these semantics available to the user via a search interface. Pundt and Bishr, (2002) outlined a process by which a user searches for data to solve a problem. This search methodology is also applicable to the methods and human experts to be used with the data. This solution fails when multiple sources are available and nothing is known of their content, structure and semantics. The use of pre-defined ontologies aids users by reducing the available search space (Pundt and Bishr, 2002). Ontological concepts relevant to a problem domain are supplied to users, allowing them to focus their query. A more advanced interface would take a user's query on its own terms and map that to an underlying domain ontology (Bishr, 1998).

As previously noted, the meaning of geo-spatial information is constructed, shaped and changed by the interaction of people and systems. Subsequently, the interaction of human experts, methods and data needs to be carefully planned. A product created as a result of these interactions is dependent on the ontology of the data and methods and on the epistemologies and ontologies of the human experts. In light of this, the knowledge framework outlined below focuses on each of the resources involved (data, methods and human experts) and the roles they play in the evolution of a new information product. In addition, the user's goal that produced the product, and any constraints placed on the process are recorded to capture aspects of intention and situation that also have an impact on meaning. This process and the impact of constraint-based searches are discussed in more detail in the following section.

3 KNOWLEDGE FRAMEWORK

The problem described in the introduction has been implemented as three components. The first, and the simplest, is the task of visualizing the network of interactions by which new information products are synthesized. The second, automating the construction of such a network for a user-defined task, is interdependent with the third, evaluating semantic change in a knowledge discovery environment, and both utilize the functionality of the first. An examination of the abstract properties of data, methods and experts is followed by an explanation of these components and their inter-relationships.

3.1   *Formal representation of components and changes*

This section explains how the abstract properties of data, methods and experts are represented, and then employed to track semantic changes as information products are produced, utilizing the tools described above. From the description in Section 2 it should be evident that such changes are a consequence of the arrangement of data, computational methods and expert interaction applied to data. At an abstract level above that of the data and methods used, we wish to represent some characteristics of these three sets of components in a formal sense, so that we can describe the effects deriving from their interaction. One strong caveat here is that our semantic description (given below) does not claim to capture all senses of meaning attached to data, methods or people. In fact, as a community of researchers we are still learning about which facets of semantics are important and how they might be described. It is not currently possible to represent all aspects of meaning and knowledge within a computer, so we aim instead to provide descriptions that are rich enough *to allow users to infer aspects of meaning* that are important for specific tasks from the visualizations or reports that we can synthesize. In this sense, our own descriptions of semantics play the role of a signifier – the focus is on conveying meaning to the reader rather than on explicitly carrying intrinsic meaning *per-se*.

The formalization of semantics based on ontologies and operated using a language capable of representing relations provides for powerful semantic modelling (Kuhn, 2002). The framework, rules and facts used in the Solution Synthesis Engine (see below) function in this way. Relationships are established between each of the entities, by calculating their membership within a set of objects capable of synthesizing a solution. We extend the approach of Kuhn by allowing the user to narrow a search for a solution based on the specific semantic attributes of entities. Using the minimal spanning tree produced from the solution synthesis it is possible to retrace the steps of the process to calculate semantic change. As each fact is asserted it contains information about the rule that created it (the method) and the data and human experts that were identified as resources required. If we are able to describe the change to the data (in terms of abstract semantic properties) imbued by each of the processes through which it passes, then it is possible to represent the change between the start state and the finish state by differencing the two.

Although the focus of our description is on semantics, there are good reasons for including syntactic and schematic information about data and methods, since methods are generally designed to work in limited circumstances, using and producing very specific types of data (*pre-conditions* and *post-conditions*). Hence, from a practical perspective it makes sense to represent and reason with these aspects in addition to semantics, since they will limit which methods can be connected together and dictate where additional conversion methods are required. Additional potentially useful properties arise when the computational and human infrastructure is distributed, for example, around a network. By encoding such properties we can extend our reasoning capabilities to address problems that arise when resources must be moved from one node to another to solve a problem (Gahegan, 1998).

3.1.1   *Description of data*
As mentioned in Section 2, datasets are described in general terms using a domain ontology drawn from generic metadata descriptions. Existing metadata descriptions hold a wealth of such practical

information that can be readily associated with datasets; for example, the FGDC (1998) defines a mix of semantic, syntactic and schematic metadata properties. These include basic semantics (abstract and purpose), syntactic (data model information and projection), and schematic (creator, theme, temporal and spatial extents, uncertainty, quality and lineage). We explicitly represent and reason with a subset of these properties in the work described here and could easily expand to represent them all, or any other given metadata description that can be expressed symbolically. Formally, we represent the set of n properties of a dataset D as: $D(p_1, p_2, ..., p_n)$ (Gahegan, 1996).

3.1.2 *Describing methods*

While standards for metadata descriptions are already mature and suit our purposes, complementary mark-up languages for methods are still in their infancy. It is straightforward to represent the signature of a method in terms of the format of data entering and leaving the method, and knowing that a method that requires data to be in a certain format will cause the system to search for and insert conversion methods automatically where they are required. Thus, for example, if a coverage must be converted from raster format to vector format before it can be used as input to a surface flow accumulation method, then the system can insert appropriate data conversion methods into the evolving query tree to connect to appropriate data resources that would otherwise not be compatible. Similarly, if an image classification method requires data at a nominal scale of 1:100,000 or a pixel size of 30 m, any data at finer scales might be generalized to meet this requirement prior to use. Although such descriptions have great practical benefit, they say nothing about the role the method plays or the transformation it imparts to the data. In short, they do not enable any kind of semantic assessment to be made.

A useful approach to representing what GIS methods do, in a conceptual sense, centres on a typology (e.g., Albrecht's 20 universal GIS operators, 1994). Here, we extend this idea to address a number of different abstract properties of a dataset, in terms of how the method invoked changes to these properties (Pascoe and Penny, 1995; Gahegan, 1996). In a general sense, the transformation performed by a method (M) can be represented by pre-conditions and post-conditions, as is the common practice with interface specification and design in software engineering. Using the notation above, our semantic description takes the form: $M: D(p_1, p_2, ..., p_n) \xrightarrow{Operation} D'(p_1', p_2', ..., p_n')$, where *Operation* is a generic description of the role or function the method provides, drawn from a typology.

For example, a cartographic generalization method changes the scale at which a dataset is most applicable, a supervised classifier transforms an array of numbers into a set of categorical labels, and an extrapolation method might produce a map for next year, based on maps of the past. Clearly, there are any number of key dimensions over which such changes might be represented. The above examples highlight *spatial scale*, *conceptual 'level'* (which at a basic syntactic level could be viewed simply as statistical scale) and *temporal applicability*, or simply *time*. Others come to light following just a cursory exploration of GIS functionality such as changes in spatial extents, e.g. windowing and buffering; and changes in uncertainty (very difficult in practice to quantify but easy to show in an abstract sense that there has been a change).

Again, we have chosen not to restrict ourselves to a specific set of properties, but rather to remain flexible in representing those that are important to specific application areas or communities. We note that as Web Services (Abel *et al.*, 1998) become more established in the GIS arena, such an enhanced description of methods will be a vital component in identifying potentially useful functionality.

3.1.3 *Describing people*

Operations may require additional configuration or expertise in order to carry out their task. People use their expertise to interact with data and methods in many ways, such as gathering, creating and interpreting data, configuring methods and interpreting results. These activities are typically structured around well-defined tasks where the desired outcome is known, although as in the case of knowledge discovery, they may sometimes be more speculative in nature. In our work we have cast the various skills that experts possess in terms of their ability to help achieve some desired goal. This, in turn, can be re-expressed as their suitability to oversee the processing of some dataset by some method, either by configuring parameters, supplying judgement or even performing the task explicitly. For example, an image interpretation method may require the identification of training examples that in turn necessitate local field knowledge. Such knowledge can also be specified as a context of applicability using the time, space, scale and theme parameters that are also used to describe datasets. As such, a given expert may be able to play a number of roles that are required by the operations described above, with each role described as: $E: \xrightarrow{Operation} (p_1,$ $p_2, \ldots, p_n)$, meaning that expert E can provide the necessary knowledge to perform $Operation$ within the context of $p_1 \ldots, p_n$. Therefore, to continue the example of image interpretation, $p_1 \ldots, p_n$ might represent (for example) the floristic mapping of Western Australia, at a scale of 1:100,000 in the present day.

At the less abstract schematic level, location parameters can also be used to express the need to move people to different locations in order to conduct an analysis, or to bring data and methods distributed throughout cyberspace to the physical location of a person.

Another possibility here, one that we have not yet implemented, is to acknowledge that a person's ability to perform a task can increase as a result of experience. Thus, it should be possible for a system to keep track of how much experience an expert has accrued by working in a specific context (described as $p_1 \ldots, p_n$). (In this case, the expert expression would also require an experience or suitability score as described for constraint management described below in Section 3.3). We could then represent a feedback from the analysis exercise to the user, modifying their experience score.

3.2 *Visualization of knowledge discovery*

The visualization of the knowledge discovery process utilizes a self-organizing graph package (TouchGraph) written in Java. TouchGraph enables users to interactively construct an ontology utilizing concepts (visually represented as shapes) and relationships (represented as links between shapes). Each of the concepts and relationships can have associated descriptions that give more details for each of the entity types (data, methods and people). A sample of the visualization environment is shown below in Figure 5.

Touchgraph supports serialization allowing the development of the information product to be recorded and shared among collaborators. Serialization is the process of storing and converting an object into a form that can be readily reused or transported. For example, an ontology can be serialized and transported over the Internet. At the other end, deserialization is the reconstruction of the object from the input stream. Information products described using this tool are stored as

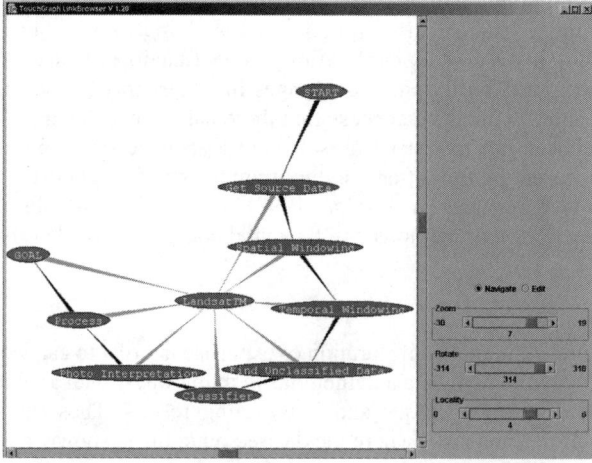

Figure 5.   A sample of the visualization environment.

DAML + OIL objects so that the interrelationships between concepts can be described semantically. The DAML + OIL architecture was chosen as the goal of the DARPA Agent Markup Language component (DAML) is used to capture meanings of terms, thereby providing a Web ontology language. The Ontology Interchange Language (OIL) contains formal semantics and efficient reasoning support, epistemologically rich modelling primitives, and a standard proposal for syntactical exchange notations (http://www.ontoknowledge.org/oil/).

3.3 *Solution synthesis engine*

The automated tool selection process or solution synthesis is more complex, relying on domain ontologies of the methods, data and human experts (resources) that are usable to solve a problem. The task of automated tool selection can be divided into a number of phases. First is the user's specification of the problem, either using a list of ontological keywords (Pundt and Bishr, 2002) or in their own terms, which are mapped to an underlying ontology (Bishr, 1997). Second, ontologies of methods, data and human experts need to be processed to determine which resources overlap with the problem ontology. Third, a description of the user's problem and any associated constraints are parsed into an expert system to define rules that describe the problem. Finally, networks of resources that satisfy the rules need to be selected and displayed.

Defining a complete set of characteristic attributes for real world entities (such as data, methods and human experts) is difficult (Bishr, 1998) due to problems selecting attributes that accurately describe the entity. Bishr's solution of using cognitive semantics to solve this problem, by referring to entities based on their function, is implemented in this framework. Methods utilize data or are utilized by human experts and are subject to conditions regarding their use such as data format, scale or a level of human knowledge. The rules describe the requirements of the methods ('if') and the output(s) of the methods ('then'). Data and human experts, specified by facts, are arguably more passive and the rules of methods are applied to or by them respectively. A set of properties governing how rules may use them are defined for data, (e.g., format, spatial and temporal extents) and human experts (e.g., roles and abilities) using an XML schema and parsed into facts.

The first stage of the solution synthesis is the user specification of the problem using concepts and keywords derived from a problem ontology. The problem ontology, derived from the methods, data and human expert ontologies, consist of concepts describing the intended uses of each of the resources. This limitation was introduced to ensure the framework had access to the necessary entities to solve a user's problem. A more advanced version of the problem specification is proposed, which uses natural language parsing to allow the user to specify a problem. This query would then be mapped to the problem ontology allowing the users to use their own semantics instead of being governed by those of the system.

The second stage of the solution synthesis process parses the rules and facts describing relationships between data, methods and human experts. The JESS rule, *compare* (Table 1), illustrates the interaction between the rule (or method) requirements and the facts (data and human experts). Sections of the rule not essential for illustrating its function have been removed. It is important to note that these rules do not perform the operations described; rather, they mimic the semantic

Table 1. JESS sample code.

```
(1)  defrule compare ;; compare two data sets
(2)  (need-comparison_result $?)
(3)  (datasource_a ?srcA)
(4)  (datasource_b ?srcB)
(5)  intersection_result ← (intersect ?srcA ?srcB)
(6)  union_result ← (union ?srcA ?srcB)
(7)  ⇒ ;; THEN
(8)  (assert (comparison_result (inputA ?srcA) (inputB ?srcB) (intersect ?intersection_result) (union
     ?union_result) ;; 'perform' the operation
```

change that would accompany such an operation. The future work section outlines the goal of running this system in tandem with a codeless programming environment to run the selected toolset automatically.

With all of the resource rules defined, the missing link is the problem to be solved using these rules. The problem ontology is parsed into JESS to create a set of facts. These facts form the 'goal' rule that mirrors the user's problem specification. Each of the facts in the 'if' component of the goal rule are in the form 'need-method_x'. The JESS engine now has the requisite components for tool selection.

Utilizing backward-chaining, JESS searches for rules that satisfy the left-hand side (LHS) of the rule. In the case of dependencies (rules preceded by 'need-') JESS searches for rules that satisfy the 'need-' request and runs them prior to running the rule generating the request. The compare rule (above) runs only when a previous rule requires a comparison_result fact to be asserted in order for that rule to be completed.

The compare rule (Table 1) has dependencies on rules that collect data sources (used for comparisons) and rules that accomplish those comparisons (intersection and union). If each of these rules can be satisfied on the 'if' side of the clause, then the results of the comparison rules are stored, together with the data sources that were used in the comparison and the products of the comparison. The results of the rule 'firing' are stored in a list that will be used to form a minimal spanning tree for graphing.

As the engine runs, each of the rules 'needed' are satisfied using backward chaining, the goal is fulfilled, and a network of resources is constructed. As each rule fires and populates the network a set of criteria is added to a JESS fact describing each user criterion that limits the network. Each criterion is used to create a minimal spanning tree of operations. User criteria are initially based upon the key spatial concepts of identity, location, direction, distance, magnitude, scale, time (Fabrikant and Buttenfield, 2001), availability, operation time and semantic change.

Users specify the initial constraints, via the user interface (Figure 6), prior to the automated selection of tools. As an example, a satellite image is required for an interpretation task, but the only available data is 30 days old and data from the next orbit over the region will not be available for another 8 hours. Is it 'better' to wait for that data to become available or is it more crucial to achieve a solution in a shorter time using potentially out-of-date data? It is possible that the user will request a set of limiting conditions that are too strict to permit a solution. In these cases all possible solutions will be displayed allowing the user to modify their constraints. The user-specified constraints are used to prune the network of resources constructed (i.e., all possible solutions to the problem) to a minimal spanning tree, which is the solution that satisfies all of the user's constraints.

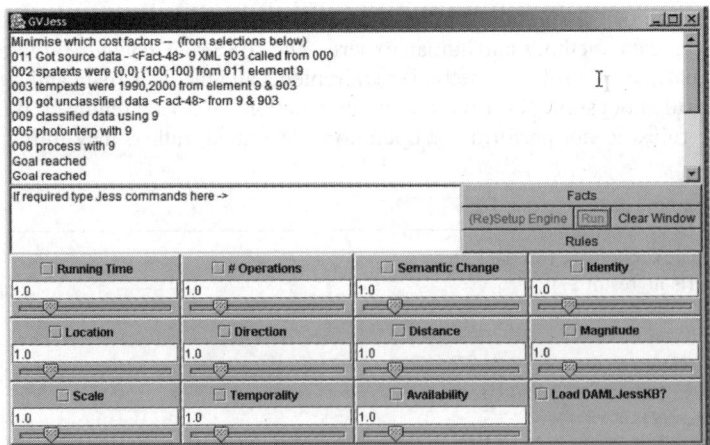

Figure 6.   Interface showing constraint selection.

4 RESULTS

This section presents the results of the framework's solution synthesis and representation of semantic change. The results of the knowledge discovery visualization are implicit in this discussion as that component is used to display the minimal spanning tree.

A sample problem, finding a home location with a sunset view is used to demonstrate the solution synthesis. In order to solve this problem, raster (DEM) and vector (road network) data need to be integrated. A raster overlay, using map algebra, followed by buffer operations is required to find suitable locations from height, slope and aspect data. The raster data of potential sites needs to be converted to a vector layer to enable a buffering operation with vector road data. Finally, a viewshed analysis is performed to determine how much of the landscape is visible from candidate sites.

The problem specification was simplified by hard-coding the user requirements into a set of facts loaded from an XML file. The user's problem specification was reduced to selecting pre-defined problems from a menu.

A user constraint of scale was set to ensure that data used by the methods in the framework was at a consistent scale and that appropriate data layers were selected based on their metadata and format. With the user requirements parsed into JESS and a problem selected, the solution engine selected the methods, data and human experts required to solve the problem. The solution engine constructed a set of all possible combinations and then determined the shortest path by summing the weighted constraints specified by the user. Utilizing the abstract notation from above, with methods specifying change as follows: $M_1: D(p_1, p_2, ..., p_n) \xrightarrow{Operation} D'(p_1', p_2', ..., p_n')$, the user weights were included and summed for all modified data sets: $\sum; D_1'(u_1 p_1', u_2 p_2', ..., u_n p_n'), ... D_n'(u_1 p_1', u_2 p_2', ..., u_n p_n')$ As a result of this process the solution set is pruned until only the optimal solution remains (based on user constraints).

5 FUTURE WORK

The ultimate goal of this project is to integrate the problem-solving environment with the codeless programming environment GEOVISTA Studio (Gahegan *et al.*, 2002) currently under development

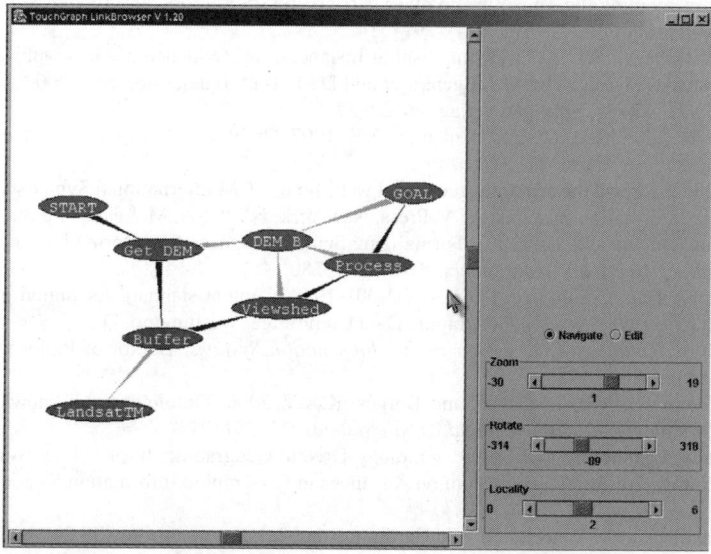

Figure 7. Diagram showing the optimal path derived from a thinned network.

at Pennsylvania State University. The possibility of supplying data to the framework and determining the types of questions that could be answered with it is also an interesting problem. A final goal is the use of natural language parsing of the user's problem specification.

6   CONCLUSIONS

This paper outlined a framework for representing, manipulating and reasoning with geographic semantics. The framework enables the visualizing of knowledge discovery, the automating of tool selection for user-defined geographic problem-solving, and the evaluating of semantic change in knowledge discovery environments. A minimal spanning tree representing the optimal (least-cost) solution was extracted from this graph, and can be displayed in real-time. The semantic change(s) that result from the interaction of data, methods and people contained within the resulting tree represents the formation history of each new information product (such as a map or overlay) and can be stored, indexed and searched as required.

ACKNOWLEDGEMENT

Our thanks go to Sachin Oswal, who helped with the customization of the TouchGraph concept visualization tool used here. This work was partly funded by NSF grants: ITR (BCS)-0219025 and the ITR Geosciences Network (GEON).

REFERENCES

Abel, D.J., Taylor, K., Ackland, R. and Hungerford, S. 1998, An Exploration of GIS Architectures for Internet Environments. *Computers, Environment and Urban Systems*, 22(1): 7–23.

Albrecht, J. 1994. Universal elementary GIS tasks- beyond low-level commands. In: Waugh, T.C. and Healey, R.G. (eds) *Sixth International Symposium on Spatial Data Handling*: 209–222.

Bishr, Y. 1997. *Semantic aspects of interoperable GIS*. PhD Dissertation Thesis, Enschede, The Netherlands, pp. 154.

Bishr, Y. 1998. Overcoming the semantic and other barriers to GIS interoperability. *International Journal of Geographical Information Science*, 12(4): 299–314.

Brodaric, B. and Gahegan, M. 2002. Distinguishing Instances and Evidence of Geographical Concepts for Geospatial Database Design. In: M.J. Egenhofer and D.M. Mark (eds), GIScience 2002. *Lecture Notes in Computing Science 2478*. Springer-Verlag, pp. 22–37.

Chandrasekaran, B., Josephson, J.R. and Benjamins, V.R. 1997. Ontology of Tasks and Methods, AAAI Spring Symposium.

Egenhofer, M. 2002. Toward the semantic geospatial web, Tenth ACM International Symposium on Advances in Geographic Information Systems. ACM Press, New York, NY, USA, McLean, Virginia, USA, pp. 1–4.

Fabrikant, S.I. and Buttenfield, B.P. 2001. Formalizing Semantic Spaces for Information Access. *Annuals of the Association of American Geographers*, 91(2): 263–280.

Federal Geographic Data Committee. FGDC-STD-001-1998. Content standard for digital geospatial metadata (revised June 1998). Federal Geographic Data Committee. Washington, D.C.

Fonseca, F.T., 2001. *Ontology-Driven Geographic Information Systems*. Doctor of Philosophy Thesis, The University of Maine, pp. 131.

Fonseca, F.T. Egenhofer, M.J. Jr. C.A.D. and Borges, K.A.V. 2000. Ontologies and knowledge sharing in urban GIS. *Computers, Environment and Urban Systems*, 24: 251–271.

Fonseca, F.T. and Egenhofer, M.J. 1999. Ontology-Driven Geographic Information Systems. In: C.B. Medeiros (Editor), 7th ACM Symposium on Advances in Geographic Information Systems, Kansas City, MO, p. 7.

Gahegan, M., Takatsuka, M., Wheeler, M. and Hardisty, F. 2002. Introducing GeoVISTA Studio: an integrated suite of visualization and computational methods for exploration and knowledge construction in geography. *Computers, Environment and Urban Systems*, 26: 267–292.

Gahegan, M.N. 1999. Characterizing the semantic content of geographic data, models, and systems. In *Interoperating Geographic Information Systems* (Eds. Goodchild, M.F., Egenhofer, M.J. Fegeas, R. and Kottman, C.A.). Boston: Kluwer Academic Publishers, pp. 71–84.

Gahegan, M.N. 1996. Specifying the transformations within and between geographic data models. *Transactions in GIS*, Vol. 1, No. 2, pp. 137–152.

Guarino, N. 1997a. Semantic Matching: Formal Ontological Distinctions for Information Organization, Extraction, and Integration. In: M.T. Pazienza (Editor), *Information Extraction: A Multidisciplinary Approach to an Emerging Information Technology*. Springer-Verlag, pp. 139–170.

Guarino, N. 1997b. Understanding, building and using ontologies. *International Journal of Human-Computer Studies*, 46: 293–310.

Hakimpour, F. and Timpf, S. 2002. A Step towards GeoData Integration using Formal Ontologies. In: M. Ruiz, M. Gould and J. Ramon (Editors), 5th AGILE Conference on Geographic Information Science. Universitat de les Illes Balears, Palma de Mallorca, Spain, p. 5.

Honda, K. and Mizoguchi, F. 1995. Constraint-based approach for automatic spatial layout planning. 11th conference on Artificial Intelligence for Applications, Los Angeles, CA. p. 38.

Kokla, M. and Kavouras, M. 2002. Theories of Concepts in Resolving Semantic Heterogeneities, 5th AGILE Conference on Geographic Information Science, Palma, Spain, p. 2.

Kuhn, W. 2002. Modelling the Semantics of Geographic Categories through Conceptual Integration. In: M.J. Egenhofer and D.M. Mark (Editors), GIScience 2002. *Lecture Notes in Computer Science*. Springer-Verlag.

MacEachren, A.M. in press. An evolving cognitive-semiotic approach to geographic visualization and knowledge construction. *Information Design Journal*.

Mark, D., Egenhofer, M., Hirtle, S. and Smith, B. 2002. Ontological Foundations for Geographic Information Science. UCGIS Emerging Resource Theme.

Pascoe, R.T. and Penny, J.P. 1995. Constructing interfaces between (and within) geographical information systems. *International Journal of Geographical Information Systems*, 9: p. 275.

Pundt, H. and Bishr, Y. 2002. Domain ontologies for data sharing – an example from environmental monitoring using field GIS. *Computers & Geosciences*, 28: 95–102.

Smith, B. and Mark, D.M. 2001. Geographical categories: an ontological investigation. *International Journal of Geographical Information Science*, 15(7): 591–612.

Sotnykova, A. 2001. *Design and Implementation of Federation of Spatio-Temporal Databases: Methods and Tools*, Centre de Recherche Public – Henri Tudor and Laboratoire de Bases de Donnees Database Laboratory.

Sowa, J.F. 2000. *Knowledge Representation: Logical, Philosophical and Computational Foundations* (USA: Brooks/Cole).

Turner, M. and Fauconnier, G. 1998. Conceptual Integration Networks. *Cognitive Science*, 22(2): 133–187.

Visser, U., Stuckenschmidt, H., Schuster, G. and Vogele, T. 2002. Ontologies for geographic information processing. *Computers & Geosciences*, 28: 103–117.

Advances in Spatial Analysis and Decision Making, Li, Zhou & Kainz (eds)
© 2004 Swets & Zeitlinger, Lisse, ISBN 90 5809 652 1

Time complexity analysis for two non-angle algorithms to determine the radial spatial adjacent relationship

Hua Qi

Department of Surveying Engineering, School of Civil Engineering, The Southwest Jiaotong University, Sichuan Province, P.R. China

Deren Li & Qing Zhu

The State Key Laboratory for Information Engineering in Surveying, Mapping and Remote Sensing, Wuhan University, Hubei Province, P.R. China

ABSTRACT: Angle is an essential parameter in describing and analysing the spatial object correlation. Several important algorithms adopt an azimuth angle or angle as the parameter in spatial analysis. How to select a credible algorithm with high efficiency is of great concern, especially for mass data processing and applications demanding real-time analysis. Taking the arc–arc topological relationship established on the nodes as an example, the time complexity for two non-angle algorithms to determine the radial spatial adjacent relationship is comparatively analysed. The effectiveness of the two algorithms is also discussed when the application range is extended to determine the spatial relationship of point sets. It is shown that, in such a spatial analysis, Qi based on $Qi(x_i, y_i)$ is an algorithm with low time complexity and high credibility. It allows people to keep using the azimuth angle in spatial analysis. At the same time, it provides an efficient approach with the same function as the azimuth angle but with lower time complexity. It can be used to improve on the time complexity for some important related algorithms. The result of this paper has a useful effect on how to select a credible algorithm with high efficiency according to the problems occurring in the related area of spatial analysis software engineering.

1 INTRODUCTION

In describing the real world, angle plays an important role in two aspects. One is its function in precise geometric computing. The other is its function in spatial analysis, where it is used to help determine the relative position relationship of spatial objects. This paper mainly focuses on the second aspect in discussing algorithms and related issues in the application of angle.

Angle is an essential parameter in analysing the correlation between spatial objects. For example, the spatial relationship for point objects is established based on the radial angle. This relationship appears in the Graham algorithm for convex hulls (Graham, 1972). The spatial relationship for point objects is also based on radial sweep. This relationship can be used in the algorithm to construct Triangulated Irregular Networks (TIN) (Mirante and Weignarten, 1982). In the process of establishing a polygon topologic relationship, the azimuth is commonly adopted to build the spatial arc–arc adjacent relationship on nodes (Huang, 1989; Qi, 1996, 1997; Wang, 2001; Gong, 2001). In the computing of the point-and-surface topologic relationship, angles may be used for summation (Guo, 2001). In addition, the azimuth position is a measurement used to describe the positioning relationship of two objects, and the azimuth position is often represented an angle (Guo, 2001). There are plenty of examples of spatial analysis using angle as the parameter.

The first step is to work out the angle or azimuth angle. The following analysis can then be carried out. The angle α_i is obtained by using $\arctan(x)$. $\arctan(x)$ is obtained by approaching the sum of the infinite items with the sum of the finite items after expanding according to the Taylor levels (Mathews and Fink, 2002). The issue of time complexity for high-precision angle computing has aroused great interest among some scholars (Li and Xu, 1991; Qi and Liu, 1996; Qi, 1997; Gao *et al.*, 2002), who have proposed improved algorithms to deal with the issue.

Li and Xu (1991) analysed the factors affecting time efficiency in automatically creating the polygon topologic relationship. They proposed that being the ordering parameter for building the arc–arc adjacent relationship, the precision requirement for an angle is not high. Traditionally, it is not effective to work out α_i by using arctan(x). The Taylor levels formula that uses arctan(x) takes α_i^0, the approximation of the first ten items in place of α_i. The experiment showed that the efficiency of computing could be increased (Li and Xu, 1991). It is a special algorithm that improves the time complexity for angle computing by decreasing the computing precision of the angle according to spatial analysis. As it is still an algorithm that uses angles, the spatial resolution would be reduced when the computing precision of the angle is decreased. Both the utilization range of the algorithm and the time efficiency after improvement are limited (Qi and Liu, 1996).

Qi (1997) took the true north as the starting direction for the Qi length computing. He worked out the length from the starting direction to the point at which the ray and the externally tangent rectangle of the unit circle intersect along the side of the externally tangent rectangle of the unit circle in the clockwise direction. He proposed the Qi algorithm instead of the azimuth angle (or angle) and gave the computing function as $Qi(x_i, y_i)$ (Qi and Liu, 1996). Any ray has only one point at which it interacts with the externally tangent rectangle of the unit circle. $Qi(x_i, y_i)$ possesses monotonicity and continuity in its domain. The Qi algorithm is a precision algorithm. It has been shown that the time complexity is lower than the direct computing. In not increasing the time complexity, it fits the demands of spatial analysis for any level of precision. The Qi algorithm has gained attention and recognition from some scholars (Wang, 1998, 2002).

Gao *et al.* (2002) proposed a new algorithm to determine the arc–arc topological relationship on the same node based on the vector product. This algorithm uses the geometric principle and directly carries out balanced sorting with the vector product of each line segment. The algorithm avoids the computing of parameter value and tedious mathematical computing. Compared with the previous algorithms, this algorithm has obviously optimized the code and improved the efficiency of implementation (Gao *et al.*, 2002). In the following context, this algorithm is called the 'Vector Product Algorithm'.

The Qi algorithm and the 'Vector Product Algorithm' do not compute the angle directly. Taking the arc–arc topological relationship established on nodes as an example, this paper makes a comparative analysis of the two algorithms. In the paper, the time complexity of the two algorithms is studied. Also discussed is the issue of how to extend the range of application of the two algorithms to the spatial relationship for determining point sets, as well as the efficiency of the algorithms. The findings of this paper may be applied to the selection of a reliable algorithm with a high level of efficiency in the area of spatial analysis software engineering.

2 TIME COMPLEXITY OF THE VECTOR PRODUCT ALGORITHM AND THE *QI* ALGORITHM

Take the arc–arc adjacent relationship established on nodes as an example.

2.1 *The algorithm based on the vector product*

Gao *et al.* (2002) proposed an algorithm based on the vector product to determine the spatial adjacent relationship between rays. This algorithm uses the idea of the orientation test between points and directed line segments and the binary sort tree.

The following are the steps for the algorithm. Use vector products to sort the arcs. First, select one arc from the original sequence to function as the base. Put the arc in the anticlockwise direction of the base arc before the arc in the sequence. Put the arc on the clockwise direction of the base arc after the base arc in the sequence. When the boundary condition appears, it will be easy to determine whether the arc and the base arc are on the same line but in a different direction. For the convenience of the test, put the arc before the base arc. Finish the sorting by comparing the vector products. In conducting the test, the process of sorting arcs by vector products coincides with the process of constructing a binary sort tree. In order to illustrate the whole sorting process, take the binary tree as an example for illustration. Construct a binary tree by interpolating the nodes step by step. Finally, traverse the binary sort tree according to the interthem. An arc sequence with adjacent spatial positions can be obtained (Gao *et al.*, 2002).

2.2 *The algorithm based on the* $Qi(x_i, y_i)$ *function*

For all of the arcs on the same node, take the node as the origin and select the points of the arcs adjacent to the node to construct the respective rays. Work out the Qi value of each ray and then sort, using Qi as the sorting keyword. Output a Qi-ordered sequence according to the sorting indexing order. In such a sequence, the arcs corresponding to the adjacent identifications are adjacent in their spatial positions. The arc corresponding to the last identification is adjacent to the arc corresponding to the first identification in the spatial position (Qi and Liu, 1996).

2.3 *Time complexity analysis of the 'Vector Product Algorithm' and the* Qi *algorithm*

2.3.1 *Orientation test between points and directed line segments*
As shown as Figure 1, the product $\vec{ab} \times \vec{ap}$ of vector $\vec{ab}$ and vector $\vec{ap}$ composes the right-hand principle. Formula (1) shows how to judge the spatial relationship between Point p and the Director-Line-Segment ab.

Set $\vec{A} = \vec{ab}$, $\vec{B} = \vec{ap}$, where

$$\vec{A} \times \vec{B} = (y_b - y_a)(x_p - x_a) - (y_p - y_a)(x_b - x_a) \quad \begin{cases} > 0 & \text{P (left of ab)} \\ = 0 & \text{P and ab: collinear} \\ < 0 & \text{P (right of ab)} \end{cases} \quad (1)$$

Vector $\vec{A}$ is defined as the directed line segment, while Vector $\vec{B}$ is defined as the vector constructed by the starting-point a of the directed line segment and a random point p.

2.3.2 *The* $Qi(x_i, y_i)$ *function*
In the right-angle coordinate shown in Figure 2, tangent ABCD is the externally tangent rectangle of the unit circle O, N is the point of intersection of the starting direction X and tangent ABCD, and p_i is the point of intersection of the ray l_i and tangent ABCD. $Qi(x_i, y_i)$, Formula (2), is the function used to compute the length from N to p_i along the side of tangent ABCD in the clockwise direction.

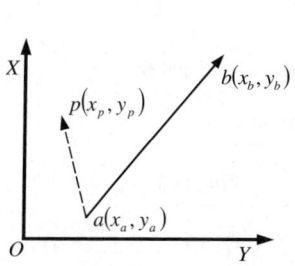

Figure 1.   The product of two vectors.

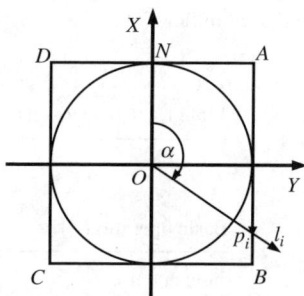

Figure 2.   The vector direction within unit rectangle.

47

$$Qi(x_i, y_i) = \begin{cases} \Delta y_i / \Delta x_i & (\Delta x_i > 0) \wedge (\Delta y_i \geq 0) \wedge (\Delta x_i > \Delta y_i) \\ 2 - \Delta x_i / \Delta y_i & (\Delta x_i > 0) \wedge (\Delta y_i > 0) \wedge (\Delta x_i \leq \Delta y_i) \\ 2 - \Delta x_i / \Delta y_i & (\Delta x_i \leq 0) \wedge (\Delta y_i > 0) \wedge (\Delta y_i > -\Delta x_i) \\ 4 + \Delta y_i / \Delta x_i & (\Delta x_i < 0) \wedge (\Delta y_i > 0) \wedge (\Delta y_i \leq -\Delta x_i) \\ 4 + \Delta y_i / \Delta x_i & (\Delta x_i < 0) \wedge (\Delta y_i \leq 0) \wedge (\Delta x_i \leq \Delta y_i) \\ 6 - \Delta x_i / \Delta y_i & (\Delta x_i \leq 0) \wedge (\Delta y_i < 0) \wedge (\Delta x_i > \Delta y_i) \\ 6 - \Delta x_i / \Delta y_i & (\Delta x_i > 0) \wedge (\Delta y_i < 0) \wedge (\Delta x_i \leq -\Delta y_i) \\ 8 + \Delta y_i / \Delta x_i & (\Delta x_i > 0) \wedge (\Delta y_i < 0) \wedge (\Delta x_i > -\Delta y_i) \end{cases} \qquad (2)$$

where $\Delta x_i = x_i - x_0$, $\Delta y_i = y_i - y_0$, $i = 1, \cdots, n$

In spatial analysis, the starting direction for $Qi(x_i, y_i)$ is the true north. The computing is carried out along the clockwise direction of the side of the externally tangent rectangle of the unit circle. The domain is $(-\infty, \infty)$, while the range of values is $(0,8)$. The $Qi(x_i, y_i)$ function possesses monotonicity and continuity. Figure 3 shows the flow of the algorithm for the $Qi(x_i, y_i)$ function.

2.3.3 Time complexity comparison between $\vec{A} \times \vec{B}$ and $Qi(x_i, y_i)$

Table 1 shows the comparison of the basic operation between $\vec{A} \times \vec{B}$ and $Qi(x_i, y_i)$.

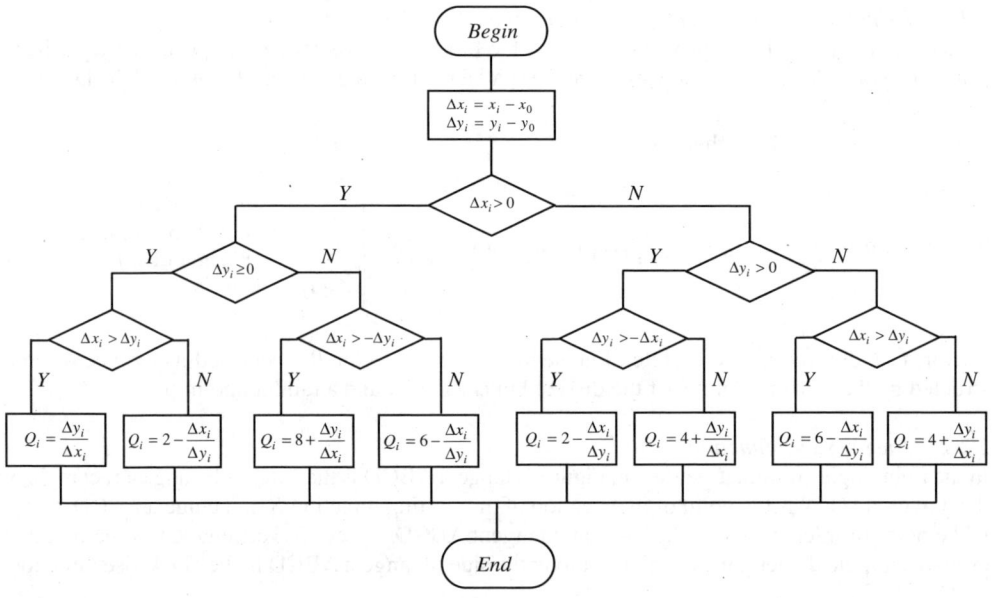

Figure 3.   Algorithm flow.

Table 1.   Comparison of basic operation between $\vec{A} \times \vec{B}$ and $Qi(x_i, y_i)$.

Basic operation	Implementation frequency	
	$\vec{A} \times \vec{B}$	$Qi(x_i, y_i)$
Judgement	0	3
+/−	5	2
×/÷	2	1

Seen from Table 1, when the problem size is n, if 'judgement' is taken as the same level as '+/−', the frequency count of $\vec{A} \times \vec{B}$ to implement '×/÷' is 2n, while the frequency count of $Qi(x_i, y_i)$ to implement '×/÷' is n. Take '×/÷' as the basic operation to measure the difference in time complexity between the two algorithms. The result is that the time complexity of $Qi(x_i, y_i)$ is $O(n)$ less than that of $\vec{A} \times \vec{B}$.

2.3.4 Process analysis for the 'Vector Product Algorithm' and the Qi algorithm to build a binary sort tree

Generally, before interpolating a new node into the binary sort tree, the search should be carried out. Starting from the root node, the searching is carried out from the top down and layer by layer. Compare the fixed value *Key* and the keyword of the root node. If *Key* is smaller, then enter the left son tree to search. If *Key* is bigger, then enter the right son tree to search. Time and again, if a *Key* node exists in the binary sort tree, the interpolation will not be implemented. If not, the new node will be interpolated as one leaf into the suitable position (Xue, 2002).

For example, interpolate the following arc rays sequence (Table 2) into the binary sort tree.

'Vector Product Algorithm'

The first interpolation uses 'a' to establish the root node. The left son and right son point at 'null'. The second interpolation will establish a new node including 'b'. Compare the relationship between Point 'b' and the directed line segment Oa. When $\vec{A} \times \vec{B} > 0$, 'b' will be interpolated as the left son tree of 'a'. The third interpolation will establish a new node that will include 'c'. Interpolate the new node as the right son tree of 'b', because when comparing 'c' with Oa, $\vec{A} \times \vec{B} > 0$; while when comparing 'c' with Ob, $\vec{A} \times \vec{B} < 0$. When all of the points in the sequence are interpolated, a binary sort tree *Tree*1 (shown as Figure 4(a)) is obtained. In the above process, the $\vec{A} \times \vec{B}$ operation is implemented 22 times. The height of *Tree*1 is 5. This means that when interpolating one node the maximal comparison time is 4. Sequence 1 of {f, b, g, h, c, a, d, i, j, e} is obtained by the inorder traversing *Tree*1.

Qi Algorithm

First of all, work out the *Qi* value of each ray. The first interpolation uses 'a' to establish the root node. The left son and right son point at 'null'. The second interpolation will establish a new node including 'b'. Compare the *Qi* value of Ob and that of Oa. When $Qi_b < Qi_a$, b will be interpolated as the left son tree of 'a'. The third interpolation will establish a new node including 'c'. Interpolate the new node as the right son tree of 'b', because $Qi_c < Qi_a$ while $Qi_c > Qi_b$. When all of the

Table 2. Coordinate of the arc rays.

ID	O	a	b	c	d	e	f	g	h	i	j
X	1.703	2.496	0.336	2.077	3.143	1.166	2.357	0.284	0.916	3.025	2.357
Y	2.388	1.153	2.954	0.939	1.940	3.800	3.793	1.792	1.145	3.035	3.800

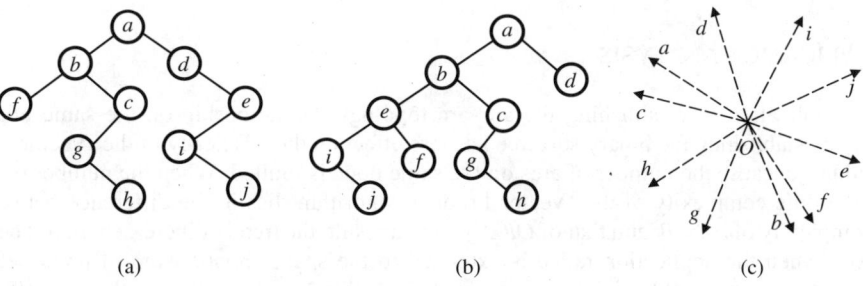

(a)   (b)   (c)

Figure 4. Illustration of the binary sort tree.

points in the sequence are interpolated, a binary sort tree *Tree*2 (shown as Figure 4(b)) is obtained. In the above process, the Qi values of 10 rays are computed. The height of *Tree*2 is 5. This means that when interpolating one node the maximal comparison time is 4. Sequence 2 of {i, j, e, f, b, g, h, c, a, d} is obtained by the inorder traversing *Tree*2.

In Sequences 1 and 2, the spatial positions of the arcs corresponding to the adjacent ray identities are adjacent. The spatial positions of the arcs corresponding to the first and last rays are adjacent, as well. Refer to Figure 4(c) for the real spatial position relationship of the arc rays.

2.3.5 *Time complexity analysis of the algorithms*

(1) In the process of establishing the binary sort tree, the 'Vector Product Algorithm' has no fixed keywords. As a result, except for the root node, the relationship between the points and directed line segments should be judged for every newly interpolated node and original node on the binary sort tree. Then, determine whether the new son node is the left or right son tree of the father node according to the result. The following is the analysis to work out the times to implement $\vec{A} \times \vec{B}$ when the n nodes are interpolated into one binary sort tree on the condition that the problem size is n.

- In a binary sort tree with n nodes (*height = h*), if the *i*th layer has *ni* nodes ($1 \leq i \leq h$), when interpolating one new node, the average comparison time is close to $O(\log_2 n)$ before the interpolation position is found.
- The first interpolation, i.e. the interpolation of the root node, does not need any comparison. The comparison time to interpolate the last node is $h - 1 = \lfloor \log_2 n \rfloor$. At the time, the binary sort tree that should be considered is the full binary sort tree, because the minimal comparison time can be obtained in this condition.
- Generally, the sum of the comparison time is $O(n\log_2 n)$. This is as same as the sum of the comparison time for establishing the binary sort tree.

(2) In the same process, the Qi algorithm uses the Qi value as the keyword for interpolating nodes. As a result, computing the Qi value of each arc for just one time is sufficient. The computing capacity is a constant that has nothing to do with the height of the tree but is only related to the sum of rays. The time to implement is $Qi(x_i, y_i)$ is $O(n)$.

(3) The time complexity of $\vec{A} \times \vec{B}$ and $Qi(x_i, y_i)$ has been analysed. For the convenience of the analysis, the difference in the time complexities of $\vec{A} \times \vec{B}$ and $Qi(x_i, y_i)$ is ignored here. Measure the time complexity of the 'Vector Product Algorithm' and the Qi algorithm based on $\vec{A} \times \vec{B}$. The time complexity to establish the same binary sort tree can be obtained for both algorithms.

- The time complexity for the Qi algorithm is $O(n)$.
- The time complexity for the 'Vector Product Algorithm' is $O(n \log_2 n)$. When *n* keeps constant, the time complexity increases when the height of the tree increases.

For the convenience of illustration, the example of establishing the binary sort tree is used for both algorithms. However, in the real operation, the Qi algorithm may directly use the Qi value as the sorting keyword and adopt any efficient sorting algorithm.

3 APPLICATION ANALYSIS

For the application of establishing the arc–arc topological relationship on the same node, the process of establishing the binary sort tree has less effect on the efficiency of the 'Vector Product Algorithm', because the number of arcs on the same node is limited. When the number of arcs is large, the time complexity of the 'Vector Product Algorithm' lies in the difference between the time complexity of $\vec{A} \times \vec{B}$ and that of $Qi(x_i, y_i)$. As a result, the trend of increase will not be great. However, when the application range is extended to the spatial relationship of point sets on a plane, the process of establishing the binary sort tree will have a big effect on the time efficiency of the 'Vector Product Algorithm'. The difference between the two algorithms will be obvious.

(1) When n is big enough, the Qi algorithm with a slow-rising function is sure to work more quickly.

(2) When the 'Vector Product Algorithm' is used to establish the arc–arc topological relationship on the same node, the node is defined as the point at which three or more arcsintersect. Before this step, the matching of nodes is proven to be efficient; i.e., the node is not in the same direction and collinear with the root node ray. As a result, only the situation of collinearity and hetero-directionality exists. $\vec{A} \times \vec{B} = 0$ is the ray that is of collinear and hetero-direction of the root node ray. In the 'Vector Product Algorithm', interpolating the ray that is of collinear and hetero-direction of the root node can solve the problem of $\vec{A} \times \vec{B} = 0$. However, when extending the algorithm to deal with the spatial relationship of point sets on the plane, the probability that collinearity will occur will increase along with an increase in the size of the problem. There is a possibility that the co-direction or collinearity and hetero-directionality of multiple rays will occur. This being the case, $\vec{A} \times \vec{B} = 0$ does not merely mean that the node is collinear and hetero-directional to the root node ray and it does not merely appear on the root node. The interpolation of multiple rays that are collinear to the root node rays is an issue for the 'Vector Product Algorithm'. Certain estimation and processing methods should be carried out to guarantee the efficiency of the algorithm.

(3) For the Qi algorithm, there is one and only one Qi value in one direction. As a result, the issue of collinearity and hetero-directionality is distinguished naturally. The efficiency of the algorithm can be assured when it is adopted to analyse the arc–arc topological relationship. When the Qi algorithm is applied in dealing with the spatial relationship of point sets on the plane, Qi is used as the keyword of the original ray sequence for *Quick Sort* or *Shell Sort*. The rays with the same Qi value are adjacent in the sorted sequence. Analyse the points in the same direction. Conceptually speaking, compute the distance between the ending point and the starting point of the ray; the spatial adjacent relationship of one point and another in the same direction can be obtained by comparing the size of the distance (Mirante and Weingarten, 1982). However, as the computation for distance is comparatively complicated, such a computation should be avoided in real operations. The author has determined that the method to establish the spatial adjacent relationship of one point and another in the same direction should be as follows:

- When $x = 0$, take $|y|$ as the keyword for the ascending sort;
- When $x \neq 0$, take $|x|$ as the keyword for the ascending sort.

As a result, the relationship of one point and another in the same direction can be obtained.

4   CONCLUSION

The paper mainly discusses the time complexity of two non-angle algorithms that are applied in spatial analysis and determine the relative positional relationship of spatial objects in two dimensions. In real applications, such as determining the spatial relationship of point sets, the time efficiency and validity of the algorithms are analysed on the condition that the problem size is big enough. The following conclusions have been obtained:

(1) When the problem size is n, the difference in the time complexity of the two algorithms is measured based on '$\times/\div$'. The time complexity of $Qi(x_i, y_i)$ is $O(n)$ less than that of the $\vec{A} \times \vec{B}$.

(2) When the problem size is n, the time complexity of 'Vector Product Algorithm' and Qi algorithm is measured to establish the arc–arc spatial adjacent relationship on the same node based on $\vec{A} \times \vec{B}$. The time complexity of the Qi algorithm is $O(n)$, while that of the 'Vector Product Algorithm' is $O(n \log_2 n)$.

(3) For the application of establishing the arc–arc topological relationship on the same node, the process of establishing the binary sort tree has less effect on the efficiency of the 'Vector Product Algorithm', because the number of arcs on the same node is limited. When the number of arc nodes is large, the time complexity of the 'Vector Product Algorithm' mainly depends on the

difference between the time complexity of $\vec{A} \times \vec{B}$ and that of $Qi(x_i, y_i)$. Thus, the trend of increase will not be great. In applications of problems of a bigger size, the time complexity of the 'Vector Product Algorithm' mainly depends on the process of establishing the binary sort tree. In the same process, the Qi algorithm has a relatively slow ascending function. Therefore, it has advantage over the 'Vector Product Algorithm' on time efficiency.

(4) When using the Qi algorithm, attention should be paid to the fact that people retain the habit of using azimuth angles for spatial analysis. At the same time, the Qi algorithm is more efficient in terms of time complexity for computation than the azimuth angle, and they have the similar functions in spatial analysis.

In conclusion, establishing the adjacent relationship of rays is an important method for spatial analysis. In analysing the spatial relationship between rays, the Qi algorithm turns the spatial adjacent relationship between rays into the adjacent relationship between the point at which the rays and the side of the externally tangent rectangle of the unit circle intersect. Then, the computation of the azimuth angle is shifted to the computation of the Qi length. The change plays a positive role in decreasing the time complexity of the algorithm.

ACKNOWLEDGEMENTS

We are very grateful to Jin Guoqing (Department of Aerial Photogrammetry and Remote Sensing, the No. 4 Surveying and Designing Institute, the Ministry of Railway) for providing the experimental data, and to Prof. Zhilin Li for his important comments.

REFERENCES

Kemighan, B.W. and Rob, P. 2000. *Program Design and Practice*. Machine Industry Publishing House, Beijing, 31–33.

Gao, Y., Xu, J. and Tang W. 2002. A new algorithm for arc-arc topological relationship on the same node. *Journal of Computer Application and Study*, (4): 58–59.

Gong, J. 2001. *Fundamental Theories for Geographic Information System*. Publishing House of Science, Beijing, 145–146.

Graham, R.L. 1972. An efficient algorithm for determining the convex hull of a finite planar set. *Information Processing Letters*, 1(4): 132–133.

Guo, R. 2001. *Spatial Analysis (the 2nd edition)*. Higher Education Publishing House, Beijing.

Huang, X. and Tang, Q. 1989. *A General Introduction to Geographic Information System*. Higher Education Publishing House, Beijing, 43–44.

Mathews, J.H. and Kurtis, D.F. 2002. *Numeric Value Methodology (MATLAB version) (the 3rd edition)*. Electronic Industry Publishing House, 140–149.

Li, Q. and Xu, Z. 1991. *Map representation algorithm based on vector processing – An Information System for Forest Resources and Environment Dynamic Monitoring in the Three Northern Areas*. Publishing House of Surveying and Mapping, Beijing, 48–60.

Mirante, A. and Weingarten, N. 1982. The Radial Sweep Algorithm for Constructing Triangulated Irregular Networks. *IEEE CGA*, 2(1): 11–21.

Qi, H. 1997. Optimization and improvement for the algorithm to automatically establish polygon topological relationship. *ACTA GEODAETICA et CARTOGRAPHICA SINICA*, 26(3): 254–260.

Qi, H. and Liu, W. 1996. *Qi* algorithm for arc-arc topological relationship built on nodes. *ACTA GEODAETICA et CARTOGRAPHICA SINICA*, 25(3): 233–235.

Wang, J. 1998. Development of cartography and GIS – 1996–1997 Commission of Cartography and GIS of Chinese Association of Surveying and Mapping. *Bulletin of Surveying and Mapping*, (11): 15–25.

Wang, J. 2001. *Principles for Spatial Information System*. Publishing House of Science, Beijing, 246–248.

Wang, J. 2002. Raster algorithm to construct the polygon topological relationship. *ACTA GEODAETICA et CARTOGRAPHICA SINICA*, 31(3): 249–254.

Xue, C. 2002. *Data Structure (the 2nd edition)*. Publishing House of Hua Zhong Scientific and Technological University, Wuhan.

Advances in Spatial Analysis and Decision Making, Li, Zhou & Kainz (eds)
© 2004 Swets & Zeitlinger, Lisse, ISBN 90 5809 652 1

Creating and analysing a constrained TIN in a geo-DBMS: integration of point heights and parcel boundaries

Jantien Stoter & Ben Gorte

Department of Geodesy, Faculty of Civil Engineering and Geosciences, Delft University of Technology, The Netherlands

ABSTRACT: In a study concerning 3-D cadastre (Stoter et al., 2002) a prototype has been built that provides insight on the vertical dimension of rights registered in the cadastral registration system. This is important when different properties are located above each other (such as tunnels, pipelines and building complexes). In the prototype, parcel boundaries need to be located in 3-D space as they have to be combined with 3-D objects such as tunnels and pipelines. To extend the spatial model of parcel boundaries into 3-D, simply assigning one z-coordinate to each parcel is not sufficient. Providing a z-coordinate to the vertices describing parcel boundaries also does not meet the requirements for a 3-D cadastre. In this paper we elaborate on how parcel boundaries can be integrated with point heights to meet the requirements for a 3-D cadastre; i.e., obtaining a height surface for parcels. A height surface for parcels is the digital terrain model (DTM) on the locations of parcels. A parcel surface changes when the terrain surface changes. We will describe and evaluate several possibilities for obtaining a height surface for parcels based on a DBMS approach.

1   INTRODUCTION

According to FIG (1995) a cadastre is 'a parcel-based, and up-to-date land information system containing records of interest in land (rights, restrictions and responsibilities)'. Cadastral systems maintain rights, limited rights and restrictions as attribute data on parcels that are described geometrically. In those systems parcel boundaries are maintained in 2-D.

In a study concerning 3-D cadastres (Stoter & Ploeger 2002) a prototype was built in which 3-D physical objects were modelled in a DBMS. In the prototype rights established for 3-D physical objects are also defined in 3-D.

An important question was how the z-coordinates of these 3-D objects should be defined: with absolute values (in the national coordinate frame) or relatively, with respect to the surface (e.g., 6 metres below the surface). Absolute z-coordinates are not influenced by surface changes; furthermore, the definition of the surface level (the reference level used for values with respect to the surface) is sometimes not clear. Finally, when using z-coordinates with respect to the surface it is complicated to define the actual geometry of 3-D objects. Therefore, the most sustainable solution is to define 3-D objects with absolute z-coordinates. In some flat urban areas it might be a better option to define z-coordinates with respect to the surface because surface levels do not differ and because DTMs based on laserscan data are complex to obtain in urban areas (see Figure 1).

In the prototype two case studies were carried out in rural areas: 1) two pipelines and 2) a railway tunnel (Stoter & Ploeger 2003). These 3-D objects are known in absolute z-coordinates (in the Netherlands National Ordnance Datum: NAP). Those 3-D descriptions do not reveal where the 3-D

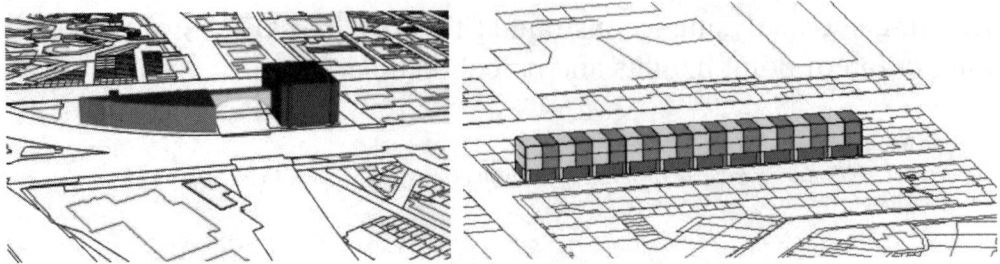

Figure 1. The heights of the buildings in these figures are defined with respect to the surface. Since the surface level in some urban areas is relatively 'flat', this is sometimes sufficient for representing the 3-D situation.

objects are located with respect to the surface and with respect to the parcels on the surface: Are the 3-D objects situated above or under the ground, and what is the depth of the tunnel and the pipelines?

To know the vertical position of 3-D objects, the surface of the parcels is also needed. Obtaining the height surface for parcels by the integration of parcel boundaries and point heights is the topic of this paper. Lenk (2001) has already performed a study on the integration of 2-D GIS data and height information.

Using one z-coordinate for each parcel is not sufficient. Assigning height to the nodes describing parcel boundaries is also not sufficient, because with this only the height on the location of the parcel boundary is known, and not the height within the parcel itself. In this research we look for a solution where the surface covering a parcel is modelled in 2.5-D.

First, we use an example to illustrate why parcel surfaces are needed in the 3-D cadastre (section 2), followed by a description of the datasets used (section 3). Then, the TINs (Triangular Irregular Networks) generated in this research are described: an unconstrained and a constrained TIN, together with the data structures (section 4). The actual integration of height information and parcel boundaries is explained by describing how the surface of a specific parcel can be extracted from the DBMS (section 5). Section 6 describes how the constrained TIN, based on the undivided parcel boundaries and height points, can be improved. In section 7 the experiments on calculating the area of a parcel in 2-D and 3-D space are described. Our conclusions are presented in section 8. Part of this research was presented in (Stoter & Gorte 2003).

2   INTEGRATING HEIGHT DATA AND PARCEL BOUNDARIES

For a 3-D cadastre we are interested in the combination of parcel boundaries and height data in order to obtain a parcel surface that can be combined with 3-D objects (such as tunnels, cables and pipelines). This indicates where the 3-D object is located with respect to the surface level.

To illustrate this, a pipeline is examined as a case study in this study on 3-D cadastres. In this case study, the NAM (Nederlandse Aardolie Maatschappij, a company that owns an important part of the natural gas network in the Netherlands) provided us with 3-D information on a pipeline located in the study area. Figure 2 is a combination of the (original) 2-D parcel boundaries and the 3-D pipeline (defined with absolute z-coordinates). It does not show where the pipeline is positioned with respect to the surface. In this specific case, the 3-D pipeline (which has absolute z-coordinates of between 5 and 10 metres) appears above the surface (the 2-D parcel boundaries are positioned in the $z = 0$ plane).

In Figure 2, on the right the dashed lined shows the projection of the 3-D pipeline on the plane where the z-coordinates equal zero (the plane where the 2-D parcel boundaries are positioned), which shows that the 3-D pipeline is located (5 to 10 metres) above the parcel boundaries. This is not correct, since it is an underground pipeline.

54

Figure 2.   A pipeline defined in 3-D combined with 2-D parcels. The dashed line in the figure on the right is the projection of the pipeline on the plane where the z-coordinate equals zero (which is the plane where the 2-D parcel boundaries are positioned).

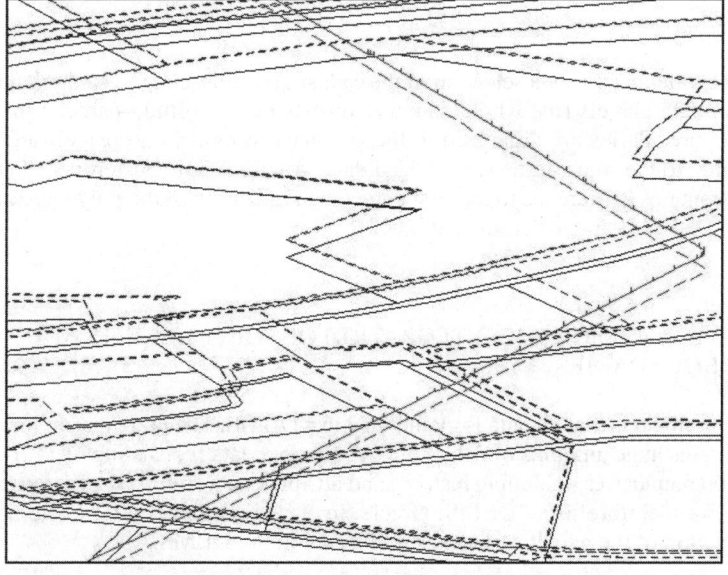

Figure 3.   Parcel boundaries defined in 3-D (dashed lines) give insight on where the pipeline is positioned with respect to the surface. Sticks indicate the distance between the pipeline and the surface. The dashed line under the pipeline is the projection of the pipeline on the $z = 0$ plane.

We described the parcel boundaries in 3-D by assigning z-coordinates to the nodes of the parcel boundaries. A DTM (Digital Terrain Model) represented by an unconstrained TIN of laser altimetry points was used to extract the z-coordinates (see further). The 3-D parcel boundaries are drawn with dashed lines in Figure 3.

55

On the locations of the parcel boundaries it is now possible to determine the depth (or height) of the pipeline. However, within the surface of one parcel this information still is not clear. Therefore, the parcel surface needs to be obtained. When the height surface of a parcel is known, this information can also be used to find the actual area of the parcel (instead of the projected area, which is currently registered in the cadastral system).

3 DESCRIPTION OF DATA SETS AND THEIR DATA STRUCTURES

For the integration of height data and cadastral data, two types of registration are used: the cadastral data maintained by the Netherlands' Kadaster and height data maintained by Rijkswaterstaat (Dutch Ministry of Transport and Public Works). In this research we combine these registrations by using a DBMS approach.

3.1 Terrain model

For the terrain model we use a data set representing the DTM of the Netherlands; i.e., AHN (Actueel Hoogtebestand Nederland) (Van Heerd et al. 2000). The AHN is a data set of points with heights obtained with laser altimetry with a density of one point per 25 square metres (with a validation accuracy of 5 cm). The data set used is a selection of the AHN covering an area of 5 km by 6 km, consisting of 1198029 data points.

3.2 Parcel boundaries

The used parcel boundaries are a selection of the cadastral database of the Netherlands. This selection contains 2,225 parcels (the whole country consists of 6.4 million parcels). In the cadastral DBMS, parcel boundaries are organized in the geometrical model and parcels are topologically stored according to the rules of the wing-edged data structure (Van Oosterom & Lemmen 2001). Every parcel contains a reference to the first edge. The realization of the polygons can also be performed in the DBMS (Van Oosterom et al. 2002).

4 GENERATING TIN FOR THE INTEGRATION OF POINT HEIGHTS AND PARCEL BOUNDARIES

An initial requirement is that all data is maintained in a DBMS. The main reason for this is that the cadastre maintains huge amounts of data, both geometrical data (consisting of 17,642,437 parcels and 45,868,700 boundaries, including history) and attribute data (rights, restrictions and subjects). In this study, we therefore use a DBMS: Oracle Spatial 9i (Oracle, 2001) to store the data sets. Also the extraction of the parcel surfaces is performed in the DBMS.

4.1 Unconstrained TIN

To be able to combine parcel boundaries with the point heights, a TIN was first generated using only the point data, by means of triangulation software called Triangle (Shewchuk 1996). The TIN is generated with Delauny triangulation (Worboys 1995) and consists of 2,393,676 triangles and 1,198,029 nodes. The triangulation was performed outside the DBMS (with Triangle software). The resulting TIN (containing x-, y-, z-coordinates on point data and the TIN defined with reference to those points) was inserted in the DBMS in a topological model. The UML model of the TIN is shown in Figure 4.

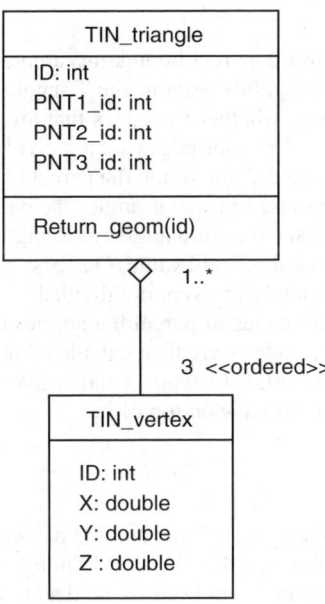

```
            TIN_triangle
      ID: int
      PNT1_id: int
      PNT2_id: int
      PNT3_id: int

      Return_geom(id)
```

Figure 4.   UML class diagramme of TIN storage.

In the topological model two tables are stored: one table ('tin') contains references to the ids of the points for every triangle (three references for every triangle):

```
SQL > describe tin
Name                                          Null?    Type
--------------------------------------------------------------------------
ID                                                     NUMBER (8)
PNT1_ID                                               NUMBER (8)
PNT2_ID                                               NUMBER (8)
PNT3_ID                                               NUMBER (8)
```

In the second table ("tin_vertex") the coordinates of the points are stored together with their ids:

```
SQL > describe tin_vertex
Name                                          Null?    Type
--------------------------------------------------------------------------
ID                                                     NUMBER (8)
X                                                     NUMBER (12,3)
Y                                                     NUMBER (12,3)
Z                                                     NUMBER (12,3)
```

In this way, every point is stored only once. A function has been written to generate ('realize') the geometry of the triangles (3-D polygons) based on the topological tables. The function returns a 3-D polygon of type mdsys.sdo_geometry (the spatial data type of Oracle). The geometry is stored as a view of the topological model by means of the following function:

```
create view tin_geom as select id, return_geom(id) shape from tin_c;
```

An advantage of having the geometries is that geometries are recognized by CAD/GIS software that can make a connection to the DBMS. In this way, it is possible to visualize the data stored in the DBMS as is done in the figures in this paper. The geometries are also needed in computations such as 'area of a polygon' or 'overlays' (where all polygons that overlap with a specific rectangle are selected).

4.2 Constrained TIN

In order to integrate the AHN with the parcel boundaries a constrained TIN was also generated, using the parcel boundaries as constraints. Again, the Triangle software was used (outside the DBMS). In Triangle you can choose whether the edges that are constraints should be divided to make more optimal triangles or not. When the edges are not divided, the original parcel boundaries exist as edges in the TIN. At first, we did not divide the parcel boundaries in order to preserve the original boundaries in the TIN. To make optimal triangles the boundaries used for the constrained TIN can be divided based on criteria; e.g., avoid angles in triangles smaller than a threshold value, avoid areas of triangles larger than a threshold value. Our first experiments were based on a constrained TIN in which the parcel boundaries were not divided.

We assigned z-coordinates to the nodes of parcel boundaries by projecting them in the unconstrained TIN. Height values for the nodes were then calculated by interpolating the z-coordinates.

The constrained TIN contains 1,500,074 triangles and 783,415 nodes, and is again stored as a topological model, with a geometrical view on top of it.

4.3 Optimal data structure

The ideal case is to simply store the point heights and the parcel boundaries in the DBMS and to generate the TIN of the area of interest within the DBMS upon the user's request, without storing the TIN in the DBMS. This is more efficient because no data transfer (and conversion) is needed from DBMS to TIN software and back. This will be a topic for future research.

5 EXTRACTING PARCEL SURFACES FROM THE DBMS

To obtain a parcel surface, all triangles that are covered by one parcel need to be selected. This can be done by either the unconstrained or the constrained TIN (see Figure 5).

In the unconstrained TIN (Figure 5, left) the selection represents an area larger than the parcel itself, since triangles cross parcel boundaries. In the constrained TIN (Figure 5, right) each triangle belongs to exactly one parcel, and therefore the selection of triangles exactly equals the area of a parcel, which offers better results.

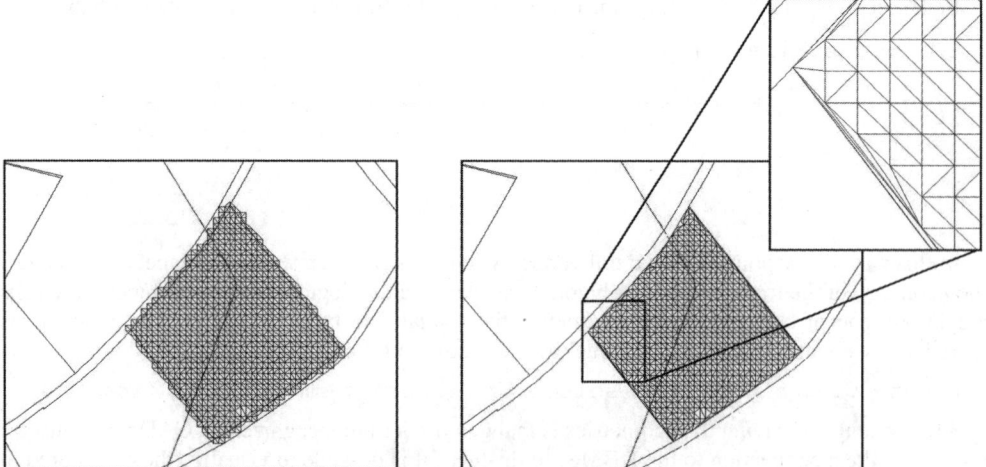

Figure 5.   A parcel surface extracted form the DBMS on an unconstrained TIN (left) and based on a constrained TIN on the right. Note that triangles do not match parcel boundaries in the case of the unconstrained TIN.

To compute the actual area in 3-D space of a parcel, the constrained TIN is also a better option, as can be concluded from Figure 5. The actual area of a parcel in 3-D can be computed by adding up all areas of the triangles covering one parcel (see section 7).

Having the surface, we can now also compute the position of the pipeline with respect to the surface (Table 1). The values of 'surface' were generated by projecting the pipeline in the TIN. The values for 'with respect to the surface' can then be computed by subtracting the 'z' value of the pipeline from the 'surface' value.

5.1 Querying

The query to obtain the triangles covering one parcel first proceeds with the realization of the geometries of triangles. The triangles that are located within a parcel are then selected. The selected realized geometries are the triangles of interest. To simplify the query, the realized polygons of parcels have been stored explicitly in the parcel table; however the geometries of triangles are computed during the query (on the fly). As was concluded before, the constrained TIN offers better possibilities. Therefore, the query to extract a parcel surface from the DBMS is performed on the constrained TIN.

To speed up the process of generating triangle-geometries (polygons) a function-based index was built on the TIN table. A function-based spatial index facilitates queries that use locational information (of type sdo_geometry) returned by a function. The spatial index is created based on the precomputed values returned by the function. The index is built based on the information stored in the metadata table:

```
insert                into            user_sdo_geom_metadata
values('TIN_C', 'stoter.return_geom(id)', mdsys.sdo_ dim_array
  (
  mdsys.sdo_dim_element('X',      0,      254330,     .5),
  mdsys.sdo_dim_element('Y',      0,      503929,     .5)), NULL
  );

create  index  tin_idx  on  tin_c(RETURN_geom(ID))  indextype  is
mdsys.spatial_index;
```

The spatial query to find all points or triangles that are located within one parcel can be performed in two ways (in Oracle Spatial terms): with the spatial operator (sdo_relate) and with the spatial function (sdo_geom.relate). The spatial operator requires and utilizes a spatial index and is therefore faster than the spatial function, which does not use an index.

Table 1.   Pipeline combined with height surface results in z-coordinates at the surface level and z-coordinates with respect to the surface level. Note that the first part of the pipeline is located above the surface.

x	y	z	surface	wrt_surface
242847.22	512941.22	10.34	9.17208	1.16792
242847.21	512941.25	10.38	9.172072	1.207928
242847.17	512941.25	10.38	9.171944	1.208056
242845.24	512942.82	10.38	9.165616	1.214384
242844.80	512943.11	10.23	9.163164	1.066836
242843.01	512944.55	8.89	9.165188	−0.275188
242841.22	512945.95	7.58	9.170208	−1.590208
242840.92	512946.21	7.42	9.170436	−1.750436
242840.47	512946.54	7.33	9.17049	−1.84049
242839.11	512947.64	7.32	9.170586	−1.850586
242835.29	512950.64	7.23	9.197065	−1.967065

Unfortunately, the current version of Oracle operators only works if the dimensions of the operands are equal to each other. It is not possible to overlay a 3-D spatial layer with a 2-D spatial layer, as in our case in which an overlay is needed between 3-D triangles and 2-D parcels. However, to illustrate the difference between the spatial operator and the spatial function, we stored the triangles of the TIN (constrained TIN) in 2-D (without the z-coordinate) and queried which 2-D triangles are inside one particular parcel (number 461, municipality: GBG00, section L):

```
/* query to obtain triangles covering one parcel

/* operator
select id, return_geom (id) shape from tin_c, parcels par
where
parcel = '461' and
municip = 'GBG00' and
section = 'L' and
sdo_relate (return_geom (id), par.geom, 'mask = COVEREDBY +
INSIDE, querytype = JOIN') = 'TRUE';
/* function
select id, return_geom (id) shape from tin_c, parcels par
where
parcel = '461' and
municip = 'GBG00' and
section = 'L' and
sdo_geom.relate (par.geom,
'COVEREDBY + INSIDE', return_geom (id),1) = 'TRUE';
```

In both cases the result is 1,680 triangles. In the operator case, the query took 2 minutes; in the function case, the query took 52 minutes.

In the constrained TIN, the maximum number of triangles in a parcel is 31,193, the minimum number is 15 and the average number is 817. For the unconstrained TIN these numbers are: 15,121, 1 and 348.

The sdo_relate operator and the sdo_geom.relate function are the implementation of the nine-intersection model of Egenhofer (Egenhofer 1992) in Oracle for finding binary topological relations between points, lines and polygons. Each spatial object has an interior, a boundary and an exterior. The boundary consists of points or lines that separate the interior from the exterior. The boundary of a line consists of its endpoints. The boundary of a polygon is the line that describes its perimeter. The interior consists of points that are in the object but not on its boundary, and the exterior consists of points that are not in the object. Some of the topological relationships of the nine-intersection model have names associated with them. The following names are used in the examples above:

– INSIDE: returns INSIDE if the first object is entirely within the second object and the object boundaries do not touch; otherwise, returns FALSE;
– COVEREDBY: returns COVEREDBY if the first object is entirely within the second object and the object boundaries touch at one or more points; otherwise, returns FALSE.

For the unconstrained TIN we used the option 'ANYINTERACT', since otherwise we will miss the triangles that cross parcel boundaries. The parameter 'ANYINTERACT' returns TRUE if the two geometries are not disjointed. Two objects are labelled 'DISJOINT' when the objects have no common boundary or interior points.

6 IMPROVING THE CONSTRAINED TIN

As was seen in the previous section, the constrained TIN offers the best characteristics for the aim of the 3-D cadastre. However, keeping the edges in the constrained TIN undivided in the triangulation process leads to elongated triangles when parcel boundaries are much longer than the average distance between DTM points (5 metres), which is usually the case. Moreover, such a parcel boundary will remain a straight line in 3-D as well even when the terrain is hilly, because there are no intermediate points on the parcel boundaries.

In principal, there are three methods to improve the shape of triangles:

(1) add additional points to the parcel boundaries by densifying the vertices describing parcel boundaries before triangulation
(2) by setting a minimum angle for triangles during the triangulation process
(3) by setting a maximum area for triangles during the triangulation process

The Triangle software offers the last two possibilities for improving the quality of TINs. We carried out experiments with the data set to evaluate the two options. First, we generated a constrained TIN by setting the minimum angle for triangles. When using the minimum angle option, Triangle uses Jim Ruppert's Delauny refinement algorithm (Ruppert 1993). The algorithm adds points to the mesh to ensure that no angles smaller than the threshold angle occur. We chose a threshold of 10 degrees. The number of triangles created during this process was 2,050,232, as compared to 1,500,074 in the first constrained TIN.

In the Triangle software it is also possible to impose a maximum triangle area. No triangle will be generated larger than the maximum triangle area. The density of the height points is one point per 25 square metres. Since the number of triangles in a TIN is usually approximately twice as large as the number of points, this means that triangles have an average area of 12.5 square metres. We therefore decided to set 25 square metres as the maximum area. The number of triangles created during this process was 1,954,939.

Figure 6 shows that the constrained TIN generated with the maximum area criterion considerably improves the shape of the triangles. Moreover, points are also added on the parcel boundaries, which makes it possible to represent more variation in height across a parcel boundary.

The disadvantage of both methods is that data points are not only inserted on the edges of the parcel boundaries, but also in the mesh itself. In this process, the density of the height points

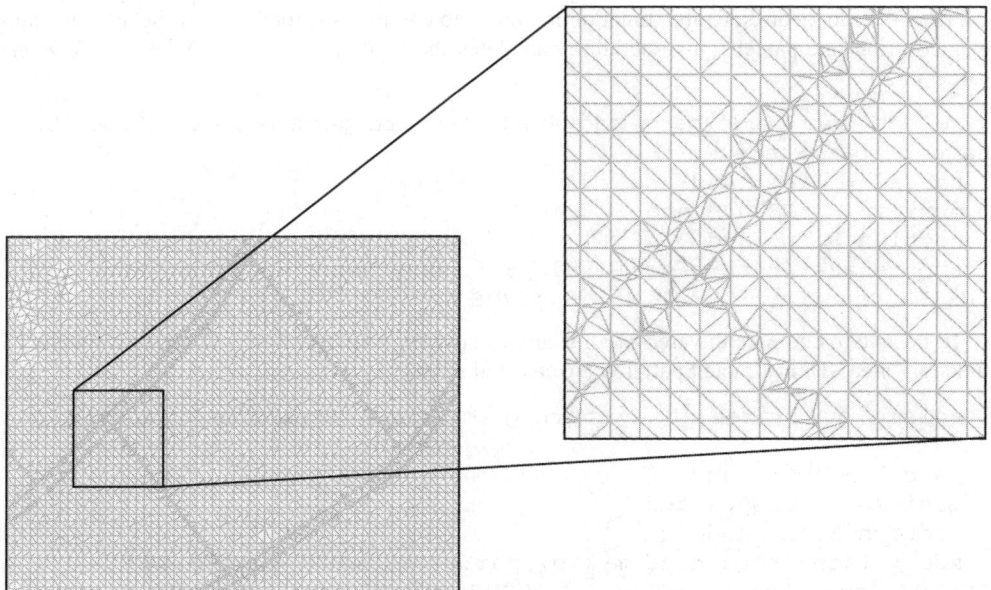

Figure 6.   Constrained TIN, generated with the criterion that the area of triangles should not be larger than 25 square metres.

becomes higher, leading to a considerable increase in redundancy. Height points are added but these height points do not contain additional information, since the heights of these added points are calculated during the triangulation process. Therefore, a better option is to add additional points only on the parcel boundaries (option 1). Future experiments will first focus on densifying the vertices on the parcel boundaries and using these edges as constraints in the triangulation process.

7   CALCULATING THE AREA OF PARCELS IN 3-D SPACE

The area of a parcel in 3-D space can be computed by summing up all of the triangles covering one parcel. All of the calculations described in this section were performed inside the DBMS. We performed the calculation of area on three TINs that were constructed in this research:

(1) unconstrained TIN
(2) constrained TIN without additional points
(3) constrained TIN with additional points based on the maximum area (25 square metres) of triangles

First, we calculated the 2-D area of the original parcel polygon:

```
select sdo_geo.sdo_area (geom, 1) from parcels where
parcel = '461' and
municip = 'GBG00' and
section = 'L';
```

Oracle Spatial 9i, as most mainstream DBMSs, does not have a function to calculate the area in 3-D. Some exceptions are PostGIS (PostGIS 2003) and the Spatialware Datablade of MapInfo (based on Informix (Mapinfo 2003; Informix 2003)) that do support calculations of geometry such as length and perimeter in 3-D. The other functions (overlap, area and distance) are also performed only in 2-D.

The area function in Oracle Spatial 9i projects the triangles defined in 3-D on the 2-D surface. We used the sdo_geom.sdo_area function in Oracle to compute the total sum of the area of triangles covering one parcel. This operation calculates the total area of all triangles covering one parcel in 2-D.

```
select sum (sdo_geom.sdo_area (return_geomID), 1) from tin,
parcels par where
  parcel = '461' and
  municip = 'GBG00' and
  section = 'L' and
  sdo_relate (return_geom (id),par.geom,'mask = COVEREDBY +
INSIDE, querytype = JOIN') = 'TRUE';
```

To be able to compute the area of all triangles covering one parcel in 3-D, we implemented a function 'area3-D' in Oracle Spatial 9i (Arens et al., 2003):

```
select sum (area3-D (return_geom (ID)) from tin, parcels par
where
  parcel 5 '461' and
  municip 5 'GBG00' and
  section 5 'L' and
  sdo_relate (return_geom (id),par.geom,'mask = COVEREDBY +
INSIDE, querytype = JOIN') = 'TRUE';
```

The results of the queries are listed in Table 2. Since we also generated a 3-D definition of the parcel boundary, as was previously described, we also computed the area of this polygon in 3-D.

Table 2. Results of the calculation of the area of one parcel in 2-D and 3-D space based on constrained and unconstrained TINs.

	Parcel polygon	Unconstrained TIN	Constrained TIN without additional points	Constrained TIN with additional points
Area 2-D	20886.47	22675	20830.95	20865.17
Area 3-D	21413.82	22694.19	20935.56	20879.67
# triangles		1810	1680	1939

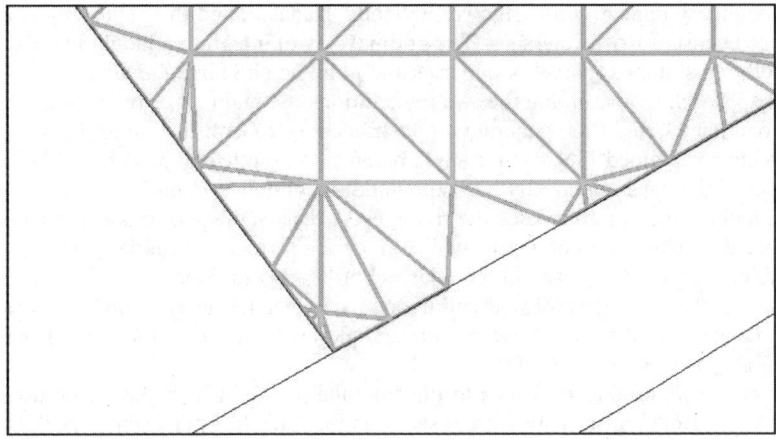

Figure 7. Zoom in on lower-left corner of parcel of interest. Triangles are selected from constrained TIN with additional points. Note that one triangle in this part is not selected (indicated by the circle) due to a too small precision number used in the overlay calculation.

7.1 *Critical remarks*

The area in 2-D based on the constrained TINs and based on the parcel polygon should be the same, since the triangles of a parcel in the constrained TINs cover the parcel exactly. However, there is 56 square metres of difference between the parcel polygon and the constrained TIN without additional points (0.27%) and 21 square metres between the parcel polygon and the constrained TIN with additional points (0.1%). This difference is due to the fact that the edges of the parcel have been slightly moved during all of the conversions that were needed. This once again emphasizes the need to compute triangles within the DBMS, to avoid conversions, which are prone to errors.

The operator "sdo_relate" uses the tolerance that has been inserted in the metadata tables of Oracle. The difference in the position of the parcel boundaries themselves and the triangles at the parcel boundaries forces the choice of a tolerance that is large enough to take care of this difference. In the case of the constrained TIN without additional points, a precision of 0.5 (units are in metres) proved to be able to select exactly all of triangles that are inside the parcel of interest (see Figure 5). Since the parcel boundaries were collected with much higher precision (1 centimetre), future experiments (when the triangulation process is incorporated within the DBMS and no conversion is needed) should strive to tolerances of 0.01. When a tolerance of 0.01 is used in this research, not all triangles of interest are selected.

An additional problem was that in the case of the constrained TIN with additional points Oracle was not able to perform the overlay between the triangles and parcel boundaries with a precision smaller than 0.1. With lower precision (a higher number) the operator that calculates the overlay crashed. This is a bug in Oracle, and shows that improvements to spatial DBMSs are still necessary.

The precision of 0.1 resulted in a selection of triangles that did not cover the whole original parcel, since three triangles were not selected (see Figure 7). However, within the current configuration we were not able to obtain more precise results. The fact that the constrained TIN with additional points has a larger area in 2-D than the constrained TIN without additional points while it covers a smaller area (because three triangles are not selected) is caused by the deterioration in the quality of the data during the calculation process, as was explained above.

8   CONCLUSION

In order to position 3-D objects (tunnels and pipelines) defined with absolute z-coordinates with respect to the surface, the surface of parcels is needed. In this research two registrations were used: the parcel boundary registration by the Netherlands' Kadaster and the height registration by the Meetkundige Dienst of Rijkswaterstaat. The optimal way of integrating height in the cadastre is to maintain both registrations (parcel boundaries and point heights) in a DBMS.

The best approach to combining the two registrations to obtain all of the height points that are located in one parcel, and then generate a constrained TIN (with the parcel boundaries as constraints). In the constrained TIN, each triangle belongs to exactly one parcel.

We improved the constrained TIN by adding additional points in the mesh based on the minimum angle and the maximum area of the triangles. A disadvantage of this is the increase in (the already dense distributed) height points, while no extra information is added. Future research will focus on adding vertices to the parcel boundaries before triangulation.

The 3-D area of all triangles located within one parcel can be summed up to get the area of the parcel in 3-D space. In the DBMS we performed queries to calculate the area of parcels in 3-D space and 2-D space, based on the TINs.

In this research we used a study area to illustrate the possibilities of the integration based on a DBMS approach. Both registrations were inserted in the DBMS. In this study, the TIN was generated outside the DBMS and stored in the DBMS in a topological model with a geometrical view on top of it. The integration of the height data and the parcel boundaries makes it possible to position 3-D objects with respect to the surface (what is the depth/height of a 3-D object at this location?). In the future, the computation of TIN should be performed inside the DBMS to avoid time-consuming conversions that lead to a decrease in the quality of the results, as was shown in this research.

One of the disadvantages of using a dense laser altimetry data set is the resulting volume of data and, with that, the poor performance of the queries. It is therefore relevant to examine how the number of TIN nodes can be reduced by removing nodes that are not significant for the TIN, taking the constraints of the parcel boundaries into account. In this process, nodes are removed based on parameters such as the maximum angle between two points and the difference between the height of the original point and the reduced TIN (does the point contributes significantly to the TIN?). In our department a study has been carried out on the reduction of iterative data in an unconstrained TIN by detecting the characteristic point heights (Penninga 2002). The reduction of data in an constrained TIN, while maintaining the quality, will considerably improve the efficiency of the TIN and will therefore be a topic of future research, which will be based on a DBMS approach.

ACKNOWLEDGEMENTS

We would like to thank the NAM and the Meetkundige Dienst of Rijkswaterstaat for the use of their data. We would also like to thank the Netherlands' Kadaster for the use of their data and for their support. Finally, we would like to thank Peter van Oosterom for his comments on a previous version of this paper.

REFERENCES

Arens, C.A., J.E. Stoter and P.J.M. van Oosterom. 2003. Modelling 3D spatial objects in a GeoDBMS using a 3D primitive, Agile 2003.

Egenhofer, M.J. and J.R. Herring. 1992. Categorising topological relations between regions, lines and point in geographic databases, *The 9-intersections: formalism and its use for natural language spatial predicates*, Technical report 94-I, NCGIA, University of California.

FIG. 1995. Fédération Internationale des Géomètres: *The FIG Statement on the Cadastre*, Publication No. 11, 1995.

Heerd, van, R.M., E.A.C. Kuijlaars, M.P. Teeuw and R.J. van't Zand. 2000. *Product specificatie AHN (in Dutch)*, March 15th 2000, Meetkundige Dienst, Rijskwaterstaat, MDTGM 2000.13.

Informix, *www.informix.com*, 2003.

Lenk, U. 2001. Strategies for integrating height information and 2D GIS data, presented at the Joint OEEPE/ISPRS Workshop, *From 2D to 3D, establishment and maintenance of national core spatial databases*. Oct. 8–10, 2001, Hannover, Germany.

MapInfo, url: http://www.mapinfo.com, 2003.

Oracle. 2001. *Oracle Spatial User's Guide and Reference*, Release 9.0.1 Part Number A88805-01, June 2001.

Oosterom, P.J.M. van, J.E. Stoter, S. Zlatanova and C.W. Quak. 2002. The balance between Geometry and Topology, *Spatial Data Handling 2002 Symposium*, 9–12 July, 2002, Ottowa, Canada.

Oosterom, P.J.M. van and C.H.J. Lemmen. 2001. *Spatial data management on a very large cadastral database*, Computers, Environments and Urban Systems (CEUS), volume 25, number 4–5: 509–528.

Penninga, F. 2002. *Detectie van kenmerkende hoogtepunten in TIN's voor iteratieve datareductie (in Dutch)*, Geo-informatiedag Nederland 2002, February 2002, Ede, the Netherlands.

PostGIS, http://postgis.refractions.net, 2003.

Ruppert, J. 1993. A Delaunay refinement algorithm for 2-dimensional mesh generation, *Proceedings 4th Symposium Discrete Algorithms*, January, 1993.

Stoter, J.E., M.A. Salzmann, P.J.M. van Oosterom and P. van der Molen. 2002. *Towards a 3D Cadastre*, FIG XXII/ACSM-ASPRS, 19–26 April, 2002, Washington, USA.

Stoter, J.E. and H.D. Ploeger. 2002. Multiple use of space: current practice of registration and development of a 3D cadastre, *UDMS 2002*, October 1–4, 2002, Prague, Czech Republic.

Stoter, J.E. and H.D. Ploeger. 2003. Registration of 3D objects crossing parcel boundaries, *FIG Working Week 2003*, April 2003, Paris.

Stoter, J.E. and B. Gorte. 2003. Height in the cadastre: integrating point heights and parcel boundaries, *FIG Working Week 2003*, April 2003, Paris, France.

Stoter, J.E. and P.J.M. van Oosterom. 2002. Incorporating 3D geo-objects into a 2D geo-DBMS, *FIG, ACSM/ASPRS*, April 19–26, 2002, Washington D.C. USA.

Shewchuk, J.R. 1996. Triangle: Engineering a 2D Quality Mesh Generator and Delaunay Triangulator, *First Workshop on Applied Computational Geometry*, May, 1996, 124–133, Philadelphia, Pennsylvania, USA, ACM.

Worboys, M.F. 1995. *GIS, a computing perspective*, Taylor and Francis, London, ISBN 0-7484-0065-6.

GIS Web services in spatial analysis and decision-making

Winnie S.M. Tang & Jan R. Selwood
ESRI China (Hong Kong) Ltd., Hong Kong

ABSTRACT: GIS Web services will be fundamental to the way in which spatial analysis and decision-making is conducted in the future. This paper briefly outlines the basis of Web service architecture and introduces some of the key protocols around which important standards are beginning to form. It then illustrates the nature of the role that Web services will increasingly play in spatial analysis and decision-making by examining the application of this technology in a number of international examples.

1 INTRODUCTION

Geographical Information System (GIS) Web Services are revolutionizing the way in which spatial datasets are presented, analysed and used. Since the introduction of the first Web-mapping application, Xerox® Palo Alto Research Centre's Map Viewer launched ten years ago in June 1993, Internet and Intranet computing architectures have evolved rapidly and in ways of great significance for spatial decision-making. Integration has always been one of the fundamental issues retarding the widespread adoption of computerized spatial analytical techniques. Spatial analysis can be found in almost every element of human endeavour. However, to date, GIS and spatial analytical tools have been treated very much as isolated, specialist applications rarely integrated directly with mainstream decision-making solutions.

In the past, this has been due in part to a divergence in the approach adopted for the electronic storage and processing of graphical and textual data. Over the last five years the widespread adoption of simple feature topology for modelling spatial features and the ability to store such models in relational or object relational databases has greatly improved integration. Increased conformance with industry standards and programming languages, and a willingness to publish Application Protocol Interfaces (APIs) and data format specifications have also been important. However, although such developments have overcome many of the technical obstacles to integration, the cost and complexity of managing spatial datasets and analytical applications remain significant issues that frequently deter potential users. This is significant. Ensuring that the spatial analytical tools are actually used must be of equal importance to the remit of this Working Group session as the increased refinement of the techniques themselves. GIS Web services offer a new paradigm for the purchase and use of spatial data and analytical routines. It will affect how small, medium and large enterprises from government departments down to local environmental groups acquire and build decision-support systems and how GIS and spatial analysis evolves in the future.

2 AN OVERVIEW OF THE WEB SERVICES ARCHITECTURE

A Web Service is simply a software component that can be accessed across the World Wide Web (WWW) for use in other applications. Web services are therefore another form of distributed computing architecture, of which, of course, there are plenty (including Common Object Request

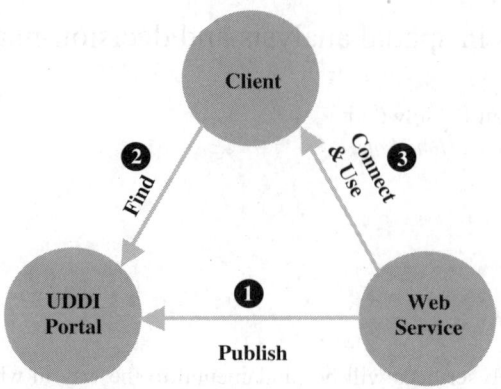

Figure 1.   Web services relationship – publish, find, bind.

Broker (COBRA), Distributed Component Object Model (DCOM) and Electronic Data Interchange (EDI)). All of these architectures have the same basic aim – to enable programs in one environment to communicate and share data and functionality with programs in another. Earlier approaches presented some technically elegant solutions; however, their impact was limited either because they were supported by a relatively small section of the IT industry and failed to garner widespread acceptance, or because they were complex and required a level of tight integration that was impractical for linking applications other than in large internal corporate Intranets. The Web services architecture differs from these earlier approaches by adopting the ubiquitous World Wide Web (WWW) as the common network backbone and focusing on retaining the independence of each application component rather than on building tight linkages between them. As a result, it has won wide and rapid support.

Fundamental to Web services are the concepts of Publish, Find and Bind that can be used to describe the basic relationship between the service provider and the client.

A service provider can *Publish* a service that it wishes to release in proscribed format to an Internet portal. Publishing the service identifies its existence to potentially entirely unknown and, at the time of publishing, undefined clients and provides details of how a client can communicate with the particular service.

Clients can search and *Find* published services in a similar way as popular Internet search engines are used to find Web pages. At present searches are more often than not undertaken manually; however, the structure permits them to be undertaken automatically, so an application realizing that it needs a particular function or dataset could search for and select remotely published services that meet its requirements.

Published information for the selected service provides the client with all of the necessary details to *Bind* to, or connect and use the service, including where and how to invoke it and definitions of parameters.

Critical in this architecture is the concentration on the messaging between services as opposed to the application providing or consuming the service. Regardless of the kind of program or software executing the service – an SQL query retrieving statistical data, a GIS application presenting a map or running a model – as long as the description of how to access it, input and output data formats all conform to a published standard, the application will be able to communicate with others. As a result of using Web services architecture, existing, legacy applications can be relatively efficiently exposed as Web services without the need for massive reprogramming.

3   WEB SERVICES TOOLS

This seamless interchange is only feasible by relying on a number of developing Web protocols including, eXtensible Markup Language (XML), Simple Object Access Protocol (SOAP),

```
Web Service:
                         LocatePlace (Placename : String):Place xy

SOAP wrapped invocation:
                         <?xml version='1.0' encoding='UTF-8'?>
                     <soap:Envelope          xmlns:xsi='http://www.w3.org/2001/XMLSchema-
                 instance'               xmlns:xsd='http://www.w3.org/2001/XMLSchema'
                 xmlns:soap='http://schemas.xmlsoap.org/soap/envelope/'
                 xmlns:soapenc='http://schemas.xmlsoap.org/soap/encoding/'
                 soap:encodingStyle='http://schemas.xmlsoap.org/soap/encoding/'>
                         <soap:Body>
                         <ns0:LocatePlace                          xmlns:ns0='http://esrichina-
                 hk.com/LocatePlaceSample'>
                             <Placename xsi:type='xsd:string'>Hong kong</Placename>
                         </ns0:LocatePlace>
                         </soap:Body>
                     </soap:Envelope>
```

Figure 2.   Typical SOAP wrapping invocation.

Universal Description Discovery and Integration (UDDI) and Web Services Description Language (WSDL).

The World Wide Web Consortium's (W3C) (www.w3.org) XML protocol has been adopted as the de-facto standard for describing data transferred in Web service applications. Now widely established throughout Web computing, it owes its success to its flexibility – defining a syntax with which data descriptions can be defined rather than attempting to describe all forms of data itself. In this way, XML has been able to be adopted by a variety of different vertical markets each agreeing their XML-compliant definitions or schema. Thus, for spatial features, the Open GIS Consortium (OGC) (www.opengis.org) has led the development of the Geographic Markup Language (GML), an XML schema designed to provide a cross-platform descriptions for spatial data. Since its launch in March 2001, this effort has been gaining wide support within the GIS community. The standard is still evolving, and work within the OGC by leading GIS and IT vendors such as ESRI, Intergraph, Oracle, Sun Microsystems and key users, is continuing, with the goal of enhancing this standard to allow for complex commands and very large datasets. OGC is working on similar Web map server specifications.

XML also forms the basis of SOAP and WSDL. SOAP is designed as a standard envelope for delivering method invocations – basically a means of wrapping an XML document so that the recipient knows what to do with it on receipt. SOAP enables an XML statement to be sent over HTTP to a Web services and provides a clear mapping between parameters and function calls, as illustrated in Figure 2 below, for a typical place locator Web service. Invoking the LocatePlace Web Service, the user would need to supply a text string with a Place name, and would receive the coordinate of the selected place.

WSDL is another W3C standard that defines a template to be used for describing a service. This tells the client what the service offers and, in detail, how to create and interpret both request and response. It defines the methods available, what their parameters are, parameter types and the nature of the output generated. WSDL is used by service providers to publish information on their services. UDDI represents a standard Web-based directory of services – in effect a Yellow Pages of Web services. Although it is not necessarily a requirement to publish WSDL documents to UDDI, doing so avoids the need to hardcode service locations and parameter details in to client applications, giving them greater flexibility in the event that a particular service is temporarily out of action.

While it is perfectly feasible to develop and host a Web service using only these components, Web services toolkits or frameworks, such as Microsoft .Net and The Mind Electric's GLUE, are often used as a means of simplifying the process. These can automatically read WSDL documents and create client code for a particular service, as well as handle standard administration services such as security and privacy and convert communications between the host and client into and out of SOAP.

4   BENEFITS AND IMPLICATIONS

The Web services architecture brings a number of general benefits as applicable to spatial analysis as they are to the IT industry. These include enabling service providers (whether these be data providers, telecoms companies, government departments or analytical research labs) to more easily host and manage a very diverse range of services that can be highly targeted to meet the needs of a particular group of clients. In addition, accessible across the Web, services can be hosted and consumed virtually anywhere, changing the dynamics of global service provision. For developers, they increase ease and speed with which complex technical applications can be delivered. Application development and testing durations are reduced as they are established by integrating existing services that are already operational and validated. Enabling access to and use of niche data and services hosted by third party specialist organizations eliminates the need to acquire and maintain these specialist skills locally. For clients, this translates to greater flexibility and choice – the range of services offered and their payment structure, and the way in which technology solutions can be purchased.

For spatial analysis, these bring a number of particularly important implications in the form of Web-based spatial data networks and Web-based analytical tools.

Web-based Data Networks promote data sharing between organizations and permit multiple organizations to locate and utilize a single copy of a dataset without the need to duplicate it. Effective management and hosting of spatial data is complex, and demands different skills from the management of other datasets held in most companies. Until Web-based data networks started appearing, if a company used spatial analysis within its decision-making process, it by default needed to manage its own spatial data – often, identical datasets needed to be individually maintained in multiple organizations. The same city plan, for example, is maintained separately by national, state and city government departments; by individual rescue services and by a host of private companies working in the area. This duplication is inefficient and increases the overall costs of spatial analysis. The data network is designed around the Web services concept of Publish, Discover and Use. Specialist data service providers often maintain their own data without needing to modify or change existing systems. The provider submits a basic description of the service offered that can be posted to a central registry. Documentation includes metadata (such as the Federal Geographic Data Committee (FGDC) or ISO standards), as well as a service description increasingly provided using the UDDI and WSDL protocols. The result is an open, multi-participant, community-based network that encourages data sharing and communication.

The implications are as follows:

- Standard datasets can be accessed directly from the supplier – clients do not need to maintain separate copies;
- More flexible data supply contracts (e.g., based on the intensity of usage and/or areal coverage);
- Economies of scale, permitting investment in data updates, maintenance and the provision of server security and backup facilities well beyond the means of most client organizations;
- Competition, ensuring quality is maintained and costs minimized.

Spatial data networks were pioneered by ESRI with the launch of the Geography Network (www.geographynetwork.com) in June 2000. The Geography Network was established as a registry and as a discovery engine for spatial data and application services offered by a large number of public, private and commercial organizations. Highly scalable, the Geography Network model has been duplicated around the world (e.g., in New Zealand, the UK, Holland and Hong Kong), with each new node linking into the wider network framework. It has also served as a model for the evolution of a number of national spatial data infrastructure projects; for example, the United States Geological Survey's (USGS's) The National Map (http://nationalmap.usgs.gov/) and the Federal government's Geospatial Onestop (www.geo-one-stop.gov). The National Map based on the data

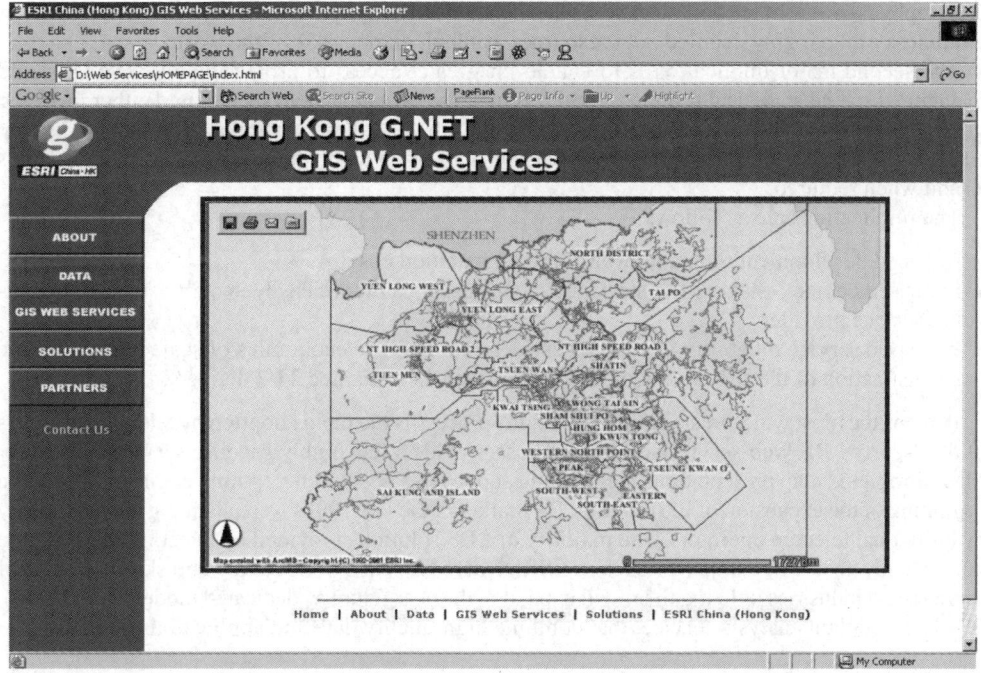

Figure 3.   Hong Kong G.Net.

network concept is designed to create a robust base map framework for the United States and uses the network concept to develop creative partnerships that will ensure that base topography data is kept current, complete and consistent throughout the country; and, secondly, to help ensure that its data is made available to a national (and potentially international) audience. The Geospatial One-Stop is a similar initiative sponsored by the US Federal Government that aims at ensuring that spatial analysis and decision-making by different parts of government is made on consistent, reliable geospatial information. Within Asia, the Hong Kong G.Net (www.gis-webservices.com) is illustrative of this trend.

The Hong Kong G.Net (HK-G.Net) offers a range of datasets covering the HK SAR, including:

- base map information and demographic and census data;
- point of interest (PoI) at various levels (public services, commercial, tourist, etc.);
- Yellow Pages location directories;
- a complete navigable road network (including turn directions);
- Ortho-photography; and
- Satellite imagery.

These datasets, derived from both commercial and public suppliers, are either hosted locally by the HK-G.Net or by the suppliers themselves. The G.Net site permits data to be viewed, searched and interrogated. As data is served in ArcXML, live data can be served to any ArcXML-compliant application. Some of the datasets (for example imagery and census data) require payment before they are made available to the user, either on a one-off purchase or subscription basis. The registry handles payment through direct linkage to third-party payment gateways. The site also provides Simple Open Access Protocol (SOAP) wrapped location and visualization functions, permitting clients to integrate maps and spatial search functionality directly into their local applications/site. Thus, for example, banks and property agents are using the service to integrate maps generated by the Hong Kong G.Net into their own Websites and internal systems, saving the additional overhead involved in the purchase and management of separate GIS software and datasets. As the Hong Kong G.Net is part of the wider

71

Geography Network framework, it exposes local datasets to the global community – providing greater awareness of the highly detailed data resources available in one area of the world.

The second major implication is to enable far greater access to professionally developed and maintained spatial analytical routines. Until recently, use of spatial analysis required either investing in skills, hardware and software to maintain a local team or obtaining external assistance to develop custom applications. Web services make it feasible to invest in highly targeted spatial analytical tools as and when required.

The implications are as follows:

- Reduced development time and overall implementation costs;
- Increased choice – standard compliance means users can more easily switch between competing service providers;
- Increased service reliability as hosted services can mobilize economies of scale to increase the sophistication of the spatial models offered, as well as guarantee 24/7 delivery.

Perhaps the most widespread illustration of this process is the rapid adoption by telecom providers of third-party LBS Web-service solutions. These provide a few highly specific services – WAP or MMS formatted map presentations, geo-coding and reverse geo-coding, routing engines – that at the beginning of the expansion of the mobile LBS market in the late 1990s were all being hosted directly by individual telecom operators. The majority of LBS solutions provided by telecom operators are now actually hosted by companies such as ESRI, Microsoft, GEODATA and Webraska. The range of services and industries utilizing them will expand in the near future as dedicated models, for example ESRI's floodplain analysis service, that combine high-quality data and applications, increase.

5 CONCLUSIONS

This paper has attempted to outline the basic architecture, structure and implications of GIS Web services for spatial analysis and presentation. GIS Web services are now permitting GIS to become seamlessly integrated with decision-making at every level. Web-based visualization and query tools have led to widespread access to existing geographic data stores. The recent focus on Web services within GIS is permitting these to be directly integrated with applications that have had no spatial functionality in themselves. While this is good for spatial analysis – enabling it to be more widely used than ever before – the trend raises a key issue. If Web services are going to address more than standard geo-processing and, for example, be used to run Web-based environmental models or coastal protection systems, the focus must turn to documentation and the ability to communicate in clear legible terms the assumptions, implications and usage of a service. Current metadata standards are evolving both in terms of content and usability, but focus on data services. It is now essential to address metadata descriptions of Web services themselves in a form that goes beyond the technical description of WSDL and helps the potential user understand the implications of the processes being undertaken. If this can be achieved, Web services will present GIS and spatial analysis in decision-making with new and exciting opportunities.

REFERENCES

ESRI, 2001, *GeoNetwork Handbook*, Redlands, ESRI Whitepaper.

Hjelm J., 2002, *Creating Location Services for the Wireless Web*, New York, John Wiley & Sons Ltd.

Kern T., Lacity M. and Willcocks L., 2002, *NetSourcing*, New Jersey, Financial Times Prentice Hall.

Kraak M-J. and Brown A. (eds), 2001, *Web Cartography: Developments and Prospects*, London, Taylor & Francis.

Longley P., Goodchild M., Maguire D. and Rhind D. (eds), 2001, *Geographical Information Systems and Science*, New York, John Wily & Sons Ltd.

Open GIS Consortium, 2002, *Open GIS Geography Markup Language (GML) Implementation Specification 2.1.2*, Open GIS Consortium.

Formation of fuzzy land cover objects from TM images

Xinming Tang
International Institute for Geo-Information Science and Earth Observation (ITC), Enschede,
The Netherlands

Wolfgang Kainz
Cartography and Geoinformation, Department of Geography and Regional Research, University of Vienna,
Vienna, Austria

ABSTRACT: Spatial features are usually simplified as crisp objects in most GIS models. However, this kind of abstraction is not accurate enough, since in many cases natural phenomena are continuous and there are no determinate boundaries between different spatial objects. The handling of these fuzzy spatial objects is one of the most important questions in GIS. The first step in modeling fuzzy spatial objects is to identify and obtain these objects. This paper discusses the general procedure for forming fuzzy spatial objects and proposes a composite method to generate fuzzy land cover from TM images. The method is developed based on a series of image analyses including supervised classification, fuzzy convolution and rule-based processing. It is verified in a test area of Sanya city of China. The result shows that the methodology is applicable for forming fuzzy land cover objects.

1 INTRODUCTION

Generally speaking, data modeling is a process for the generalization and analysis of various phenomena for a certain application. Defining, identifying and generating entities or objects are its primary steps. GIS is a tool that can handle spatial features. Currently, most GISs handle spatial phenomena in either vector or raster data models. No matter which model is adopted, the spatial features are generally abstracted in a crisp way. In the vector data model, the features are represented by crisp polygons, arcs or points. The boundary of a polygon is an arc, the boundaries of an arc are nodes (the points at the end of an arc), which are crisp. The raster data model represents spatial features in pixels with a specific value. The group of pixels of the same value represents a certain feature, where the boundary of the feature is also crisp. Both models represent spatial features with one common characteristic: all spatial features have clear boundaries.

However, natural phenomena are not always as crisp as the objects represented in GIS data models. In many cases spatial features are continuous such as downtown areas, mountains, soils, grasslands and forests. They are difficult to describe in a crisp way. These features have another characteristic in common – they have indeterminate boundaries. Currently, most applications simplify these spatial features with a crisp boundary, without taking continuous characteristics into consideration. This may lead to the loss of some useful information. Take a land cover classification as an example. When a TM image is classified into types of land cover, every pixel will be assigned by a single land cover type. However, on TM images, some pixels are composed of different types of land cover, which are called mixed pixels. On the other hand, land cover types usually cannot be clearly defined. Under such cases, it is difficult to determine that these pixels definitely belong only to a single class. If we analyze changes in land cover in a certain area, the transition from one type to another will be definite and abrupt. That is, the change is qualitative. In nature, land cover changes gradually. A forest may slowly degrade into bush when the natural environment degenerates, if there is no artificial interference. During this evolution, the land cover usually changes from forest, then to mixed forest and bush, and finally, to bush. If TM images are classified into crisp land cover types, then the classified results will indicate that the change from forest to bush occurred suddenly. The loss of information causes the imprecision of the analysis.

This problem leads to the requirement for fuzzy spatial objects. Fuzzy objects are normally defined as those with indeterminate boundaries. A great deal of research has been conducted in this field. Most of the studies, however, emphasize how to adopt fuzzy mathematics for better classifications and decisions based on crisp objects. Several focus on the characteristics of fuzzy spatial objects, including the definition and types (Schneider 1999), the relationships (Cohn and Gotts 1996, Clementini and Di Felice 1996, Molenaar 1996, Tang and Kainz 2002), and identification of fuzzy spatial objects (Cheng et al. 1997).

In order to model fuzzy spatial objects in spatial databases, the first step is to form them. The key problem is how to represent the fuzziness and then how to calculate membership functions for fuzzy spatial objects. According to Fisher (1996) and Burrough (1996), two kinds of fuzziness exist in spatial objects: fuzziness in spatial extent and fuzziness in object definition. Understanding the fuzziness helps us to design membership functions and compute membership values for a fuzzy object. In general, membership values are calculated by two kinds of methods: active and passive. The active method derives the membership function and values by experts or based on some knowledge. The passive method calculates the fuzzy membership functions according to the data itself. Cheng et al. (2001) discussed the general processes of identification of fuzzy spatial objects. They also proposed three models for forming fuzzy objects using active and passive methods. However, the passive method, which was applied in the classification of TM images needs to be discussed further since the extent of fuzzy spatial objects covers a very broad area.

This paper puts forward a composite method for identifying and forming fuzzy spatial objects. The paper first introduces the general procedure for forming fuzzy spatial objects using the passive method. A composite approach is then proposed for computing the membership values of the fuzzy spatial objects. The method involves many steps such as fuzzy classification, fuzzy convolution, rule-based processing and more. The method is then applied to form fuzzy land cover objects from TM images in Sanya city, China. The results show that the method is applicable for the formation of data-oriented fuzzy objects. Finally, the conclusions and discussions are summarized.

2 GENERAL PROCEDURE FOR FORMING FUZZY SPATIAL OBJECTS

2.1 *Procedures for forming fuzzy spatial objects*

The general procedures to identify fuzzy spatial objects include three steps (Figure 1, refer to Cheng et al. 1997): the analysis of fuzzy types of spatial objects, the computation of membership values and the evaluation of the accuracy of fuzzy spatial objects.

2.2 *Analysis of fuzzy types*

Understanding fuzzy types is the starting point for forming fuzzy spatial objects. At this stage, the aspects that cause the fuzziness of spatial objects should be interpreted, so as to make sure they are related to certain applications.

In general, two types of fuzziness exist in spatial objects: spatial extent fuzziness and thematic fuzziness (object definition). Thematic fuzziness exists when we cannot clearly define an object. For example, when we define a land cover type, although we try to determine each piece of land cover clearly, fuzziness is inevitable in each class. When we browse the definition of forest in the USGS land cover standard (Anderson 1976), it is defined as an area characterized by tree cover (natural or semi-natural woody vegetation, generally more than 6 metres tall); where tree canopy accounts for 25–100 percent of the cover. In this definition, 'natural', 'semi-natural' and 'generally greater than 6 metres tall' are fuzzy terms. In the definition, the size of the area is also not specified. Another kind of

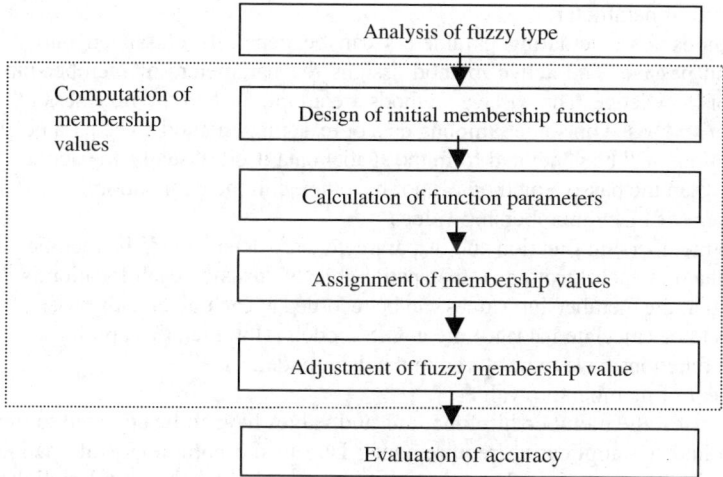

Figure 1.   General procedure of formation of fuzzy spatial objects.

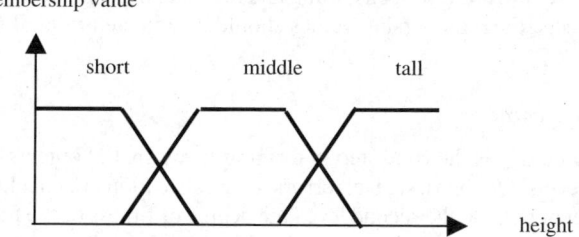

Figure 2.   Membership function for human stature *short, middle* and *tall.*

fuzziness is spatial extent fuzziness. Sometimes, we can clearly define an object, but we cannot clearly obtain it. When a TM image is classified into land cover classes such as grassland, we will immediately find that some pixels are a mixture of grassland with some trees or dry land; some pixels have some grassland at one side and other types of land cover at the other side; and some pixels contain both of the above cases.

In some applications, the location can be measured precisely, leaving the object definition fuzzy. In other applications, the definition is clear but the location cannot be measured precisely. In some cases, both definitions and locations contain fuzziness. In general, we should bear in mind which fuzziness is mainly concerned in the applications.

2.3   *Computation of membership values*

The computation of membership values should be done for all fuzzy spatial objects. Generally, it contains the following steps:

– Design of initial membership functions
   After understanding which type of fuzziness of objects is concerned, the initial membership functions should be designed. The design is usually obtained by selecting one function from a list of existing membership functions such as triangular, trapezoidal, bell-shaped, or normal distribution functions, based on the fitness of the fuzziness of the spatial objects with those functions. For example, if we subdivide the human stature into three fuzzy classes: *short, middle* and *tall*, the trapezoidal membership function (Figure 2) can be adopted as the initial function for *short, middle* and *tall*.

75

– Calculation of parameters

The methods for calculating parameters can be generally classified into two categories: active and passive. The active method assigns the parameters of membership functions by experts or knowledge. The passive methods are adopted when the fuzziness of spatial objects can be derived based on some sampling data or reasoning methods. The parameters of membership functions will be generated from the spatial data itself. Usually, the active method will be selected when the passive method cannot be adopted in the application.

– Assignment of fuzzy membership values

After the membership function and its parameters are determined, the membership values can be calculated at each location of the spatial objects. Usually, each location is described by a pixel, so that the membership values can be recorded at each pixel. Otherwise, the membership value has to be calculated at run-time in GIS models. However, this approach is rarely adopted since the functions have to be stored in the data model.

– Adjustment of membership values

In many cases, the membership functions and values have to be adjusted to meet the factual situation and the application requirements. Due to the complexity of spatial features and problems in data sampling, there could be errors in the membership functions or values. In some situations, although the membership values can reflect the factual situation, they are too complicated for applications. For example, the extent of spatial objects may be too small. Thus, the analyses are too time-consuming and the visualization is very poor. To minimize the above side effects, the membership values should be adjusted to facilitate the analysis.

2.4 *Evaluation of accuracy*

The evaluation of accuracy is the final step in forming fuzzy spatial objects. The evaluation can be conducted on two levels. On the first level, errors in classification are checked; that is, whether the object type is correct or not. On the second level, the degree of fitness of the fuzziness is verified with the factual situation. Normally, a field survey should be conducted to verify the accuracy of fuzzy spatial objects on both levels.

3 FORMATION OF FUZZY LAND COVER TYPES

3.1 *Understanding of land cover type*

The importance of land cover needs no further explanation since it plays a fundamental role in many fields such as land use planning, urban construction and natural resource exploitation. We address the method how to form fuzzy land cover from TM images.

On TM images the pixel value is a reflection of all spatial features per pixel. One pixel may contain different features. Therefore, the fuzziness of a land cover object is raised by both thematic and spatial resolution. Since the fuzziness in object definition and object extent cannot be differentiated from the value itself, the result of classification contains fuzziness in both the thematic and spatial aspects of land cover objects.

3.2 *Method to form fuzzy land cover objects*

Following the general procedure discussed in Section 2, a method for forming fuzzy land cover objects from TM images is proposed. The procedure is illustrated in Figure 3.

The method includes seven steps: analysis of fuzziness in land covers, selection of appropriate initial membership functions, computation of parameters of membership functions, fuzzy convolution for adjusting the membership values according to the land cover texture, rule-based processing for finalizing membership values, representation of fuzzy land cover objects and testing of accuracy.

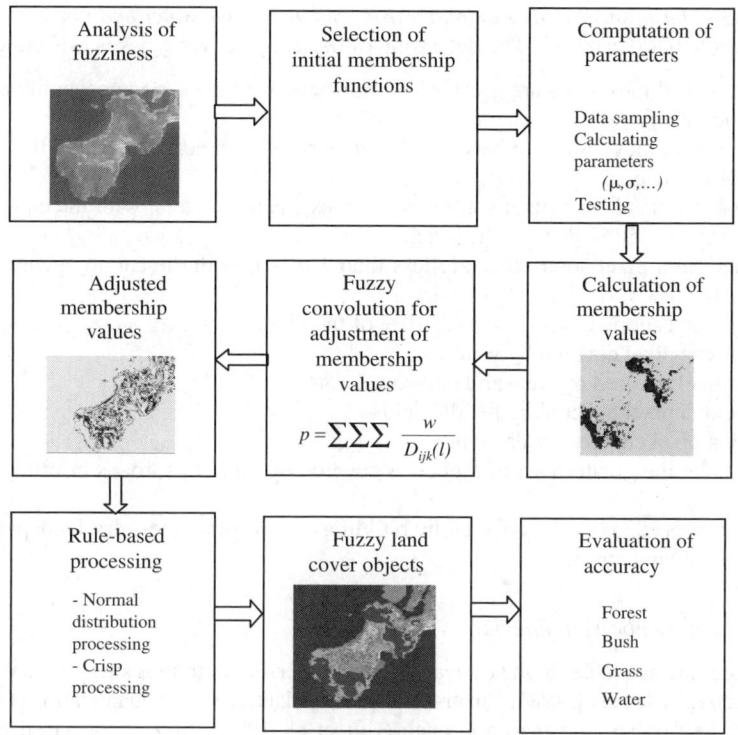

Figure 3. Procedure for forming fuzzy land cover objects from TM images.

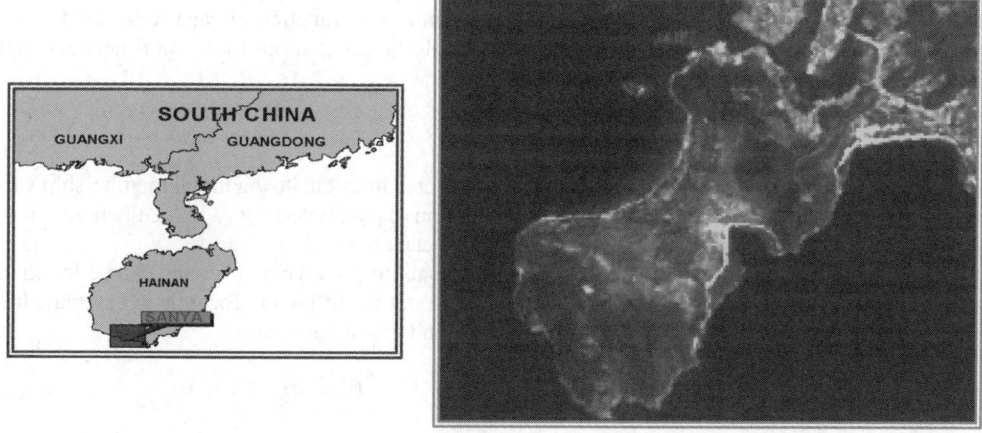

Figure 4. Location and TM image of test area.

3.3 *Test area*

Sanya city is selected for the test of the proposed method. The city is located on Hainan island, in south China (Figure 4). The TM image was obtained on 18 April 1990. The study area ranges from E108.97257° N18.160917° to E109.550161° N18.5904583°, covering the city and rural areas, with 1960 * 2350 pixels on the TM image. Eleven types of land cover will be classified from the seven-band

TM image: *forest, bush, shrub and grassland, waste land, bare land, water body, beach, built-up area, rural area, paddy field, dry land.* The definition of these land cover types is as follows:

1. Forest: in a pixel most trees are greater than 6 meters, with the canopy generally covering over 80% of the pixel.
2. Bush: in a pixel most trees are between 2–6 meters, with the canopy generally covering over 50%–80% of the pixel.
3. Shrub and grassland: in a pixel some trees are less than 2 meters, with the canopy generally covering between 50%–80% of the pixel.
4. Waste land: in a pixel some trees are less than 1 meter, with the canopy generally covering between 10–50%.
5. Bare land: the canopy covers less than 10% of the pixel.
6. Water body: a pixel covered by water.
7. Beach: a pixel covered by wet sand and some water.
8. Paddy field: a pixel covered by paddy fields.
9. Dry land: a pixel covered by dry land.
10. Built-up area: the greater part of a pixel is covered by buildings, roads or other construction material.
11. Rural area: a pixel generally covered by buildings, with some trees, dry land, paddy fields or other types of land cover.

3.4 Selection of membership function

Several methods are available for the derivation of membership functions such as the C-means clustering approach (Bezdek et al. 1984, Chi and Yan 1995), adaptive vector quantization (Dickerson and Kosko 1993), self-organizing map approach (Chi et al. 1995), fuzzy supervised classification approach (Wang 1990, Mannan et al. 1998), and neural network approach (Sun and Jang 1993). The first three are clustering methods that generate the cluster center and variance. They are useful when the center should be calculated using all of the data. The fuzzy supervised classification approach is mainly used to derive crisp classes. In this application, the conventional supervised classification is selected, since we assume that the membership function of each fuzzy land cover type *basically* follows the normal distribution. It should be pointed out that neural network and other classification approaches are also applicable.

3.5 Derivation of membership functions and values

The conventional maximum likelihood classifier is selected to calculate the initial membership values. This process includes data sampling, the computation of parameters for each membership function, and the calculation of the membership values for each type of land cover.

The notation of maximum likelihood is a classic probability-based classifier and can be found in many remote sensing textbooks (Richards 2000). We adopt the following formula to calculate the weighted distance of pixel values belonging to a certain type of land cover:

$$D_c = \ln(a_c) - 0.5\ln(|Cov(c)|) - 0.5(X - M_c)^T (Cov(c))^{-1} (X - M_c) \tag{1}$$

where
D_c = the weighted distance (the distance of a pixel belonging to class c);
c = a particular class;
X = the measurement vector of the candidate pixel;
M_c = the mean vector of the sample of class c;
a_c = percent possibility that any candidate pixel is a member of class c, (defaults to 1.0, or is entered from *a priori* knowledge);
$Cov(c)$ = the covariance matrix of the pixels in the sample of class c;
$|Cov(c)|$ = determinant of $Cov(c)$;

$Cov(c)^{-1}$ = inverse of $Cov(c)$;
T = transposition function.

The membership value of a pixel belonging to class c can be calculated by the following formula:

$$p_c = \frac{g_c}{\sum_{i=1}^{m} g_i} \quad (2)$$

where

$$g_i = e^{D_i - min(D_k)}, k = 1, 2, ..., m \quad (3)$$

m is the number of land cover types.

To compute the parameters, every land cover type is trained by supervised sampling. The maximum likelihood classifier classifies the land cover types when the pixel values of each type follow a normal distribution. In some cases, a certain class has to be sampled by sub-classes, such that each of the samples follows the normal distribution. After classification, they can be merged together.

In the Sanya application, the paddy fields are split into two sub-classes: paddy field with water and paddy field with a lot of canopy of rice leaves since there is a big difference between the pixel values of these sub-classes. Therefore, in the classification, 12 types are initially sampled and classified.

The maximum likelihood classification adopts sample data to calculate the parameters of the membership function. It is well known that the sample data has a great effect on the results of the classification. Although several sampling methods are taken to check the correctness of the classification, it should be pointed out that the final result is obtained by sampling data in small polygons conventionally. The samples cover all types of land cover. The classification is carried out using ERDAS Imagine.

The prior possibilities of all land cover types are assigned to 1. We assume that each pixel may contain a maximum of 4 different types of land cover. There are four layers of membership values at each pixel. The largest membership values are stored on the first layer, which shows the maximal membership value belonging to a class. The second layer stores the second-largest membership values. The first layer of the classification result is partly shown in Figure 5. It also represents the crisp result of classification.

3.6 *Fuzzy convolution*

After the classification, the weighted distances of each pixel belonging to every class are calculated. After checking the results, several errors can be observed. For example, some dry land was

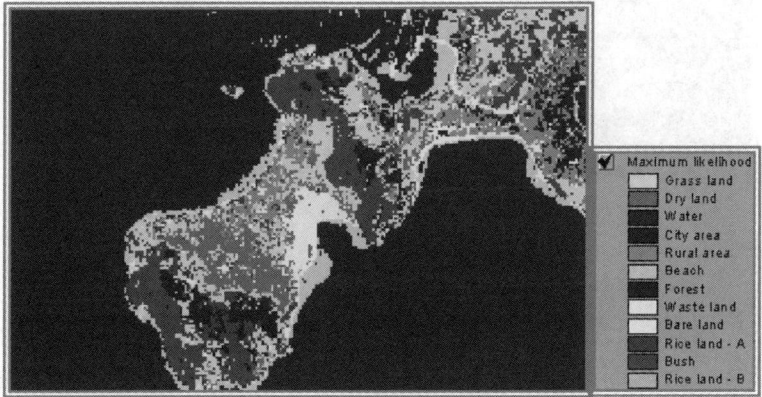

Figure 5.   The class at the first layer after the maximum likelihood classification.

classified as rural land; some land cover objects are too small. This is because the maximum likelihood classifier classifies images pixel by pixel. In order to derive a better result, the fuzzy convolution is applied to adjust the weighted distances of the pixel belonging to all classes, because it considers the membership values of neighboring pixels.

The basic formula of fuzzy convolution is:

$$T(c) = \sum_{i=1}^{x}\sum_{j=1}^{y}\sum_{l=1}^{m}\frac{w_{ij}}{D_{ijl}(c)} \qquad (4)$$

where

$T(c)$ = the distance after fuzzy convolution to class c;
w_{ij} = the weight of pixel (i,j);
c = a particular land cover class;
m = land cover classes;
$D_{ijl}(c)$ = the weighted distance of pixel (i,j) belonging to class c at layer l;
x,y = the number of neighborhood pixels.

After the fuzzy convolution, the membership value of a pixel belonging to class c can be revised from formula (3) to the following:

$$g_i = e^{\frac{\sum_{i=1}^{x}\sum_{j=1}^{y}w_{ij}}{T(c)}\frac{\sum_{i=1}^{x}\sum_{j=1}^{y}w_{ij}}{min(T(k))}}, k=1,2,...,m \qquad (5)$$

In practice, a 3*3 matrix is adopted to adjust the membership values of land cover types. The matrix is:

0.5	0.646	0.5
0.646	1.000	0.646
0.5	0.646	0.5

The sum of weight $\sum_{i=1}^{x}\sum_{j=1}^{y}w_{ij} = 5.584$. The results of the fuzzy convolution are shown in Figure 6. After fuzzy convolution, the membership values are more continuous for the same class. In the Sanya application, the class of rice with a good canopy and the class of rice with a normal canopy are then merged together. The class *paddy field* is derived simply by adding these two membership values belonging to two sub-classes at each pixel.

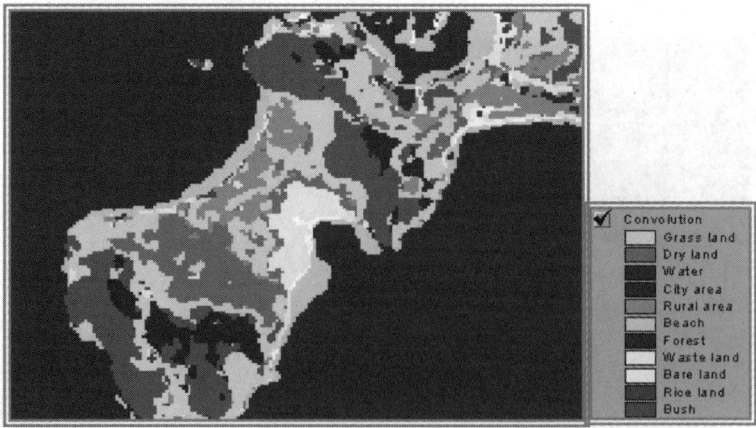

Figure 6.   The first layer after fuzzy convolution.

80

3.7   *Additional adjustment of membership values*

The next step in forming fuzzy land cover objects is to determine the boundary and core of each class. According to the definition (Warren 1977, Tang and Kainz, 2002), the *boundary* of a fuzzy set is a subset, such that the membership value at each of its points is between 0 and 1, and the *core* is the subset whose membership value at each point is 1. The simplest way is to take the pixels with membership values of greater than 0 as the extent of that type, and take the areas of pixels where the membership values are equal to 1 as the core. The difference between the core and the extent forms the boundary. Figure 7 shows the extent of the bush class.

However, the above result cannot be directly adopted to form the core and the boundary. In Figure 7 there are some pixels (marked in pink) with membership values for bush that cannot reflect the real situations. In general, three problems can be detected:

1. On some pixels the membership values for bush are very small. These pixels definitely belong to other classes, for example, to forest.
2. There are also some pixels with membership values for bush that are very large, but less than 1. These pixels actually must belong to bush.
3. Some pixels definitely belong to a single class; however the membership values for that class are far less than 1. For example, a pixel that must be dry land may simply have a not very large membership value for grassland. However, the membership value is only 0.6 for dry land.

The first two problems are caused by the maximum likelihood classifier. When this method is adopted, it is actually assumed that the membership function of a fuzzy land cover follows absolutely the normal distribution and can be calculated by the trained data. In the general situation, this assumption is correct since most pixels follow the normal distribution. However, this does not hold in extreme situations. Take a simple example. If there is a pixel with a value of 255, the membership values for all types of land cover will be 10% when 10 classes are classified (Figure 8). However, this does not mean the membership values are 10% in practice, which explains why we assume that the membership values follow the normal distribution in a *basic* sense only. The large and small values should be refined.

In order to solve the first two problems, we can define thresholds to cut off small membership values and enlarge the large membership values according to the variance of the distribution. In the normal distribution, if the membership value for a certain class equals 1, then the pixel covers all characteristics of that class. If the membership value of a pixel for a class is greater than 0.69, then the pixel falls within the interval of one variance (σ) (Figure 9). We can cut off the two tails

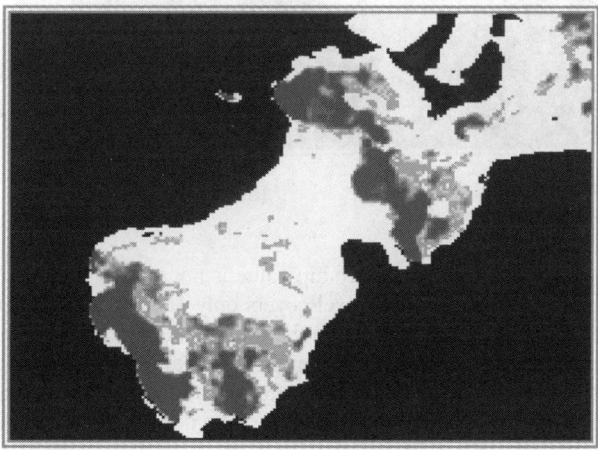

Figure 7.   Spatial extent of land cover objects bush.

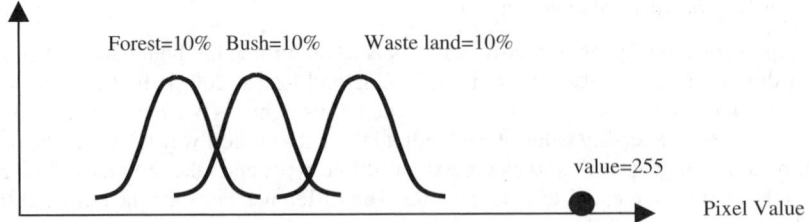

Figure 8.   The pixel value (255) has the possibilities 10% for all types.

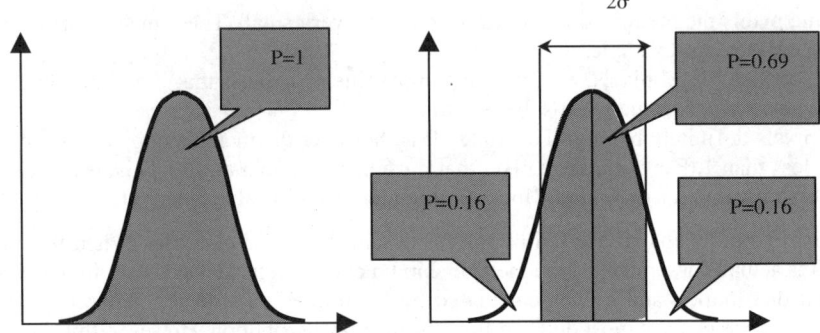

Figure 9.   Possibility distribution of a land cover for pixels.

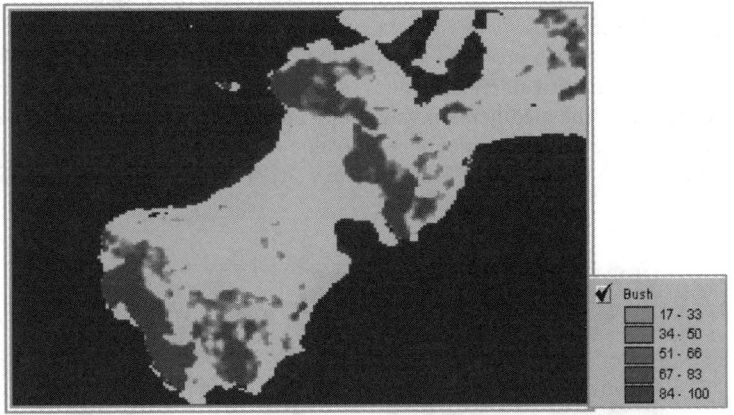

Figure 10.

according to Figure 9. That is, if the membership value is not greater than 0.16, then it is set to 0. This means that after classification, if the pixel covers only 16% of the characteristics of a certain class, the membership value is changed to 0. On the other hand, if a pixel covers 84% (= 100% − 16%) = 0.84, of the characteristics, then the membership value is assigned to 1, showing that it definitely belongs to that class. Part of the result after the adjustment is shown in Figure 10.

The membership values between 0.16 and 0.84 are then normalized by a linear transformation:

$$p_n = p_o / 0.69 - 0.16 / 0.69 \qquad (6)$$

where p_0 is the previous membership value and p_n is the new membership value. Parameters $1/0.69$ and $-0.16/0.69$ are calculated according to two assumptions: if $p_0 = 0.85$, then $p_n = 1$, and if $p_0 = 0.16$, then $p_n = 0$.

3.8 *Rule-based processing*

In order to solve the third problem, more should be considered regarding the issue of definitions of land cover types. The membership values calculated from the above procedures are derived from the pixel value, where a normal distribution is assumed for each class. However, when a certain type of land cover is clearly defined, then there is no fuzziness in this class. Therefore, the fuzziness of a pixel belonging to this class is only caused by the spatial resolution. For example, a pixel has a membership value for dry land and a membership value for grassland. If the dry land is clearly defined, then fuzziness is caused by the pixel size. That is, in one pixel, there is a part that is dry land and a part that is grassland. The maximum likelihood classifier actually calculates the membership values of a pixel for different classes.

In reality, some crops grow on dry land, which causes the pixel values to be similar between dry land and grassland or waste land. If we take the value of dry land and the value of grassland (for example) as the membership values (this means the pixel is composed of dry land and grassland), then many errors can be detected by a comparison with the reality, since many of these pixels are actually either dry land or grassland. Due to the similarity of the pixel values between grassland and dry land, even if 300 clusters are classified by unsupervised clustering, these two classes can still not be differentiated. Therefore, it is better to neglect the mixed pixels that are composed of grassland and dry land, since usually the dry land covers larger areas and is seldom mixed with the grassland in reality. According to this assumption, it is reasonable to enlarge the dominant membership values to 1, and assign 0 to the others. This strategy is useful when a lot of pixels that actually belong to one class, have values of two classes that are difficult to differentiate according to membership values.

In practice, we assume that the definition of dry land and paddy fields are crisp, and neglect the mixed pixels between these two classes and other classes. Two rules can then be stated:

1. If a pixel has membership values belonging to dry land and other classes, then it belongs to the class whose membership value is the maximum of all membership values, and is assigned membership 1.
2. If a pixel has membership values belonging to paddy field and other classes, then it belongs to the class whose membership value is the maximum of all membership values, and is assigned membership 1.

The membership value at the boundary of dry land and rice land is calculated into crisp forms according to the maximum value of pixels. The final results are shown in Figures 11 and 12, respectively.

3.9 *Representation of fuzzy land cover objects*

After the above procedures, the fuzzy and crisp land cover objects are formed. The land cover objects can be visualized in the following two ways:

1. Represent the land cover objects by different classes in different layers. On each layer, the change in colour represents the membership values. Figure 11 shows the membership values belonging to grassland. By this method, the distribution of any class can be clearly visualized; however, the overall classes on a pixel cannot be shown in one layer. This is useful for visualizing the distribution of a single class.
2. Show the dominant classes in the first layer, using a brightness representing the membership values for that class (Figure 12). The second layer shows the second class to which the pixel belongs, also using the brightness representing the membership values of that class. When there is no second class on the second layer (in the case of crisp objects), then the pixel will contain no data. The third layer can be treated in the same way. In fact, the layer is an extension of the

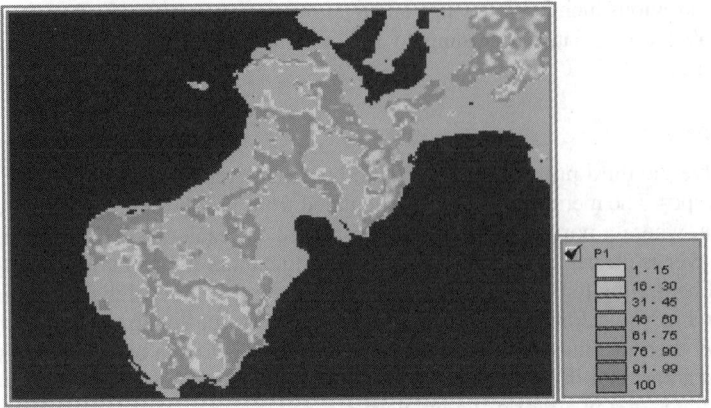

Figure 11.   Representation of fuzzy land cover object grassland.

Figure 12.   Representation of dominant land cover objects.

visualization used in showing a crisp classification result, with the difference that the membership value is shown on each pixel. The method allows for a better understanding of the overall situation of spatial objects. However, the secondary classes cannot be shown in the same layer.

3.10   *Evaluation of accuracy*

In comparison with the methods for evaluating the accuracy of crisp objects, an accuracy test for fuzzy objects is still under discussion. We adopted air-photos to verify the results, which were also checked by field work. 162 sample points were selected from areas that were difficult to classify. The truth value of fuzzy land cover objects were estimated based on expert knowledge. At each point, membership values belonging to two land cover classes were simulated. Table 1 shows the results of the test for accuracy.

Several comparisons have been made based on testing data. In Table 1, the first row compares the results between membership values of the dominant class of the classification result and the expert estimation at each point. 143 points are classified correctly; i.e., 88.2% of the 162 points. This also represents the accuracy of the crisp classification. The average of the differences is 10

Table 1.   Evaluation of the accuracy of fuzzy land cover objects.

	Number	Percentage	Average of difference between membership values	Standard deviation between membership values
C1 = T1	143	88.2%	10	14
C1 = T2	3	1.8%	77	6
C2 = T1	1	0.6%	64	0
C2 = T2	121	74.7%	2	6
C1 = T1 or C2 = T1	144	88.8%	10	14
C1 = T2 or C2 = T2	124	76.5%	2	6
C1 = T1 or C1 = T2	147	90.7%	11	14
C2 = T1 or C2 = T2	122	75.3%	3	6
C1 = T1 or C2 = T1 or C1 = T2 or C2 = T2	160	98.8%		
C1 = T1 and C2 = T2	108	66.7%		

T1 = Cover type 1 from air-photo (ground truth), C1 = Cover type 1 from TM image, T2 = Cover type 2 from air-photo (ground truth), C2 = Cover type 2 from TM image.

between the membership values. This means that if the pixel is correctly classified at the first layer, then the accuracy of the membership value is 90%.

The second row is a comparison between the dominant class after classification and the secondary class of the ground truth. It indicates that the dominant class should be the secondary class of the true value. The average of the differences is 77 between two membership values.

The third row shows the comparison between the secondary class and the dominant class of the true value. It indicates that the secondary class should be the dominant class of the ground truth. The average of the differences is 64 between two membership values.

The fourth row is a comparison between the secondary class with the true values. It shows that 121 points are correctly classified in the secondary layer; i.e., 74.7%. The mean difference is 2, which denotes that the accuracy of the membership values is 98% in the correct classes.

The fifth row is a comparison between the classes in both the dominant and secondary layer and the dominant class of the true value. This means that if any classified land cover type is equal to the true class, then the pixel is correctly represented. This comparison will add the number of pixels that are correctly classified, but enlarge the difference of the average of the membership values.

Rows six to eight have a similar meaning as the fifth row. The ninth row shows the percentage by which any one of the classes equals any of the true classes.

The tenth row denotes the percentage when both the dominant class and the secondary class are correctly classified. It indicates the accuracy of the classification in both the dominant and secondary layers. The total overall accuracy is 66.7%.

From above statistics, it can be seen that if the land cover class is correctly classified, then the differences between the classified results and the true classes are very small. That is, the accuracy of the classification is the key that affects the accuracy of the membership values of each land cover object.

4 CONCLUSIONS AND DISCUSSIONS

4.1 *Conclusions*

The paper discusses a general procedure to form fuzzy spatial objects. Three steps are involved in this procedure. Based on this procedure, the paper addresses a method of forming fuzzy land cover objects. The procedures are further explained by classifying a TM image into fuzzy land cover types. While the paper focuses on the method and its procedures, it also proposes some ways of representing fuzzy spatial objects, and discusses some aspects of evaluating the accuracy of fuzzy spatial objects.

The paper proposes a composite method for forming fuzzy land cover objects. Seven steps are included: the design of the initial membership functions for land cover types, the generation of initial parameters by supervised classification, the adjustment of membership values by fuzzy convolution and membership distribution, the determination of membership values by rule-based processing, the representation of fuzzy land cover objects, and the evaluation of accuracy of these objects. It can be seen that the key issue is how to derive the membership values of fuzzy land cover objects. According to the results of the test area and evaluation of the accuracy, it is shown that the proposed method is suitable for forming fuzzy land cover objects.

4.2 *Discussions*

The paper proposes a method of forming fuzzy land cover objects. The method can be revised in several aspects.

1. The limitation on minimum membership values can be put at the classification stage. That is, small membership values can be filtered out by setting some confidence value during classification.
2. Different weighted values can be adopted in the fuzzy convolution for different applications.
3. The fuzzy convolution can also be done after filtering out the small membership values.
4. The core and boundary of fuzzy spatial objects are formed using normal distribution theory; they can also be tuned using a neural network approach.

Less is considered on different methods of classification. Actually, many methods of classification are available, such as the knowledge-based classifier and neural-network classification, which can also generate membership values for fuzzy spatial objects. It is necessary to discuss them to compare the accuracy of land cover classes and membership values.

The accuracy of different types of land cover is not presented in this paper. In fact, how to investigate the accuracy of fuzzy objects, as well as how to represent fuzzy spatial objects are two different aspects of research in the modeling of fuzzy spatial objects. Although some efforts have been made in this paper to examine these aspects, more overall and systematic research is still needed.

REFERENCES

Anderson, J. R., Hardy, E. E., Roach, J. T. and Witmer, R. E., 1976. A Land Use and Land Cover Classification System for Use with Remote Sensor Data. *U.S. Geological Survey, Professional Paper 964*, Reston, VA.

Bezdek, J. C., Ehrlich, R. and Full, W., 1984. FCM: The Fuzzy C-Mean Clustering Algorithm'. *Computers and Geosciences*, 10, pp. 191–203.

Burrough, P. A., 1996. Natural Objects with Indeterminate Boundaries. In: P. Burrough and A. U. Frank (eds), *Geographic Objects with Indeterminate Boundaries*, Taylor & Francis.

Cheng, T., Molenaar, M. and Bouloucos, T., 1997. Identification of Fuzzy Objects from Field Objects. *Spatial Information Theory, A theoretical Basis for GIS, Lecture Notes in Computer Sciences 1327*, COSIT'97, pp. 241–259, Springer-Verlag, Germany.

Cheng, T., Molenaar, M. and Lin, H., 2001. Formalizing Fuzzy Objects from Uncertain Classification Results. *International Journal of Geographical Information Science*, 15(1): 27–42.

Chi, Z. and Yan, H., 1995. Image Segmentation using Fuzzy Rules derived from K-means Clusters. *Journal of Electronic Imaging*, 4(2): 199–206.

Chi, Z. Wu, J. and Yan, H., 1995. Handwritten Numeral Recognition using Self-organizing Maps and Fuzzy Rules. *Pattern Recognition*, 28(1): 59–66.

Clementini, E. and Di Felice, P., 1996. An Algebraic Model for Spatial Objects with Indeterminate Boundaries. In: P. Burrough and A.U. Frank (eds), *Geographic Objects with Indeterminate Boundaries*, Taylor & Francis.

Cohn, A. G. and Gotts, N. M., 1996. The 'Egg-Yolk' Representation of Regions with Indeterminate Boundaries. In: P. Burrough and A. U. Frank (eds), *Geographic Objects with Indeterminate Boundaries*, Taylor & Francis.

Dickerson, J. and Kosko, B., 1993. Fuzzy Function Learning with Covariance Ellipsoids. In *Proceeding of IEEE International Conference on Neural Networks*, vol. III, pp. 1162–1167, San Francisco, USA.

Fisher, P., 1996. Boolean and Fuzzy Region. In: P. Burrough and A. U. Frank (eds), *Geographic Objects with Indeterminate Boundaries*, Taylor & Francis.

Mannan, B., Roy, J. and Ray, A. K., 1998. Fuzzy ARTMAP Supervised Classification of Multi-spectral Remotely-sensed Images. *International Journal of Remote Sensing*, 19(4): 767–774.

Molenaar, M., 1996. A Syntactic Approach for Handling the Semantics of Fuzzy Spatial Objects. In: P. Burrough and A.U. Frank (eds), *Geographic Objects with Indeterminate Boundaries*, Taylor & Francis.

Richards, J. A., 2000. *An Introduction of Remote Sensing Digital Image Analysis*, Springer-Verlag.

Schneider, M., 1999, Uncertainty Management for Spatial Data in Databases: Fuzzy Spatial Data Types. *The 6th Int. Symp. on Advances in Spatial Databases* (*SSD*), LNCS 1651, Springer Verlag, 330–351.

Sun, C. T. and Jang, J. S., 1993. A Neuro-fuzzy Classifier and its Applications. In *Proceedings of the Second IEEE International Conference on Fuzzy Systems*, pp. 94–98. San Francisco, USA.

Tang, X. M. and Kainz, W., 2002. Analysis of Topological Relations between Fuzzy Regions in General Fuzzy Topological Space. In Proceeding of the SDH conference' 02, Ottawa, Canada.

Wang, F., 1990. Fuzzy Supervised Classification of Remote Sensing Images. *IEEE Transactions on Geosciences and Remote Sensing*, 28(2): 194–201.

Warren, H. R., 1977. Boundary of a Fuzzy Set. *Indiana University Mathematics Journal*, 26(2): 191–197.

Advances in Spatial Analysis and Decision Making, Li, Zhou & Kainz (eds)
© *2004 Swets & Zeitlinger, Lisse, ISBN 90 5809 652 1*

Satellite image classification using Modified Counter-Propagation

Mitsuyoshi Tomiya & Seitaro Kikuchi
Department of Applied Physics, Seikei University, Musasino-shi, Tokyo, Japan

ABSTRACT: A supervised classifier for satellite images by the Modified Counter-Propagation
(MCP) is proposed. The MCP is a neural network for competitive learning. It is the modified
version of the Counter-Propagation, whose competitive layer has been replaced by the Self-Organizing
map (SOM). The Landsat image data are adopted as the input data of the MCP, and the output layer
consists of the pixel values, which represent categories to be classified. Our results shows that the
MCP can classify the data more accurately and stably than can the SOM only, especially in the
case of classification of vegetation, farm and wood.

1   INTRODUCTION

Remotely sensed data by sensors on satellites are not so 'clean' as the data taken in laboratories,
where various elements of the environment can be controlled, such as temperature, humidity, pres-
sure and various physical parameters. On the other hand, the atmosphere, the ocean, etc. are completely beyond control. Consequently, the usual methods of analysis for multi-dimensional data are
not always useful for remotely sensed data. For instance, the methods usually suppose that the statistical distributions of the data are almost Gaussian. On the contrary, the remotely sensed data are
affected by a variety of natural phenomena and human activities, and cannot follow the clear
Gaussian distribution. The neural networks are expected to be available even for such complicated
data, which cannot be easily well-parameterized.

The Modified Counter-Propagation (MCP) (Kohonen, 1984; Nielsen, 1988; Kohonen, 1995) is the
neural network that has been proposed as the supervised classifier. Connection vectors between the
input layer and the competition layer are multi-dimensional. An element of connection vectors
between the competition layer and the output layer represents the frequency information for each
category. The MCP compresses the category information in the multi-dimensional input data space
to the lower-dimensional competition layer. The MCP can also visualize the category distribution in
the competition layer.

2   MCP FOR COMPETITIVE LEARNING

The MCP can also be considered the supervised version of the Self-Organizing Map (SOM)
(Kohonen, 1984; Nielsen, 1987, 1988; Kohonen, 1997; Tomiya, 2000), which is essentially an unsu-
pervised and self-organizing classifier. Adding the third layer as the output layer for the Counter-
Propagation to the SOM, the MCP has the supervising data for the classification. The prototype of the
MCP is the Counter-Propagation (CP). The CP is a neural network consisting of three layers: the input
layer, the competition layer and the output layer (Kohonen, 1984; Nielsen, 1988; Kohonen, 1995). The
MCP is also a competitive learning network, and the one-dimensional competitive layer is substi-
tuted by a two-dimensional SOM (Nielsen, 1987; Kohonen, 1997; Tomiya, 2000). Figure 1 shows the

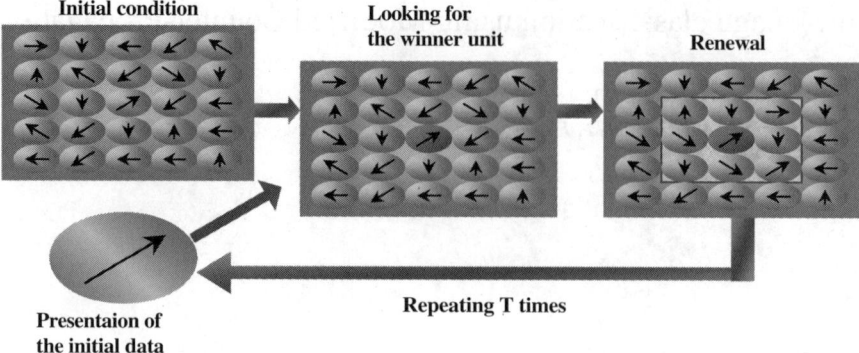

Initial condition

**Looking for
the winner unit**

Renewal

**Presentaion of
the initial data**

Repeating T times

Figure 1.   Schematic diagram of SOM.

block diagram of the MCP that we adopt in this work. The input and the competition layers form the SOM (Nielsen, 1987; Kohonen, 1997; Tomiya, 2000). The connections of the CP (Kohonen, 1984; Nielsen, 1988; Kohonen, 1995) from the competition layer are extended to the output layer.

The SOM is the most familiar neural network for competitive learning (Nielsen, 1987; Kohonen, 1997). It makes a map of the resemblance of multi-dimensional data by Kohonen's competitive learning. The SOM is composed of the input layer, to which the input data space $\mathfrak{R}_I^N$ is adopted, and the competition layer, which is the M-dimensional array of units as Figure 1, where $M < N$ and usually the shape of the array is set as $M = 2$. In this work we make the array rectangular. A parametric connection vector

$$W_j = \left[w_{j1}, w_{j2}, \cdots, w_{jN}\right] \in \mathfrak{R}^N$$

is defined on the connection from the input layer to every node j of the competition layer. The SOM needs no supervising data and the resemblance of the input data can be mapped onto the connection vectors W_j.

The competitive learning of the SOM part is an iterative method (Figure 2). The competitive layer has random data as the initial data. The first layer has N units, and N-dimensional input vector data $X_i = [x_{i1}, x_{i2}, \ldots, x_{iN}] \in \mathfrak{R}_I^N (i = 1, 2, \ldots, I, I$ represents the number of input data) determines the winning unit j, if its distance

$$D_{ij} = \sqrt{|X_i - W_j|^2} \tag{1}$$

becomes the smallest among $\forall$j for fixed i, where $\mathfrak{R}_I^N$ is the input data space. Not only the unit that wins but also units in its square neighbourhood, whose size is determined as

$$N(t) = N(0)\left(1 - \frac{t}{T}\right) \tag{2}$$

learn at the same time by the regression method

$$W_j(t+1) = W_j(t) + \alpha(t)(X_i(t) - W_j(t)), \tag{3}$$

where

$$\alpha(t) = \alpha(0)\left[1 - \frac{t}{T}\right], \tag{4}$$

90

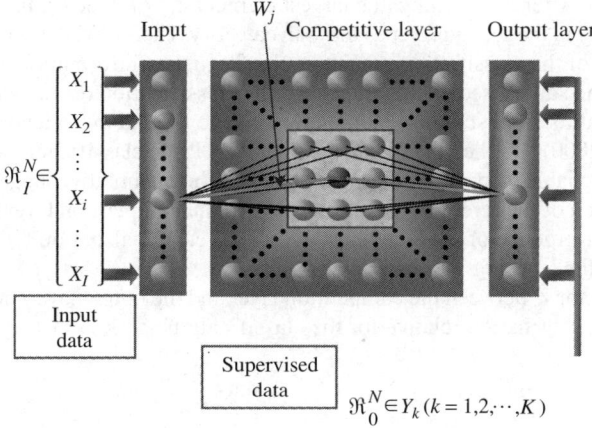

Figure 2. Competitive learning.

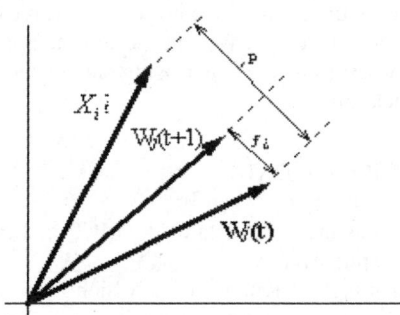

Figure 3. The correction of the connection weight.

and $W_j(t)$ is a reference vector at the t-th iteration and $W_j(t + 1)$ is the updated one (Figure 3). Note that ($t = 1, 2, \ldots, T$) denotes the learning time coordinate. Each discrete time t and whole input data ($\forall i$) are used for competitive learning.

The output layer of the MCP has the category information Y_k ($k = 1, 2, \ldots K$) for the classification of the supervising N-dimensional signals, where K is the number of the categories into which the data are classified. A frequency vector on the connections

$$Z_j = \left[z_{j1}, z_{j2}, \cdots, z_{jK} \right] \in \mathfrak{R}^K$$

between the output data space and the unit j in the array has a zero vector as the initial value. Only one element z_{jk} of Z_j is incremented by

$$z_{jk}(t+1) = z_{jk}(t) + \beta(t),\tag{5}$$

where $\beta(t) = \beta(0) + (1 - \beta(0))t/T$ and $\beta(0) < 1$, if the distance

$$D_{kj}^* = \sqrt{|Y_k - W_j|^2}\tag{6}$$

is the minimum of $\forall k$ during the Kohonen's learning. The k-th elements of the frequency vectors in the square neighbourhood of the size

$$N^*(t) = N^*(T)\,{}^t\!/_T\tag{7}$$

91

are also incremented. After the learning, the largest element $Z_{j\tilde{k}}$ of Z_j determines the most probable category of each unit j in the second layer as the $\tilde{k}$-th category. Therefore, the MCP can quantitatively judge the reliability of the classification result by the frequency information Z_j.

The SOM needs no supervised data Y_k; however, the classification result usually depends on the 'visual' judgement at its final stage. We somehow manage to read the category information from the SOM (Tomiya, 2000). On the other hand, using the MCP, the classification result of the MCP is more objective and reliable than that of the SOM, because the quantitative judgement by the relative frequency information of the reference vector Z_j between the competition layer and the output layer is available. The convergence of the competitive learning can be also monitored by the frequency information during the learning.

The frequency vector Z_j between the competition layer and the output layer must be normalized to make the frequency information relative for the classification as

$$ h_{jk} = \frac{Z_{jk}}{A_{\max}} \quad , \quad A_{\max} = \max_{j} \sum_{k=1}^{K} Z_{jk} \quad , \tag{8} $$

$h_j = [h_{j1}, h_{j2}, \ldots , h_{jk}]$ is the relative frequency vector of the unit j. Finally, eq. (8) can make the visual map of the category frequency information. Of course the maximum $h_{j\tilde{k}}$ specifies the most probable category of each unit j in the competitive layer. If the value of $h_{j\tilde{k}}$ is larger, it is more certain that the unit j belongs to the category $\tilde{k}$. We can see the category information distribution on the SOM. It is then necessary to examine the supervising data or 'code words' carefully for higher reliability of the MCP classification.

3   RESULT FOR LANDSAT IMAGE DATA

Landsat image data are adopted as the input data of the MCP, and the RGB value that represents each category is given to the output layer.

The Landsat natural-colour image of Kitaura lake, which is the large shallow lake at Ibaraki prefecture, Japan, is shown in Figure 4. This means that if $N = 3$ and we put $M = 2$, the size of the image is 512×400 ($= 204800 = I$). The initial condition of the SOM is between the input layer and the competition layer, whose size is 30×20 is shown in Figure 5. We chose seven categories ($K = 7$) in the image: 1. Sea, 2. Lake, 3. Vegetation, 4. Woods, 5. Farm, 6. Soil, 7. City Area. The initial values of the parameters of the MCP are set at $\alpha(0) = 0.50$, $\beta = 0.15$, $N(0) = 10$ and $N^*(0) = 2$. These values were essentially selected after repeated trial and error.

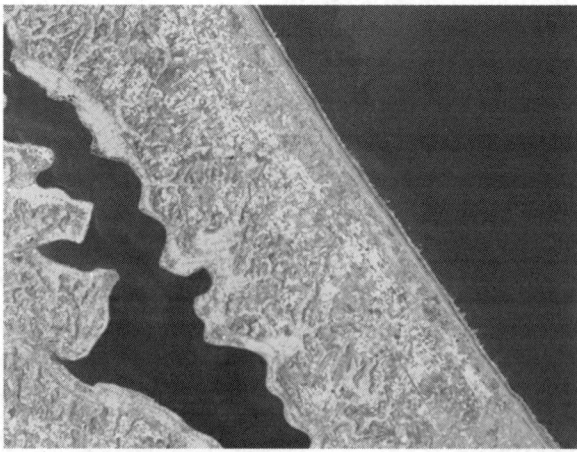

Figure 4.   Pseudocolour image of Kitaura lake.

The result of $T = 100$ is shown in Figure 6. The frequency information for each category can be obtained by eq. (8), as seen in Figure 7. We apply the grey scale to represent the information for eq. (8). For example, if $h_{jk} = 1(0)$, the colour of the category k on the unit j becomes completely black (white).

Figure 8 shows the categories of all units in the competition layer. They are determined by the maximum elements of the relative frequency vectors in eq. (8). The portion of the area of each category corresponds to the ratio of its land cover in the natural image; however, it is not proportional to the ratio. For example, the Pacific Ocean has a larger land cover than Kitaura lake (Figure 4). On the other hand, the lake acquires the largest area in the category information map. This implies that the category of lake actually consists of various categories. Of course, the main contribution is the water body, and some parts consist of vegetation, because Kitaura lake is notorious for outbreaks of the phytoplankton, 'aoko'. Owing to its shallow water, the contribution of soil cannot be ignored. Therefore, the cluster of the lake is located in the middle of the category distribution map (Figure 7(b)). It extends toward the other clusters, i.e. sea, vegetation, soil, etc. and shares the boundaries with them. The category distribution often does not seem to be simply connected. It may have several islands of high frequency. This would be due to the effect of the embedding three-dimensional input space into two-dimensional competitive layer.

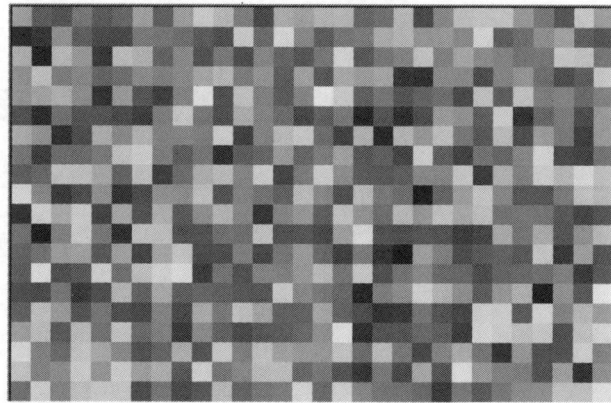

Figure 5.   SOM with the initial condition.

Figure 6.   SOM after Kitaura $T = 100$ times learning.

93

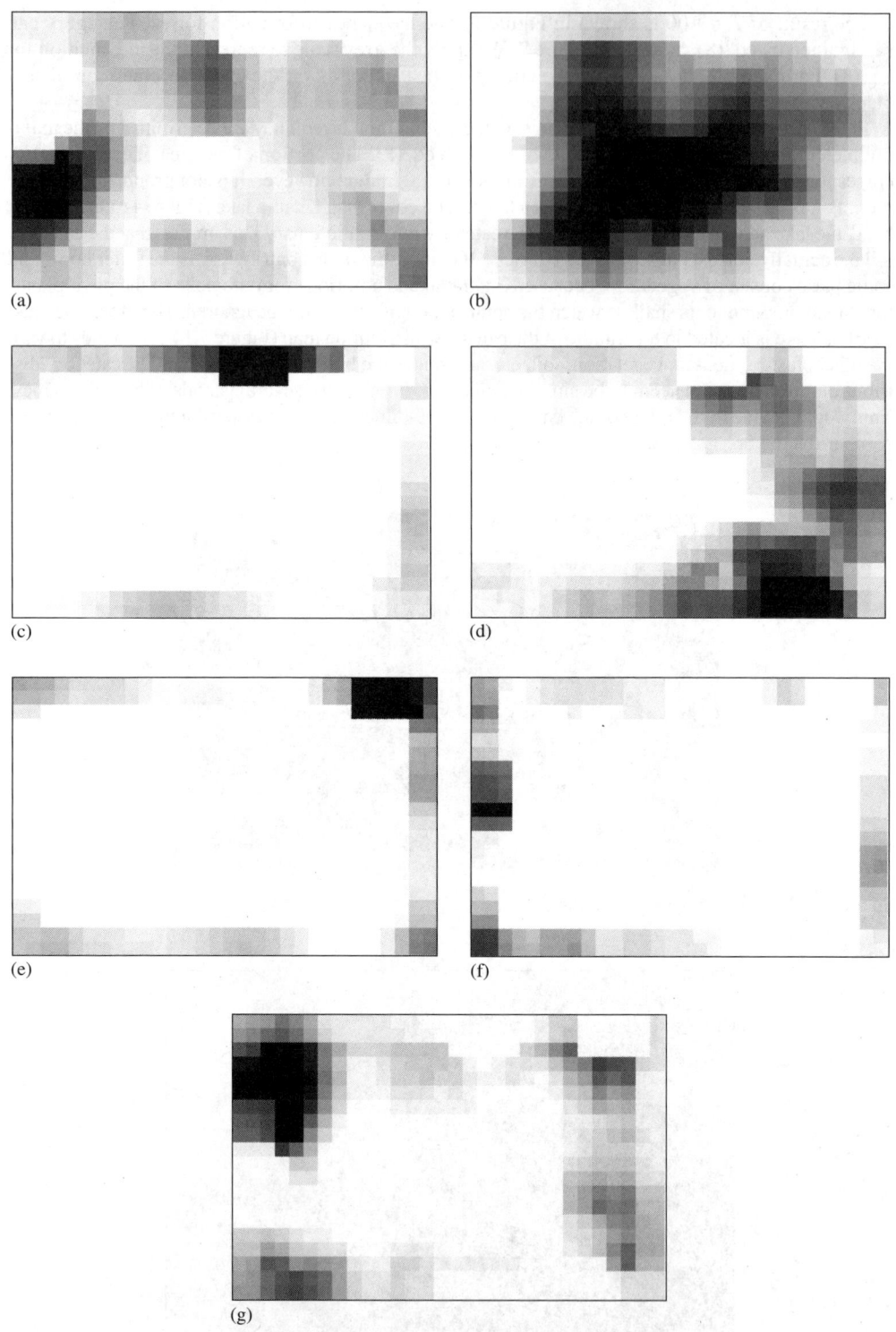

Figure 7. Frequency information of categories to categorize the data of Kitaura lake: (a) sea, (b) lake, (c) vegetation, (d) woods, (e) farm, (f) soil and (g) city area.

4   CONCLUSION AND DISCUSSIONS

The classification result is shown in Figure 9. The result is composed from the category information of Figure 8. To evaluate the classification result, we independently make a 'ground truth' map. We classify all of the pixels (204800 pixels) on the image by hand, making full use of the land cover map distributed by the Japan Geological Institute and several maps and different kinds of software. This took about a month, because of huge number of the pixels involved.

From a comparison with the 'ground truth' map, the accuracy of our method can be estimated. If the category of each pixel that the MCP identifies is the same on the 'ground truth' map, the result is considered to be accurate. The accuracy is defined as the ratio (the number of accurately classified pixels)/(the number of all pixels = 204800). During the learning, it increases rapidly at first and it steadily saturates at around $t \approx 75$ (Table 1). We can achieve high accuracy by the MCP.

Some guiding principle is always necessary for the SOM at the final stage of the classification, because the SOM itself is simply a gradation pattern of data (Kohonen, 1995). The region growing is applied to determine the category to which each unit in the self-organized competition layer belongs. If the difference between adjacent units after the learning is smaller than some threshold

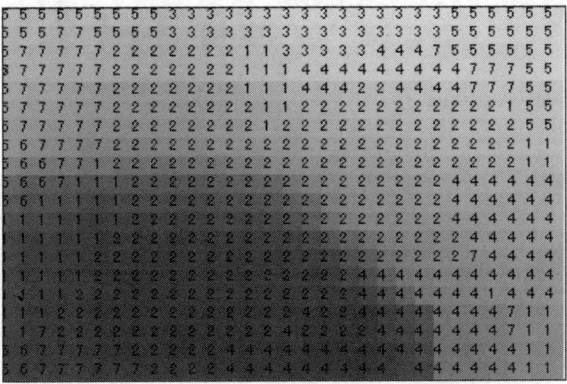

Figure 8.   Categories of all units in the competition layer determined by the relative frequency vectors eq. (8).

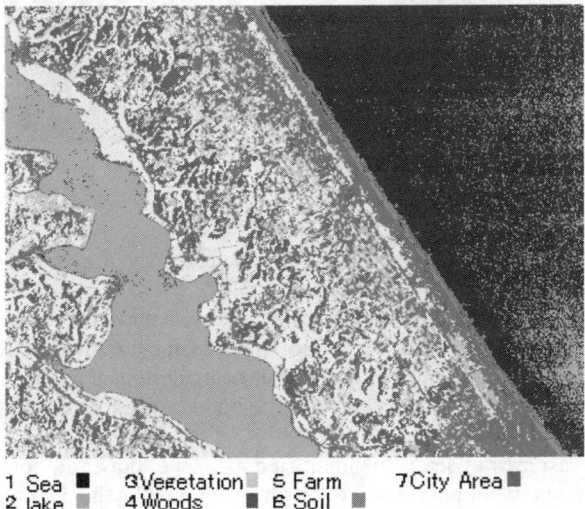

Figure 9.   Classified result 1 of Kitaura lake.

95

Table 1. Time evolution of the accuracy by the MCP.

Learning time (t)	25	50	75	$100 = T$
Accuracy (%)	92.46	95.21	96.19	96.25
CPU time (min)	48.5	95.5	142.5	189.0

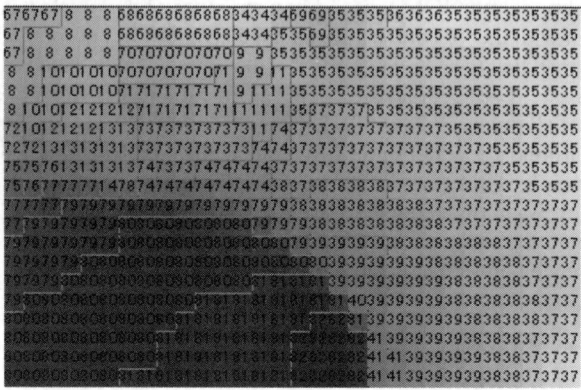

Figure 10. SOM after region growing.

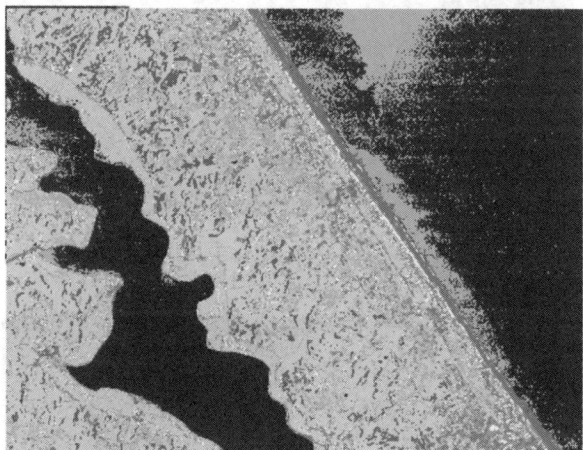

Figure 11. Classified result of SOM 1.

value, it is determined that the units belong to the same category. Figure 10 is the result of region growing as applied to the SOM.

As seen in Figure 10, the boundaries often become open and make the region too small to classify. We have to 'fix' the boundaries, usually by using 'the human eyes' after the region growing. Both Figure 11 and Figure 12 are the classified results after being corrected by the human eyes. Owing to 'the correction', even if the same person corrects the boundaries, the result can be different every time, and are generally unstable.

In Figure 11, the vast area of sea is misidentified as a lake; however, both are bodies of water and a further 'correction' would not be so difficult. In Figure 12 the area that is actually land is classified as a sea or a lake. Some of the areas that should be soil and city areas are also classified as vegetation, and the mis-recognition of the woods are more disastrous. On the other hand, the

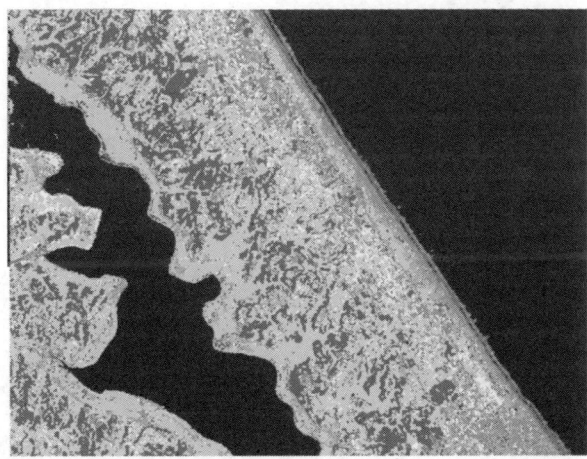

Figure 12.   Classified result of SOM 2.

MCP considerably reduces such mis-recognition. This is especially important for the classification of vegetation, farms and woodland (Figure 9). Therefore, the MCP is superior to the SOM
owing to the quantitative method for classification.

The classification result using the MCP with the supervised data and the results using the SOM
with no supervised data were compared. A more definite, quantitative and stable classification
result can be derived using the MCP. The result acquires reliability when the category frequency
information is added on the output layer.

REFERENCES

Kohonen, T. 1984. Self-Organization and associative memory, Springer Series in Information Sciences, Vol. 8.
Kohonen, T. 1995. Self-Organizing Maps, Springer Series in Information Sciences, Vol. 30.
Kohonen, T. 1997. Self-Organizing Maps, Second edition. Springer-Verlag.
Nielsen, R.H. 1988. Applications of Counter-propagation Networks, *Neural Networks*, 1: 131–139.
Nielsen, R.H. 1987. Counterpropagation Networks. Proceedings of IEEE first international conference on
    Neural Networls: 19–32.
Toniya, M. Classifier for Remotely Sensed Imagery Using Kohonen's Self-Organizing Feature Map with Region
    Growing. International Archives of Photogrammetry and Remote Sensing, Vol. 33, part B7: 1518–1523.

Advances in Spatial Analysis and Decision Making, Li, Zhou & Kainz (eds)
© 2004 Swets & Zeitlinger, Lisse, ISBN 90 5809 652 1

An efficient algorithm for the extraction of topographic structures from large scale grid DEMs

Qing Zhu, Jie Zhao, Zheng Zhong & Hai-gang Sui
National Key Laboratory for Information Enigineering in Surverying, Mapping and Remote Sensing, Wuhan University, Wuhan, P.R.China

ABSTRACT: As a fundamental problem in terrain analysis, the extraction of topographic structures plays a role in many applications such as hydrographic analysis, mineral deposition, land erosion pollution diffusion analysis, and so forth. The inherent problems in the efficiency and accuracy of existing approaches to extraction have hindered their application in the processing of large-scale Digital Elevation Models (DEMs). In this paper, an efficient approach to the extraction of topographic structures based on grid DEMs is proposed. The authors make two significant improvements to the extraction of topographic structures. The first is to fill depressions by adopting vector processes combined with traditional neighbourhood raster processes, and the second is to assign flow direction over flat by applying a neighbour-grouping scan method. After the experimental analysis using various volumes of real DEMs, the results show a significantly improved level of efficiency in the extraction of topographic structures, especially when applied to large-scale DEMs. The accuracy of the results is also better than existing methods. Thus, these improvements make the digital terrain analysis method more practical for large regional spatial decision-making.

1   INTRODUCTION

With the rapidly increasing availability of DEMs of various types and scales, the derivation of information on topographic structures (especially on ridges and valleys) from DEMs has attracted an increasing amount of attention in late years. The extraction of topographic structures is a fundamental problem in many applications such as hydrographic analysis, mineral deposition, land erosion, and pollution diffusion analysis (Wolock et al., 1989, 1990; Chen, 1991; Freeman, 1991; Moore et al., 1994; Li and Zhu, 2000). Information on topographic structures is essential in simulating floods and assessing flood plain risk, which can help decision-makers make the correct decisions to limit and control flood damage. It is also significant in studies of sustainable development in large drainage areas. In our applications, combined with other GIS information, information on ridges and valleys derived from the DEMs played a key role in accurately confirming the borderline between China and Vietnam.

   According to different terrain models, many extraction approaches have been proposed: O'Callaghan and Mark (1984), Jenson and Domingue (1988), and Tarboton et al. (1991) developed methods for elevation grids. Briggs (1989) developed a method based on TIN models. Moore et al. (1988) and Chen (1991) developed methods based on elevation contours. In existing grid DEMs based topographic structure extraction approaches, the hydrological flow modelling method is widely applied (O'Callaghan and Mark, 1984; Jenson and Domingue, 1988; Tarboton et al., 1991; Moore, 1994). The inherent problems of the efficiency and accuracy of these approaches, especially when handling depressions and flat surfaces, has hindered their application in the processing of large-scale DEMs. Concerned with these problems, an extraction approach based on grid DEMs is proposed in this paper. In this approach depressions are filled by adopting vector processes combined with traditional neighbourhood raster processes, and the direction of the flow over flat surface is assigned by applying a neighbour-grouping scan method. The approach was tested with different volumes of DEMs. The results indicate that its efficiency and accuracy are all better than existing methods.

2 EXTRACTION APPROACH BASED ON THE HYDROLOGICAL FLOW MODELLING METHOD

The main principle of the hydrological flow modelling method is to compute the catchment amount of each terrain cell according to the natural rule that water flows from high to low. Those cells with catchment amounts greater than a threshold are considered to be located in a valley line, which can be extracted by linking these cells together. The ridges are extracted by calculating the watershed of each valley line and taking the boundaries of these watersheds as ridge lines (Li and Zhu, 2000). In such an approach the main task is to derive three sets of data from the original DEM: a depression-less elevation matrix, flow direction matrix and flow accumulation matrix. Upon the completion of the generation of these three matrixes, a series of specific applications including the extraction of the topographic structure can be carried out. Practically, an initial flow direction matrix is calculated first because information on the direction of the flow is required in the procedure to fill the depression. The method of calculating the flow direction matrix is as follows: For each cell in the grid the direction to the lowest of its neighbours (steepest descent direction) is determined, allowing for a $1/\sqrt{2}$ factor for diagonal ones. The direction is then encoded in powers of two and stored in the flow direction matrix (O'Callaghan and Mark, 1984; Jenson and Domingue, 1988; Freeman, 1991).

Depressions, especially looping depressions (adjacent depressions spilling into each other) have been recognized to be one of the chief obstacles in the extraction of topographic structures (Jenson and Domingue, 1988; Freeman, 1991; Tarboton, 1991; Moore et al., 1994). In this paper the bottom cells of the depression (a cell with an elevation lower than all of its eight neighbours) are marked out while calculating the initial flow direction matrix. Starting from these bottom cells, the procedures to detect and fill the depression are carried out by combing vector processes with traditional neighbourhood raster processes. The efficiency of handling depressions (especially looping depressions) is improved much compared to traditional methods. Assigning the direction of flow over flat surfaces is another stubborn problem in extracting topographic structures from grid DEMs. The main disadvantages of existing methods include ineffective parallel flow and sawtooth phenomena (Jenson and Domingue, 1988; Martz and de Jong, 1988; Freeman, 1991; Tribe, 1992; Moore et al., 1994; Martz and Garbrecht, 1995). But, here, the direction of the flow is assigned over a flat surface by adopting a neighbour-grouping scan method: the neighbour cells without flow directions are grouped according to their locality (cardinal and diagonal), and scanned in sequence (cardinal prior to diagonal) to assign flow directions, and the scan sequence within the same group is determined randomly. The approach can handle flat surfaces with multiple outlets and leads to much less parallel flow and sawtooth phenomena. In generating a flow accumulation matrix, a recursion procedure is used. Upon deriving the accumulation matrix, a threshold is applied to extract valley lines using a tree-generating algorithm. The ridges are then extracted by simply applying the same approach used to extract valleys on an upside-down elevation matrix of the original DEM. In the following sections the above-mentioned algorithm will be discussed in detail.

3 TO DETECT DEPRESSIONS

Most depressions in grid DEMs are mistakes in the data brought about during the process of collecting data, while some represent real terrain features; e.g., quarries or grottoes, etc. O'Callaghan and Mark (1984) have investigated the causes of depressions while generating DEMs using a digital photogrammetry workstation (DPW) and interpolations from contour maps. There are three kinds of depressions in grid DEMs: single-point depression, standalone depression and compound depression, of which compound depressions are dominant in most grid DEMs.

All the bottom cells of a depression are marked out while generating the initial flow direction matrix. The single-point depressions can be filled by simply raising the elevation of each bottom cell to the lowest of its neighbour's. After such a step, the steepest descent value of all cells is no less than zero. Since only the cells with steepest descent value equal to zero can form a mutual-pointing phenomena (adjacent cells with the steepest descent directions point to each other), and the mutual-pointing can serve as an evidence of the existence of a depression, these cells then can be used as clues for detecting depressions. Here, we adopt a stack-based seed-filling algorithm. A flag matrix (a flow accumulation matrix can be utilized for the present) is used to mark the detected depressions. The basic work flow includes following steps:

1. Push an unmarked cell with a steepest descent value equal to zero (used as seed) into the stack, and mark its correspondent cell in the flag matrix;
2. Pop up a cell from the stack top and mark its correspondent cell in the flag matrix;
3. Scan the cell's eight neighbours in sequence; if a neighbour's flow direction points to it and the correspondent flag cell of this neighbour is not yet marked, push this neighbour into the stack;
4. Execute steps (2) and (3) until the stack is empty.

After undergoing such processing, all depressions can be detected. A data structure is defined to store the information on the depression, after which the following filling process can be carried out. The C++ language description of the structure is presented below:

```
class CDepression
{
    int          nID;            // Identification of depression
    int          nPointedID;     // Identification of spilled depression
    CRect        rectExterior;   // Depression's minimum outer-enclosed rectangle
    CPointArray  pntsOutlet;     // Depression's flow outlets
    CPolygon*    pVergin;        // Depression's rim
}
```

The *nID* is the unique identification of a depression in a flag matrix. The minimum outer-enclosed rectangle can be determined while detecting depressions. Searching along the upper side of a depression a point in its rim can be located directly. Starting from this point the depression's rim can be traced out in the flag matrix according to the following steps:

1. Starting from the known rim point, scan its neighbours clockwise from an initial scan direction until a new rim point is located;
2. Starting from this new rim, point and rotate the current scan direction 90° counter-clockwise. Scan its neighbours clockwise until a new rim point is located;
3. Repeat step (2) until the first known rim point is located;
4. Correct any point omission problems arising from the concave corner in the rim (see Figure 1).

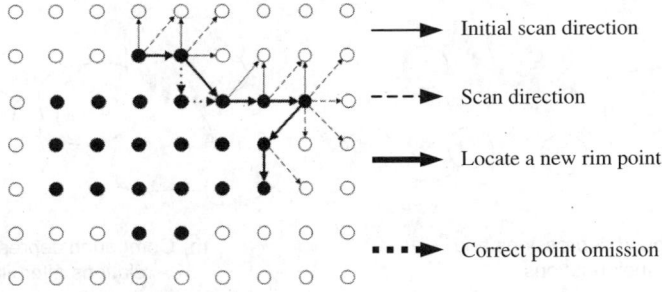

Figure 1.   Tracing the rim of the depression.

All of the outlets can be searched along the rim. The problem of selecting wrong outlets encountered by Martz and de Jong (1988) will not arise here because the searching is based on the vector characteristic of the depression. Moreover, the outlets can be determined correctly whether the depression is standalone or compound. If its outlet is located in another depression's rim, then it belongs to a compound depression. In this paper we call such a relation among adjacent depressions as a depression pointing to another (its outflow spills into another), and represent it as nPointID. The filling of these depressions is discussed in the next section.

3.1 *To fill depressions based on vector processes*

The dominant depressions in a grid DEM are compound ones, and quite a few of them are represented as looping depressions (which frequently occur when many depressions are located on a relatively flat area), as shown in Figure 2(a). In such a case, a new depression may emerge after the adjacent one had been filled; hence the procedures for detecting and filling need to be performed recursively until there is no new depression to be created. Herein lies the inefficiency of the existing approaches. The depression-filling procedure developed by Mark et al. (1984) does not include the logic for finding looping depressions. Rather, one depression is detected at a time, and a flat area is assigned entirely to the first depression that touches it. The procedure developed by Jenson and Domingue (1988) is an improvement in some respects in that it can handle multiple depressions in an execution loop and flat areas can be subdivided if they have more than one adjacent depression. However, detecting and filling procedures are still required to deal with new depressions after the compound depressions have been filled.

In this paper, the depressions are filled based on their vector characteristics such as outer-enclosed rectangles, rims, flow outlets and point relations by adopting vector processes combined with traditional neighbourhood raster processes. The filling procedure can be summarized as follows:

1. Based on the point relations recorded during the detecting of depressions, the looping depressions in compound depressions can be identified (depicted as depressions I, II and III in Figure 2 (a));
2. Merge the looping depressions in a new depression (depicted as depression V in Figure 2 (b)), the vector characteristics and point relations of which have been calculated. If the depression is a standalone depression (in that its outflow does not spill into other) it will be filled by raising the cells contained in it to the elevation value of the outlets;
3. Repeat steps (1) and (2) until all depressions in the DEM have been handled.

3.2 *Assigning flow direction over flat surfaces based on a neighbour-grouping scan*

The reasons why flat surfaces can arise in a grid DEM include: data truncation, DEM generation from contours, some real terrain features, and the result of filling depressions (O'Callaghan and Mark, 1984; Freeman, 1991). To date, many methods of assigning the direction of the flow over flat surfaces have been devised. Jenson and Domingue (1988)'s method starts from the outlets to

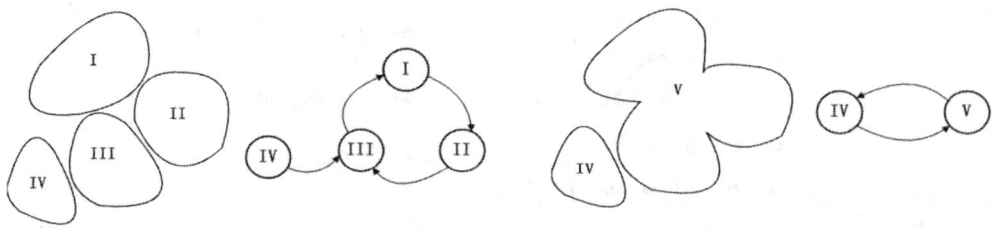

(a) Compound depressions and
their relations

(b) Compound depressions and their
relations after filling one time

Figure 2.   Processing compound depressions.

assign the direction of the flow to the adjacent cells in flat surfaces that have not been assigned. Their own neighbours in the flat surface are then scanned to give a direction. The procedure will be repeated until the whole flat area has been involved. Though this method can handle flat surfaces with multiple outlets and can be implemented easily, it introduces many parallel flow and sawtooth phenomena. Freeman (1991) developed a method that first indexes all of the cells in a flat area according their distances to outlets, and then assigns flow directions to these points according to their indices by applying a shortest route algorithm. This method leads to less parallel flow and sawtooth phenomena, but has a low efficiency because of its two global traverse procedures over the whole flat area. The method adopted by Martz and de Jong (1988) set the catchment of every point in the flat area to the total catchment of the whole flat area, as the task of assigning the direction of the flow does not need to be carried out explicitly. The efficiency is greater than those of previous methods, yet there are too many parallel flows in the result. Concerned about this deficiency, Tribe (1992) developed a method that decreases the parallel flow but brings more sawtooth phenomena (Martz and Garbrecht, 1995).

As in the detection of depressions, the vector characteristics of a flat area (minimum outer-enclosed rectangle and rim) are also calculated. All of the outlets can be searched out along the rim (see Figure 3(a)). The basic method of assigning the direction of the flow over a flat area based on the neighbour-grouping scan method is as follows:

1. Suppose there are N outlets, N first-in-first-out queues are constructed, and each outlet is separately put in a queue. The N queues are sorted by the steepest descent of the outlet placed in them;
2. Get the head from the queue as the current cell, scan its eight neighbours (cardinal ones prior to diagonal ones). If one of the neighbours is in the flat area and has not been assigned a flow direction, then set the direction to be pointing to the current cell (depicted as Figure 3(b), (c)), and put this neighbour into the queue;
3. For each queue, execute step (2) once if it is not empty.

3.3 *Generating a flow accumulation matrix and extracting valley lines*

The catchment area of each cell in a DEM is represented as a flow accumulation matrix, in which each cell is assigned a value equal to the number of cells that flow into it. After filling depressions and assigning flow directions over a flat area, the depressionless elevation matrix and flow direction matrix can be obtained. Based on these two matrixes, the flow accumulation matrix can be calculated according to the following main principle: Give a unit flow to each cell and compute the total outflow of each cell based on the flow direction matrix (Freeman, 1991; Li and Zhu, 2000). In this paper a recursive routine is applied to compute the total catchment area of each cell. The number of the local variable in the routine is kept to a minimum to save on the cost of dynamic memory. The execution time is proportional to the size of the DEM.

The extraction of valley lines is based on the flow direction matrix and flow accumulation matrix. Given a flow accumulation threshold, the cells with catchment amounts greater than this

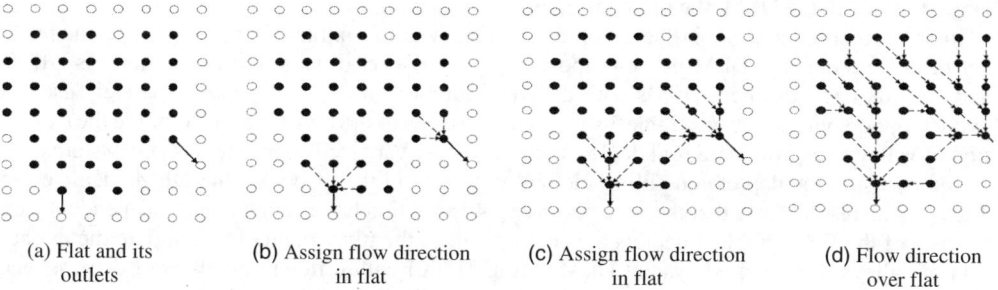

(a) Flat and its outlets   (b) Assign flow direction in flat   (c) Assign flow direction in flat   (d) Flow direction over flat

Figure 3.   Assigning the direction of the flow over flat surfaces.

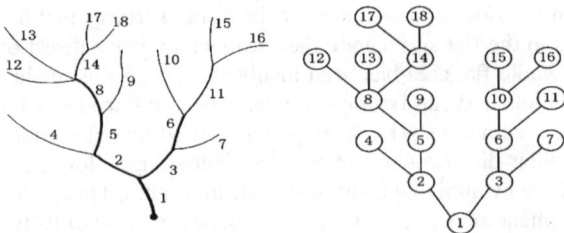

Figure 4.   Valleys represented by a tree-structure.

threshold will be linked together and vectorized to valley lines. In a depressionless DEM the outflow of each cell will flow along a decreasing path that will eventually lead to a watershed outlet located at the edge of the DEMs. The watershed has a maximal catchment. Thus, starting from this cell all of the valley lines in the corresponding watershed can be extracted by applying a tree-generating algorithm. These valley lines can be represented by a tree structure depicted as shown in Figure 4.

In some existing approaches the ridges are extracted by calculating the watershed of each valley line and taking the boundaries of these watersheds as ridges lines. They are not consistent with natural topography due to the closed watershed boundaries (Li and Zhu, 2000). In this paper the ridges are extracted by applying the same approach used to extract valleys on an upside-down elevation matrix of the original DEM; thus, the problem of closed lines can be avoided. The level of detail of the extracted topographic structure is in inverse proportion to the magnitude of the flow accumulation threshold; generally when it is set to the mean value of the accumulation of all of the cells, the ideal result can be obtained (Tang, 2000).

4   EXPERIMENTS AND RESULTS

The above proposed approach was implemented using C++ and tested with real DEMs of various sizes. The results are compared with those of existing methods (the Spatial Analysis Module of ARC/INFO V8.0 developed by ESRI Inc.). The main purpose of the experiment was to investigate the efficiency and accuracy of the approach. The data sets used in the experiment were the square grid DEMs (with a 5-metre grid cell size) obtained from aerial photogrammetry by a digital photogrammetric workstation. A detailed description of various DEMs is listed in Table 1. Most of these DEMs are mountain regions. There are also a few rivers, along which a scattered some large flood plains.

The test environment is as follows: the CPU P4 is 1.7 GHz, the Memory is 256 M, and the operating system is Windows 2000 Professional. To start with, the extraction efficiencies when applied on DEMs of varying sizes are compared (see Table 2). The comparison shows that the proposed approach has a significantly improved level of efficiency compared to existing approaches: the larger the size of the DEM, the more efficient the extraction.

In order to investigate the influence on the efficiency of the extraction due to the size and number of depressions in a DEM in the proposed approach, a set of varisized depressionless DEMs and a set of size-equivalent DEMs with different numbers of depressions are separately used to extract topographic structures. Table 3 shows comparisons of the time taken to compute the extraction of valley lines from varisized depressionless DEMs. Varisized depressionless DEMs are generated by using the depression filling method proposed in this paper on the DEMs depicted in Table 1. The result shows that the time required for extraction has a rough linearly dependency on the size of the DEM. Table 4 depicts the comparison of the time required to compute the extraction of valleys from size-equivalent DEMs (1000*1200) with different numbers of depressions. The results show that the time needed to execute the extraction also has a rough linear dependency on the number of depressions.

104

Table 1.   The basic descriptive information of varisized DEMs.

DEM size	500* 600	1000* 600	1000* 1200	1000* 2400	2000* 2400	3000* 2400	3000* 3600
Min elevation (m)	702.4	702.4	646.8	646.7	608.3	566.9	571.2
Max elevation (m)	958.8	1053.7	1055.4	1080.1	1080.1	1134.7	1134.7
Mean elevation (m)	820.4	863.3	824.5	833.2	796.3	775.9	789.2
Standard deviation	45.9	64.5	81.9	83.1	91.4	113.7	109.2

Table 2.   The comparison of computation times for the extraction of valleys.

DEM size	500* 600	1000* 600	1000* 1200	1000* 2400	2000* 2400	3000* 2400	3000* 3600
Amount of depressions and flats	108	197	821	1295	3187	5209	6173
Consumming time of proposed approach (s)	1	2	3	5	13	31	34
Consumming time of ARC/INFO (s)	5	7	28	73	212	397	510

Table 3.   The time required to compute the extraction of valleys from varisized depressionless DEMs.

DEM size	500* 600	1000* 600	1000* 1200	1000* 2400	2000* 2400	3000* 2400	3000* 3600
Consumming time (s)	1	1	2	3	6	13	21

Table 4.   The time required to compute the extraction of valleys from size-equivalent DEMs with different numbers of depressions.

Amount of depressions and flats	254	509	623	677	810	1245	1272
Consumming time (s)	1	2	2	4	4	5	5

A comparison of the accuracy of the results is also carried out between the proposed approach and ARC/INFO. Figure 5 shows the results derived using the two approaches. In mountainous regions, there is no apparent difference, but in the region dominated by flood plains (the upper-left corner), the proposed approach has more a proper convergence of flow and results in far fewer parallel lines than ARC/INFO (see the flat regions marked by circles).

5   CONCLUSION

The topographic structure extraction approach proposed in this paper fills depressions by adopting vector processes combined with traditional neighbourhood raster processes and assigns the direction of the flow over flat surfaces by applying a neighbour-grouping scan method. The approach has been tested with real DEMs, and the results show that its efficiency and accuracy are better than existing methods. This makes it more appropriate for the processing of large-scale DEMs, which can be used for spatial decision-making with regard to large regional sustainable development projects, such as the planning of drainage areas, formulating responses to flood disasters, determining borderlines, and so on.

Extensions to this approach are required to make it more suitable for the extraction of hydrologic structures and foruse in hydrologic decision support systems; e.g., the introduction of a

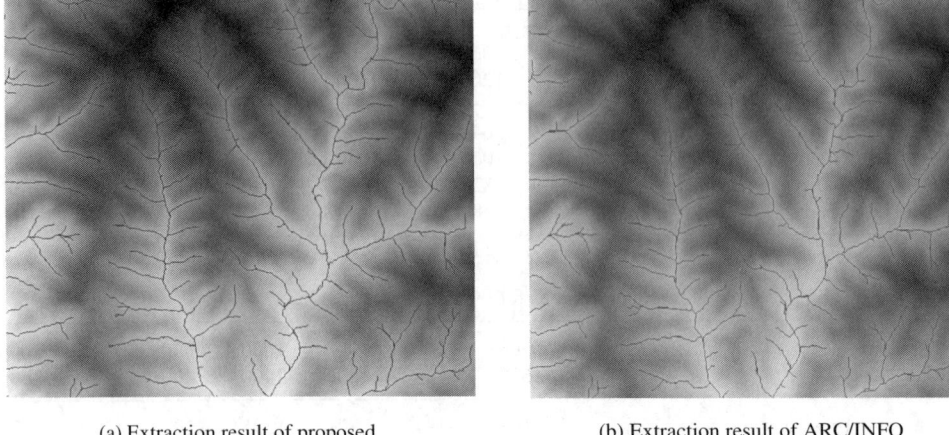

(a) Extraction result of proposed   (b) Extraction result of ARC/INFO

Figure 5.   Comparison of extraction on a mountainous region dominated DEM.

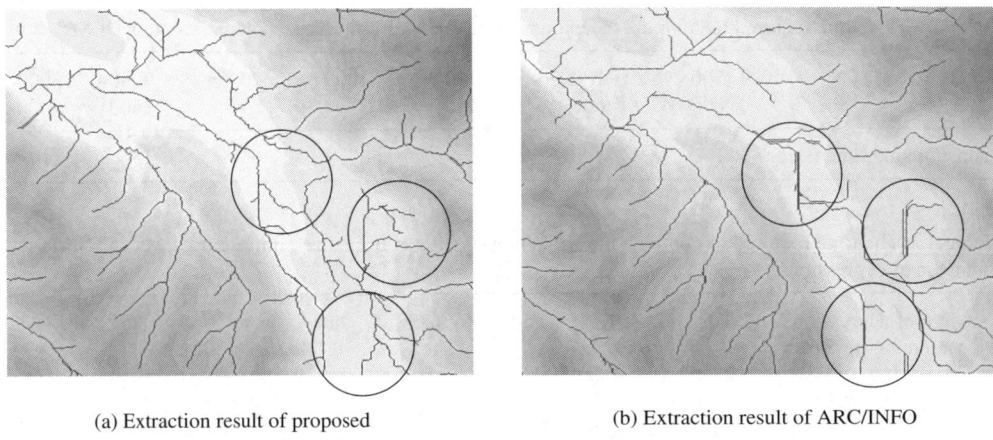

(a) Extraction result of proposed   (b) Extraction result of ARC/INFO

Figure 6.   Comparison of extraction on a plain region dominated DEM.

divergent flow method for more accurate extraction results (Freeman, 1991; Wolock and McCabe, 1995; Pilesjo et al., 1998; Li and Zhu, 2000). In addition, information on depressions is required in some hydrological analysis, and depressions that represent real terrain features should be reserved for later use. In such cases, determining the cause of the depressions should be done first.

ACKNOWLEDGEMENTS

The work described in this paper was supported by the NSFC (40001017) and by the National Key Basic Research and Development Program (2002CB312101).

REFERENCES

Briggs I., 1989, Water Flow Using Density of Area, CSIRO Division of Information Technology, Technical Report pp. 1–11.

Chen Xiao-Yong., 1991, Mathematical Morphology and Image Analysis, Beijing: Survey & Mapping Press (in Chinese) .

Freeman T G., 1991, Calculating Catchment Area with Divergent Flow Based on A Regular Grid, Computers and Geosciences, 17(3): 413–422.

Jenson S K., Domingue J O., 1988, Extraction Topographic Structure from Digital Elevation Data for Geographic Information System Analysis, Photogrammetirc Engineering and Remote Sensing, 54(11): 1593–1600.

Li Zhilin, Zhu Qing, 2000, Digital Elevation Model, Wuhan: Wuhan Technology University of Survey and Mapping Press (in Chinese).

Marks D., Dozier J, Frew J., 1984, Automated Basin Delineation from Digital Elevation Data, Geographic Processing, 2: 299–311.

Martz L W., de Jong. CATCH, 1988, A FORTRAN Program for Measuring Catchment Area from Digital Elevation Models, Computers and Geosciences, 14(5): 627–640.

Martz L W., Garbrecht J., 1995, Automated Recognition of Valley Lines and Drainage Networks from Grid Digital Elevation Models: A Review and A New Method – Comment, Journal of Hydrology, 167(5): 393–396.

Moore I D., O'Loughlin E M., Burch G J., 1988, A Contour-Based Topographic Model for Hydrological and Ecological Application, Earth Surface Processing and Landforms, 13(3): 305–320.

Moore I D., Grayson R B., 1994, Ladson A R. Digital Terrain Modelling: A Review of Hydrological, Geomorphological, and Biological Applications, Terrain Analysis and Distributed Modelling in Hydrology (Eds: Beven K J and Moore I D.), Chichester, UK: John Wiley & Sons, pp. 7–34.

O'Callaghan J F., Mark D M., 1984, The Extraction of Drainage Networks from Digital Elevation Data, Computer Vision, Graphics, and Image Processing, 28(4): 323–344.

Pilesjö P., Zhou Q M, Harrie L., 1998, Estimating Flow Distribution over Digital Elevation Models Using A Form-based Algorithm, Geographic Information Sciences, 4(1): 44–51.

Tang G A., 2000, A Research on The Accuracy of Digital Elevation Models, Beijing New York: Science Press, pp. 132–148.

Tarboton D G., Bras R L., Rodriquez I I., 1991, On The Extraction of Channel Networks from Digital Elevation Data, Hydrologic Processes, 5(1): 81–100.

Tribe A., 1992, Automated Recognition of Valley Lines and Drainage Networks from Grid Digital Elevation Models: A Review and A New Method, Journal of Hydrology, 139(3): 263–293.

Wolock D M., Hornberger G M., Beven K J., et al., 1989, The Relationship of Catchment Topography and Soil Hydraulic Characteristics to Lake Alkalinity in the Northeastern United States, Water Resources Research, 25(5): 829–837.

Wolock D M., Hornberger G M., MusGrove T J., et al., 1990, Topographic Effect on Flow Path and Surface Water Chemistry of the Llyn Brianne Catchments in Wales, Journal of Hydrology, 115(3): 243–259.

Wolock D M., McCabe J G J., 1995, Comparison of Single and Multi Flow Direction Algorithms for Computing Topographic Parameters in TOPMODEL, Water Resources Research, 31(5): 1315–1324.

Integrated techniques and application-oriented models

Advances in Spatial Analysis and Decision Making, Li, Zhou & Kainz (eds)
© 2004 Swets & Zeitlinger, Lisse, ISBN 90 5809 652 1

Integrating spatial statistics into a multi-resolution classification framework to improve the mapping of urban land use/cover

DongMei Chen
Department of Geography, Queen's University, Kingston, Ontario, Canada

ABSTRACT: The fundamental problem involved in producing accurate urban land use maps from remotely sensed data is that urban areas comprise a complex spatial assemblage of land cover types. With the recent increasing availability of 1–5 m high-resolution satellite and airborne digital images, the problem of spatial variability in each urban land use/cover type is more acute, making it more difficult to use the traditional approach of single-resolution classification. This paper presents a multi-resolution analysis and classification framework for selecting and integrating suitable information from different resolutions and analytical techniques into classification routines. Information on the spatial structure of the image should be acquired from spatial statistics for each land use/cover class prior to and during the classification. The multi-resolution approaches were tested using simulated multi-resolution images for a portion of the rural-urban fringe of the San Diego Metropolitan area. We demonstrate that the multi-resolution classification approaches integrating techniques of spatial analysis can significantly improve the accuracy of urban land use/cover classifications as compared with single-resolution approaches.

1   INTRODUCTION

Remotely sensed data have been the major source for generating land use/cover information used by many governments and private organizations at the local, regional, and national levels for different applications, such as environmental monitoring and planning, land use/cover change modelling, transportation planning, urban development planning, and urban modelling. There are two major ways of generating land use/cover maps through remotely sensed data. One is by visually interpreting image features and manually digitizing land use/cover unit boundaries in the image. This approach can generally produce relatively accurate land use/cover maps. Unfortunately, it is expensive and usually unaffordable, especially when applied to a large area. The other approach is by conducting semi-automated image classifications that provide a low-cost approach to generate information on land use/cover, but with a loss of precision and accuracy.

Over the past forty years, many semi-automated methods of classifying images of remotely sensed data have been established to improve the accuracy of land use/cover classifications (Jensen 1996). Different algorithms and strategies have been developed and tested for various environments. They include supervised, unsupervised and hybrid training approaches (Richards 1986); parametric and nonparametric classifiers; segmentation (Conner et al. 1984); artificial neural networks (ANN) (Civco 1993); fuzzy sets (Wang 1990, Foody 1996) and knowledge-based systems (Kontoes & Rokos 1996).

Previous studies of conventional semi-automated image classifications have indicated that a decrease in the overall accuracy of land use/cover classifications may occur as the spatial resolution of images is increased while other sensor characteristics remain the same (Townshend & Justice 1981, Toll 1985, Latty et al. 1985, Marceat et al. 1994a). Due to the complexity of urban environments and the increasing use of high/fine resolution images, the desired image classes often do not have homogenous spectral characteristics within each land use/cover parcel (Gong and Howarth 1990, Martin et al. 1988). These land use/cover objects have different sizes and patterns, which require different scales of analysis according to the scene model as suggested by Woodcock and Strahler (1987). At a particular level of classification, some categories (such as trees) are better classified at finer resolutions while others (such as commercial buildings) require coarser resolutions. This implies that classification routines based on a single resolution may not improve the accuracy of all categories.

It is commonly agreed that the spatial resolution that can best represent each land use/cover class is not arbitrary, but tends to concentrate in a relatively narrow range as a function of the physical laws or human activities that dominate each level (O'Neill et al. 1989). Efforts have been made to select appropriate resolutions in remote sensing applications by examining variations in the brightness of images as a function of changes in analytical scale and resolution using exploratory data analysis or geo-statistics (Woodcock & Strahler 1987, Woodcock et al. 1988a and 1988b, Jupp et al. 1989a, Marceau et al. 1994b, Atkinson & Curran 1997, Collins & Woodcock 1999). The previous studies were limited to the selection of a single optimal resolution at which an overall higher accuracy of mapping could be achieved, rather than a suite of resolutions at which each individual class could be better mapped.

The problem of selecting appropriate resolutions is complex. The appropriate resolution is a function of the type of environment, the kind of information desired, and the techniques used to extract information. The available spatial scale of image data, methods of analysis, environmental situations, and main questions about these environments make the enumeration of their combination a difficult task. A classification framework to assess the effects of resolution and to incorporate appropriate routines from different resolutions to improve the accuracy of image analysis and classification has yet to be developed.

In this paper, the problems in urban mapping using high-resolution remotely sensed images and the limitations of conventional single-resolution approaches are discussed. The different statistics used in different stages of image classification are then reviewed and theoretical background on multi-resolution analysis and classification frameworks for remotely sensed data is presented. The multi-resolution framework is based on the development of spatial analytical techniques and strategies to select and integrate suitable information from different resolution into classification routines.

2   PROBLEMS IN URBAN MAPPING WITH HIGH-RESOLUTION IMAGES

An urban landscape is a mosaic of highly heterogeneous land use/cover objects. Different land use/cover classes such as residential, commercial and recreational areas have different sizes and patterns, such that a range of within-class variances of image brightness occurs at different spatial resolutions for different classes. In high-resolution imagery different urban classes have a wider range of spatial variation than in low-resolution imagery. In traditional classification procedures, the selection of training sites is based on the assumption that specific ground or surface features will be represented as distinct pixel regions with relatively homogenous spectral reflectance values in an image. This assumption indicates a scene-model characterized by a high degree of spatial autocorrelation between image elements and low overall variance (Woodcock & Strahler 1987). When the ground sampling distance is large relative to ground objects, individual pixels often represent parts of two or more objects of interest; i.e., the L-resolution scene model. The L-resolution model is particularly amenable to traditional classification approaches, as noted by Woodcock and Strahler (1987) and many of the earlier classification decision algorithms using low-resolution data such as 80 m Landsat Multi-Spectral Scanner (MSS) imagery were training-site based. Similarly, when the pixels become very small relative to objects (H-resolution scene), the internal variance of the objects results in difficulty in delimiting homogenous training sites. In general, the assumption of the commonly-used per-pixel classifiers does not hold for most urban land use classes in high-resolution images. This, in turn, makes the training process more complex and adversely affects the results of the commonly used per-pixel classifiers (Markham & Townshend 1981, Gong & Howard 1990, Barnsley et al. 1996).

Previous studies of multi-spectral classifications using higher spatial resolution images, such as the Landsat Thematic Mapper (TM) and SPOT High Resolution Visible (HRV) data, have shown that higher accuracy may not be obtained from lower spatial resolution data such as Landsat MSS data. Townshend and Justice (1988) found a marked decrease in the accuracy of classifications of residential land use with higher resolution. The fundamental problem involved in producing accurate urban land use maps from remotely sensed data is that urban areas comprise a complex spatial assemblage of land cover types, each of which may have different spectral reflectance characteristics (Wharton 1982, Gong & Howard 1992, Barnsley et al. 1996). Unfortunately, per-pixel spectral classification algorithms are poorly equipped to deal with this type of spatial variability (Woodcock & Strahler 1987, Barnsley et al. 1996). A further problem for per-pixel spectral classification is the difficulty mentioned above of defining suitable training sets for many categories of urban land use, due to spatial variations in the spectral response of their component land cover types (Gong & Howarth 1990, Barnsley & Barr 1996).

Several approaches have been developed to overcome some of the above problems. They include:

(1) Improving the quality of the training statistics by refining the supervised training samples (Toll 1985a, Gong & Howard 1990, Chen & Stow 2002) and testing unsupervised training strategies;
(2) Using preclassification filtering, such as median filters (Cushnie & Atkinson 1985), and texture or structural measures (Baraldi & Parmiggiani 1994, Franklin & Peddle 1990, Gong & Howarth 1990);
(3) Incorporating ancillary data into the classification procedure (Ehlers et al. 1991);
(4) Using enhanced or advanced classification algorithms, ranging from context-based classification (Wharton 1982, Gong et al. 1992, Sharman & Sarkar 1998), through knowledge-based classification and expert systems (Wharton 1989, Moeller-Jensen 1990, Kontoes & Rokos 1996), to artificial neural networks (Civco 1993);
(5) Post-classification spatial processing, ranging from simple majority filters to spatial (or contextual) reclassification procedures (Wharton 1982, Gong et al. 1992, Barnsley & Barr 1992, 1996); and
(6) Image segmentation (Conners et al. 1984, Woodcock & Harward 1992).

Most of the above procedures involve incorporating spatial information on high-resolution images from window-based or region-based analyses (Gong & Howard 1992). Not all of these techniques directly address the problem of inferring land use from a complex spatial mixture of spectrally distinct land-cover types (Barnsley & Barr 1996). Up to now most of the above methods have been tested with Landsat TM or SPOT HRV images and few have been conducted with images with a ground sample distance size in the range of 1 m to 5 m of resolution. With the recent increasing availability of 1–5 m high-resolution satellite and airborne digital images such as Earthwatch, Space Imaging (IKONOS), Orbimage, IRS, ADAR, DOQQ, etc., the problem of spatial variability in each land use/cover type has become more acute, making it more difficult to use the traditional per-pixel classifier.

It is evident that most of the information on the spatial structure of the image should be required to be given prior to and during the stage of classification training. The accuracy of the classification decision rules is primarily dependent on the training data sets that are supplied. Knowledge-based classification and segmentation routines based on multi-level nested hierarchical scene models can successfully establish and incorporate spatial structures at different levels of the classification process (Woodcock & Harward 1992). However, because most commercially available classification routines are spectrally based and rely on the extraction of supervised or unsupervised training data, the effective use of these approaches requires that additional methods be employed to establish the spatial structure and characteristics of imagery.

The precise specification of suitable training algorithms and their input parameters may be estimated from spatial statistics that identify the scale and form of the spatial autocorrelation in each image class and band. The following section introduces the multi-resolution framework of image

classification and reviews spatial statistical measures to provide the spatial information requirements for classification.

3 MULTI-RESOLUTION IMAGE ANALYSIS AND CLASSIFICATION FRAMEWORK

The question of how one might choose scales that best represent each land use/cover type is the
basis for proposing multi-resolution classification approaches. Such approaches should incorporate information obtained from multi-resolution data and exploratory data analysis to improve the
accuracy and effectiveness of urban land use/cover classification procedures. A proposed multi-
resolution image analysis and classification framework for integrating information from multiple
resolutions is illustrated in Figure 1.

The framework proposed here, focuses on examining image patterns/structures/autocorrelations
using different spatial analytical techniques in order to select appropriate methods in different stages
of classification such as training strategy, feature extraction, scene models, and accuracy of classification. Multi-resolution images can be generated by aggregating fine resolution images into different
levels of coarse resolution (i.e., image pyramids). Several methods can be used in aggregation, such
as simple aggregation (Bian & Butler 1999), geo-statistical methods (Collins & Woodcock 1999),
Sacel-space transformation (Lindeberg 1994), and wavelet transformation (Mohanty 1997).

Techniques of spatial analysis for measuring the size of patterns and degree of autocorrelation are
described in section 3.1. Techniques for assessing appropriate spatial resolutions are discussed in
section 3.2. Spatial statistics are computed for each training class to guide the selection of training data
and of high-resolution or low-resolution classification models, and to ensure the selection of appropriate methods and parameters and the range of spatial resolutions used for classification. The different strategies of incorporating information from different resolutions are explained in section 3.3.

3.1 *Spatial statistics in image analysis and classification*

Several variables play a role in the success of image classification: the type and level of the classification system; spatial, spectral and radiometric resolution of imagery, landscape variability

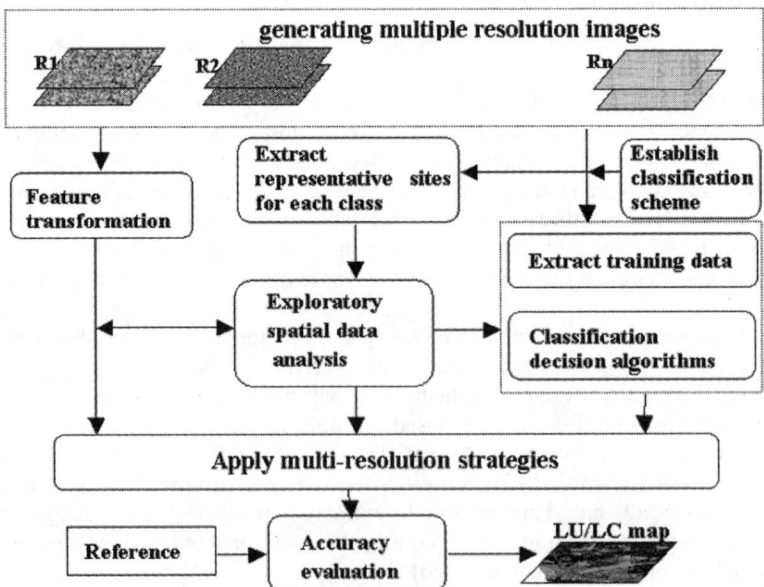

Figure 1.   A proposed multi-resolution image analysis and classification framework. Rn stands for variable
image spatial resolution.

and the classification algorithm. Proper application of image classification procedures requires knowledge of the above variables of the data to determine the appropriate classification methodology and parameters to use. Values for these attributes are often selected on an *ad-hoc* basis, ignoring the spatial characteristics of the image being considered (Woodcock & Strahler 1987, Stein et al. 1988, Woodcock & Harward 1992). Selecting an appropriate classification routine and optimally specifying its required parameters are paramount issues to facilitate the accurate use of high spatial resolution and hyper-spectral imagery, as noted by Ryerson (1989) and Lunetta et al. (1991). A framework for establishing the spatial characteristics of imagery and selecting the analysis resolution range in classification analysis has yet to be established.

Prior to classifying the images, an exploratory spectral-radiometric data analysis and a visual assessment of the spatial characteristics of each image band is normally conducted (Jensen 1996, Gong & Howard 1990). Exploratory spatial data analysis may be used to determine the required information on image spatial characteristics and to ensure that appropriate methods and parameters are used.

Since Haralick (1979) demonstrated the potential spatial information content of remotely sensed imagery based upon the interrelationships of pixels, these pixel interrelationships were subsequently identified as a possible problem in the form of spatial autocorrelation (Campbell 1981). As a result, a number of techniques have been developed for assessing and using image spatial autocorrelation in pre-processing, classification and post-classification evaluation. Table 1 summarizes the spatial statistical techniques derived from a limited review of related literature on the application of spatial techniques in remote sensing applications.

Table 1. Summary of spatial techniques utilized in different stages of image classification.

Phase of classification process	Techniques	References
Pre-classification		
Registration, filtering and enhancement	Wavelet	
Training	Spatial autocorrelation	Mohanty (1997),
	Local Variance	Labovitz and Masuoka (1984),
		Woodcock and Strahler (1987)
Deciding appropriate resolutions	Semi-variance	Curran (1988)
		Jupp et al. (1988, 1989)
		Woodcock et al. (1988)
		Atkinson (1997)
Sampling	Fractal	Weiler and Stow (1991)
	Spectral analysis	
Pattern analysis		
Classification		
Creating texture features	Simple statistical transformation	Hsu, S. (1978)
	Matrix of pixel co-occurrence (GLCM)	Gong et al. (1992)
	Texture spectrum	Haralick (1979)
		Wulder and Boots (1998)
	Fraction	Flygare (1997)
	Semi-variogram	
	Wavelet	Bian (2003)
Evaluating errors of classification	Spatial autocorrelation	Campbell (1981)
	Join count statistics	Congalton (1988)
	Semi-variance	Atkinson (1997)

3.2   *Techniques for assessing appropriate spatial resolutions*

To select an image with appropriate spatial resolution for a study, one must examine the characteristics of scene content, especially the variable patterns of a scene as a function of difference in analytical scale and resolution. Most studies on selecting optimal spatial resolutions have focused on examining the accuracy of estimating some property at the ground with remotely sensed imagery of different spatial resolutions. For example, Markham & Townshend (1981) evaluated the effect of decreasing spatial resolution on land cover classification on a per-pixel basis. Airborne multi-spectral scanner images were aggregated from 5 m resolution to 40 m resolution, each data set was classified and each classification was evaluated within the classifier training pixel areas. In an urban-suburban test site, the accuracy of land cover classifications increased as the spatial resolution was degraded.

Considering that improvements in image analysis and classification could be achieved if there is prior knowledge about the optimal spatial resolution, efforts have been made to develop methods for examining image characteristics at different resolutions. Woodcock & Strahler (1987) used the local image variance to help in selecting an appropriate image scale for forested, agricultural and urban/suburban environments. A measure of local variance at different spatial resolutions was computed and plotted, and the location and height of the peak of the curve was used to select an optimal spatial resolution for the scene. An optimal resolution of 15 to 20 m was determined for a scanned colour infrared photograph of suburban southern California. The drawback of the local variance method is that it is strongly related to the global variance of the image studied. Therefore, the measure is image-dependent and comparisons cannot be made between images.

Townshed & Justice (1988) applied scale variance and Fourier analyses on spatially degraded Multi-spectral Scanner data, from 125 m to 4 km, to investigate the required spatial resolution for the global monitoring of land transformations. They advised an average resolution of 500 m as the best compromise between the detail of the changes detected and the resulting volume of data.

For a forested environment, Marceau et al. (1994b) tested an approach to identifying the optimal spatial resolutions for the detection and discrimination of coniferous classes. The minimal intra-class variance was used as the indicator of the optimal spatial resolution. MEIS-II data with 0.5 m resolution was degraded to 29.5 m with an increment of 1 m, using an averaging window algorithm. They concluded that the optimal spatial resolution is primarily affected by the spatial and structural parameters of forest stands.

Fractal analysis is another technique used for assessing scale effects. The key concept underlying fractals is self-similarity (Lam & Quattrochi 1992). This means that a data series or surface is made up of copies of itself at reduced scales. Although true fractals with self-similarity at all scales are rare, self-similarity often exists over a limited range of scales. Several studies such as those by Bian (1997) and Lam & Quattrochi (1992) have demonstrated that fractals are a promising tool for providing insights on spatial scale and resolution.

Spatial autocorrelation is an important factor in selecting appropriate spatial resolutions and image models (L- or H-resolution). Studies by Dana (1982) with Landsat MSS images have indicated that the radiance of one pixel can affect the radiance of pixels 4–6 pixels apart. With higher resolution systems the positive autocorrelation is even higher (Labovitz & Masuoka 1984). The concept of spatial autocorrelation has been introduced as a basis for understanding the effect of scale. Several studies have explored spatial autocorrelation measures to examine the autocorrelation of pixel DNs and to determine the optimal spatial resolution of a remotely sensed application.

The semi-variogram is the most commonly used measure of autocorrelation. Since Jupp et al. (1989a, 1989b) and Curran (1988) introduced the basic methods of autocorrelation and regularization in digital images, the variogram has been widely adopted for modelling the scale variation of remote-sensing applications. In a forested landscape Hyppanen (1996) applied semi-variograms to measure the autocorrelation of pixels and local variance to define the spatial resolution that maximizes the variance between neighbouring pixels. He found that the optimal spatial resolution in his application was finer than the resolution provided by widely used natural-resource satellites. The semi-variogram was also suggested by Atkinson & Curran (1997) as a tool in choosing an appropriate spatial resolution in an agricultural application of remote sensing.

In the following sections, the formula and theory of the semi-variogram will be described briefly.

The semi-variogram of a digital image is a plot of the semi-variance of DN values for pixels at different distances of separation (often called lags). The variogram is the core tool of geo-statistics. A more detailed theoretical and mathematical exploration of variograms is presented in Cressie (1991). Mathematically, two assumptions are required to use variograms: spatial stationarity, which assumes that the mean and variance do not vary with spatial location, and ergodicity, which assumes that spatial statistics taken over the area of the images as a whole are unbiased estimates of those parameters. The semi-variance represents the average of the squared difference in values separated by a specific lag distance. The formula of the semi-variogram is:

$$\gamma(h) = \frac{\sum_{i=1}^{n-h} \sum (X(i) - X(i+h))^2}{2n} \tag{1}$$

where n is the number of pixel pairs separated at distance h, $X(i)$ and $X(i + h)$ are the pixel value at i, $i + h$, respectively.

The semi-variance is often normalized by the global variance. The shape of a semi-variogram may be fitted with a model (such as linear, exponential, spherical or guassian). Typically, the range and sill are the two parameters of semi-variograms used to describe a data set (Figure 2). If there is no nugget effect which, when it occurs, is expressed as a finite limit of the variogram at the distance of zero, the semi-variance is zero when the lag is zero. In most cases, the semi-variance tends to increase with spatial intervals. As the distance increases, the difference between pixel pairs becomes larger. After it reaches a maximum value, the semi-variogram develops a flat region called the sill. The distance (or lag) at which the sill is reached is called the range. At the peak spatial scale, the image presents the greatest variance. The range generally indicates the extent to which values sampled from a spatial process are similar. The height of the sill indicates the variance of the entire image.

The semi-variogram is an effective tool for studying the effect of scale on image analysis, because the variance of images is treated as a function of scale. The range where spatial dependence is present and the general form of the spatial variation of an image can be readily visualized from the semi-variogram.

3.3   *Different strategies for incorporating information from multiple resolutions*

Three strategies were developed by Chen & Stow (2003) to integrate information obtained from different resolutions and, thus, to improve the classification results based on the Gaussian maximum-likelihood (GML) classifier. The ML classifier is based on an estimated probability density function

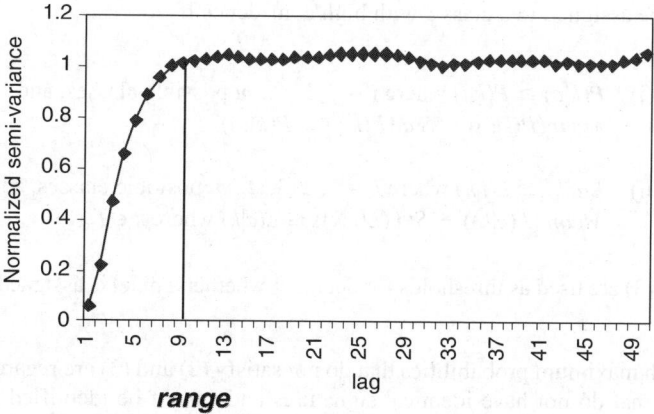

Figure 2.   An example plot of a normalized semi-variogram.

for each of the classes under consideration. The class statistics are obtained from the training data. Pixels are allocated to their most likely class of membership.

A simple means of using information from multiple resolutions is to incorporate them simultaneously in a classification routine. In this way, the feature measures obtained from coarse resolutions are merged with those from finer resolutions. Then, the ML classifier is applied. This method is easy to understand and the information from multiple resolutions is incorporated together. No other algorithms are needed to organize the data. All of the features from coarse resolutions need to be mapped back to the finest resolution using pixel replication. The major drawback is that the cost of computation may be high.

The second strategy is to compare posterior probabilities calculated from multiple resolutions. For this approach, the ML classifier is applied at each resolution to obtain the probability $P(k/i)$ for each pixel k as a member of class i ($i = 1, 2, …, m$ possible classes) and converted to the *a posteriori* probabilities $L(I/k)$ of class membership, which are assessed as the probability density of a case for a class relative to the sum of the densities. For each pixel, the *a posteriori* probabilities total 1.0. At each resolution, the highest *a posteriori* probability and its related class were outputs for each pixel, and k is assigned to the class with the highest maximum *a posteriori* probability.

With the above approach, the feature layers obtained from coarse resolutions do not have to map back to the finest resolution specified and the computation cost is lower than that in the first strategy. However, there is a requirement for classifying the entire image at all resolutions, so this method may still not be very efficient. Pixels with the highest confidence at one resolution can be removed (e.g., masked out) at subsequent finer resolutions, and this leads to the third strategy, a top-down procedure.

A top-down multi-resolution procedure starts with the coarser resolution image. The finer resolution images are used only when necessary. The basic steps involved in the top-down, per-pixel classification process can be stated explicitly as follows:

- At each level of resolution l, select groups of training pixels for each class through the image. Use $T_l(i)$ to denote the training data set for class i at resolution level l. These training data are used for calibrating the classification routine.
- Classify the image at the coarsest resolution R with the maximum likelihood (*ML*) classifier. For each pixel k, calculate the probability $P(k/i)$ and *a posterior* probability $L(k/i)$ as a member of class i, respectively, and then obtain the maximum probability and the maximum a posterior probability for all classes.
- Calculate the mean and standard deviations of maximum probabilities ($Mean(P(k/i))$ and $Std.(P(k/i))$), and the maximum *a posteriori* probability ($Mean(L(i/k))$, $Std.(L(i/k))$) for each class i based on the training data $T_R(i)$. If the training data for each class are unbiased samples, a pixel k can be assigned to a class c with high confidence if

(i) $P(k|c) \geq P(k|i)$ where $i = 1, 2, …, m$ possible classes, and
$$Mean(P(k|c)) - Std.(P(k|c) \geq P(k|c)) \tag{2}$$

(ii) $L(c|k) \geq L(i|k)$ where $i = 1, 2, 3, …, m$ posibble classes, and
$$Mean (L(c|k)) - Std.(L(c|k)) \leq L(c|k) \text{ where } k \in T_R(c) \tag{3}$$

Thus, (2) and (3) are used as thresholds for deciding whether a pixel is assigned to the maximum likelihood class.

- All pixels with maximum probabilities that do not satisfy (2) and (3) are regarded as mixed pixels, or pixels that do not have identical signatures and cannot be identified at this resolution level. For these pixels the *a posteriori* probability $L_R(i/k)$ is calculated, where $L_R(i/k)$ is the *a posteriori* probability $L(i/k)$ at resolution level R for pixel k as a member of class i.

– All pixels that are already assigned to a class are excluded (masked) for subsequent processing. For other pixels, the process goes to finer resolution images. Repeat step 2.
– Repeat steps 3–5.

The above sequential process stops once a) all the pixels are assigned to a class, or b) the finest resolution is reached. In case b) if there are still pixels unassigned to a class, the rule in equation (2) is used. That rule is to compare the *a posteriori* probability at all resolution levels and assign a pixel to the class with the highest *a posteriori* probability.

4 CASE STUDY

The multi-resolution framework was tested using simulated multi-resolution images derived from 1 m USGS CIR Digital Ortho-photo Quarter Quads (DOQQ) data for a portion of the rural-urban fringe of Del Mar of San Diego County, California. The image data have a spatial resolution of 1 m with three spectral bands (Green, Red, and NIR). The study area covers about 3.5 Km^2.

DOQQ images with 1 m resolution were aggregated progressively into six nominal resolution levels (2 m, 4 m, 8 m, 12 m, 16 m and 20 m) by an averaging method. Eight land use/cover classes were used, including single-family residential, multi-family residential, industrial/commercial, irrigated grassland, high-density vegetation, cleared land, undeveloped land and agricultural land. Figure 3 shows a DOQQ subset of the study area.

An initial exploratory data analysis was carried out as three trials. The first trial used histograms to determine the types of distribution exhibited by each band and resolution, since MLC assumes normally distributed data. The second trial included the mean and standard deviations to assess

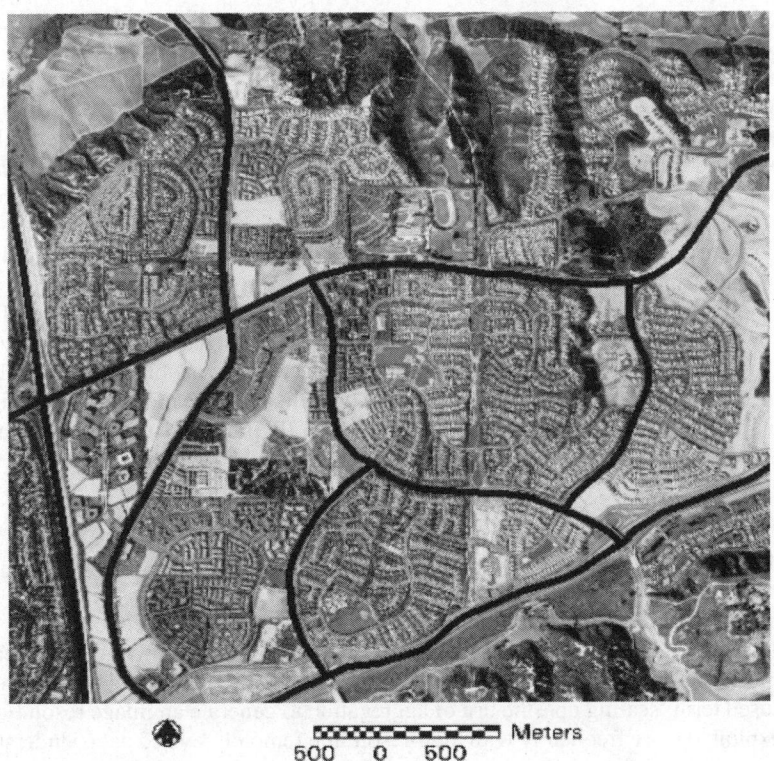

Figure 3.   Subset of a USGS DOQQ for an area of DelMar, CA. The black lines mask out the major roads.

119

Table 2. Summary of classification accuracies derived from single-resolution and multi-resolution strategies. Accuracy is expressed as Kappa values. A high Kappa value indicates better classification accuracy.

Approach	Number of classification maps	Average Kappa	Maximum Kappa	Minimum Kappa
4 m	1	0.4789	0.4789	0.4789
8 m	1	0.5181	0.5181	0.5181
12 m	1	0.5435	0.5435	0.5435
16 m	1	0.5526	0.5526	0.5526
20 m	1	0.5664	0.5664	0.5664
Strategy I	24	0.569	0.601	0.5145
Strategy II	15	0.5822	0.6342	0.5387
Strategy III	14	0.6078	0.6483	0.5557

distribution properties. The final trial established whether each band offers or similar or different information; i.e., whether they are correlated.

The non-directional or isotropic semi-variogram was calculated and plotted for a set of training data to assess the degree of spatial autocorrelation in respective bands. The ranges were determined by visual examination and through a comparison of piecewise slopes. The shape and range of each semi-variogram were useful for determining suitable sizes for training data, sampling intervals, resolutions or window sizes used for the extraction of spatial features. Based on the discussion in a previous section, when image resolution is close to or coarser than the range, an L-resolution scene model is generally most appropriate. Otherwise, spatial features that incorporate information on texture/contexture should be generated at H-resolution.

Both single-resolution and multi-resolution classifications were conducted. The single-resolution classification was used as a benchmark for evaluating various multi-resolution approaches.

The results were evaluated and analysed based on their classification accuracies in discriminating between eight land use/cover classes. The reference data were derived with the aid of an extant land use GIS layer, aerial photographs and field reconnaissance. In each classification exactly the same training pixels and reference data were used. The Kappa coefficients (Jensen 1996) were reported for the general study area for a series of classification maps to evaluate the agreement between the classification results and the reference data.

Table 2 presents the summarized results obtained from single-resolution and multi-resolution approaches. The Kappa values obtained from a classification using three strategies are greater than those from a single-resolution image input. When compared to the accuracy results obtained from three strategies, Strategy III resulted in the highest classification accuracy (0.6483). The improvement in the classification from the best classification results obtained from three strategies is at the 0.95 confidence level when compared with the classification results obtained from a single resolution. Among the three strategies, Strategy III produced the highest classification accuracy while Strategy I showed the least improvement in classification accuracy.

5   SUMMARY

The information content of remotely sensed imagery is strongly dependent on spatial resolution. The spatial resolution of an image can substantially affect the results of image classifications. The characteristic properties of image classes are not the same at different spatial resolutions. Through the widely used term 'scaling up', the use of aggregation to generate an image resolution pyramid should be exploited more frequently with high-resolution remotely sensed data. Understanding the change in variation and the relationship between different resolutions is important in image processing and classification.

Several techniques have been employed to assess appropriate (or optimal) spatial resolutions. Among them, the semi-variogram is the most commonly used. Although a particular class can achieve the best results from a single resolution appropriate to the class, no single resolution would give the best results from all classes (Marceau et al. 1994b). Clearly, landscape objects (land cover/use types) are not the same size; therefore, the scale of analysis corresponding to one object may not be applicable to others.

The correct application of image classification procedures for mapping land use/cover requires knowledge of certain spatial attributes of the data to determine the appropriate classification methodology and parameters to use. In general, traditional single-resolution classification procedures are inadequate for understanding the effects of the chosen spatial resolution. They have difficulty discriminating between land use/cover classes with complex spectral/spatial features and patterns. Although a number of different approaches have been developed for classifying highly heterogeneous landscapes, current research focuses on contextual, knowledge-based and segmentation routines using spatial and spectral information. Most approaches developed mainly for Landsat TM and SPOT HRV images are often scene-specific and untested on high-resolution images (1 m to 10 m).

The multi-resolution framework recognizes that the selection of an image classification procedure should be cognizant of the spatial structure of images to minimize errors, increase efficiency and extract information from the classification process. The selection of the training scheme and classification decision rules should be guided by specifications of the type of scene model (H- and L- resolution) and level of spatial variance represented by the image to be classified.

The advantage of integrating spatial analytical techniques into the classification routine was introduced and discussed. A variety of methods of spatial analysis can provide the above information to allow the effects of resolution on individual classes to be examined. The case study illustrated the potential of integrating spatial statistics into the multi-resolution classification framework. Using a simulated multi-resolution data set, it was demonstrated that multi-resolution classification approaches could significantly improve the accuracy of land use/cover classifications compared with single-resolution approaches. Exploratory data analysis can provide useful information to ensure that subsequent classification methods and parameters are suited to the spatial characteristics of the features (or classes). The results confirm the validity and efficiency of the proposed framework.

REFERENCES

Atkinson, P. M. 1997. Selecting the spatial resolution of airborne MSS imagery for small-scale agricultural mapping. *International Journal of Remote Sensing,* 18(9): 1903–1917.

Atkinson, P. M., and P. J. Curran 1997. Choosing an appropriate spatial resolution for remote sensing investigations. *Photogrammetric Engineering and Remote Sensing,* 63(12): 1345–1351.

Baraldi, A., and F. Parmiggiani. 1994. A Nagao-Matsuyama approach to high-resolution satellite image classification, IEEE Transactions on Geoscience and Remote Sensing, 32(4): 749–758.

Barnsley, M. J., and S. L. Barr. 1996. Inferring urban land use from satellite sensor images using kernel-based spatial reclassification. *Photogrammetric Engineering and Remote Sensing,* 62(8): 949–958.

Bian, L., and R. Butler. 1999. Comparing effects of aggregation methods on statistical and spatial properties of simulated spatial data. *Photogrammetric Engineering and Remote Sensing,* 65(1): 73–84.

Bian, L. 2003. Retrieving urban objects using a wavelet transform approach. *Photogrammetric Engineering and Remote Sensing,* 69(2): 133–142.

Campbell, J. B. 1981. Spatial correlation effects upon accuracy of supervised classification of land cover, *Photogrammetric Engineering and Remote Sensing,* 47(3): 355–363.

Chen, D., and D. A. Stow. 2002. The effect of training strategies on supervised classification at different spatial resolutions. *Photogrammetric Engineering and Remote Sensing,* 68(11): 1155–1161.

Chen, D., and D. A. Stow. 2003. Strategies for integrating information from multiple spatial resolutions into land use/cover classification routines. Photogrammetric Engineering and Remote Sensing (Forthcoming).

Civco, D. L. 1993. Artificial neural networks for land-cover classification and mapping. *International Journal of Geographical Information Systems,* 7(2): 173–186.

Congalton, R. G. 1988b. Using spatial autocorrelation analysis to explore the errors in maps generated from remotely sensed data, *Photogrammetric Engineering and Remote Sensing,* 54(5): 587–592.

Conners, R. W., M. M. Trivedi, and C. A. Harlow. 1984. Segmentation of a high-resolution urban scene using texture operators, *Computer Vision, Graphics, and Image Processing*, 25: 273–310.

Collins, J. B., and C. E. Woodcock. 1999. Geostatistical estimation of resolution-dependent variance in remotely sensed image. *Photogrammetric Engineering and Remote Sensing*, 65(1): 41–51.

Cressie, N. 1991. *Statistics for spatial data*. Chichester: John Wiley.

Curran, P. J. 1988. The semi-variogram in remote sensing: an introduction. *Remote Sensing of Environment*, 24: 493–507.

Cushine, J. L., and P. Atkinson. 1985. Effect of spatial filtering on scene noise and boundary detail in TM imagery. *Photogrammetric Engineering and Remote Sensing*, 51(9): 1183–1193.

Flygare, A-M. 1997. A comparison of contextual classification methods using Landsat TM, *International Journal of Remote Sensing*, 18(18): 3835–3842.

Foody, G. M. 1996. Approaches for the production and evaluation of fuzzy land cover classification from remotely sensed data, *International Journal of Remote Sensing*, 17: 1317–1340.

Gong, P., and P. J. Howarth. 1990. An assessment of some factors influencing multispectral land-cover classification, *Photogrammetric Engineering and Remote Sensing*, 56(5): 597–603.

Gong, P., and P. J. Howarth. 1992. Frequency-based contextual classification and gray-level vector reduction for land-use identification, *Photogrammetric Engineering and Remote Sensing*, 58(4): 423–437.

Haralick, R. M. 1979. Statistical and Structural Approaches to Texture. *Proceedings of the IEEE* 67(5): 786–803.

Hlavka, C. A., and G. P. Livingston. 1997. Statistical models of fragmented land cover and the effects of coarse spatial resolution on the estimation of area with satellite sensor imagery, *International Journal of Remote Sensing*, 18(10): 2253–2259.

Hsu, S. 1978. Texture-tone analysis for automated landuse mapping. *Photogrammetric Engineering and Remote Sensing*, 44(11): 1393–1404.

Hyppanen, H. 1996. Spatial autocorrelation and optimal spatial resolution of optical remote sensing data in boreal forest environment. *International Journal of Remote Sensing*, 17(17): 3441–3452.

Irons, J. R., B. L. Markham, R. F. Nelson, D. L. Toll, D. L. Williams, S. Latty, and M. L. Staufer. 1985. The effects of spatial resolution on the classification of TM data, *International Journal of Remote Sensing*, 6(8): 1385–1403.

Jensen, J. R. 1996. *Introductory Digital Image Processing: A Remote Sensing Perspective*. Prentice Hall Series in Geographic Information Science. Prentice Hall, Upper Saddle River, New Jersey, pp.316.

Johnson, D. D., and P. J. Howarth. 1987. The effects of spatial resolution on land cover/use theme extraction from airborne digital data. *Canadian Journal of Remote Sensing*, 13(2): 68–74.

Jupp, D. L. B., A. H. Strahler, and C. E. Woodcock. 1989a. Autocorrelation and regularization in digital images I: Basic Theory, *IEEE Trans. on Geoscience and Remote Sensing*, 26(4): 463–473.

Jupp, D. L. B., A. H. Strahler, and C. E. Woodcock. 1989b. Autocorrelation and regularization in digital images II: Simple image models, *IEEE Trans. on Geoscience and Remote Sensing*, 27(3): 247–258.

Kontoes, C. C., and D. Rokos. 1996. The integration of spatial context information in an experimental knowledge-based system and the supervised relaxation algorithms – two successful approaches to improving SPOT-XS classification, *International Journal of Remote Sensing*, 17(16): 3093–3106.

Labovitz, M. L., and E. J. Masuoka. 1984. The influence of autocorrelation in signature extraction – an example from a geobotanical investigation of Cotter Basin, Montana. *International Journal of Remote Sensing*, 5(2): 315–332.

Latty, R. S., R. F. Nelson, B. L. Markham, D. L. Williams, D. L. Toll, and J. R. Irons. 1985. Performance comparisons between information extraction techniques using variable spatial resolution data. *Photogrammetric Engineering and Remote Sensing*, 51(9): 1159–1170.

Lam, N. S-N., and D. A. Quattrochi. 1992. On the issues of scale, resolution, and fractal analysis in the mapping sciences. *Professional Geographer*, 44(1): 88–98.

Lindeberg, T. 1994. Scale-space theory: A basic tool for analysing structures at different scales. *Journal of Applied Statistics*, 21(2): 225–270.

Lunetta, R. S., R. G. Congalton, L. K. Fenstermaker, J. R. Jensen, K. C. McGwire, and L. R. Tinney. 1991. Remote sensing and geographic information system data integration: error sources and research issues. *Photogrammetric Engineering and Remote Sensing*, 57(6): 677–687.

Marceau, D. J., P. J. Howarth, and D. J. Gratton. 1994a. Remote sensing and the measurement of geographical entities in a forest environment 1: The scale and spatial aggregation problem, *Remote Sensing of Environment* 49: 93–104.

Marceau, D. J., P. J. Howarth, and D. J. Gratton. 1994b. Remote sensing and the measurement of geographical entities in a forest environment 2: The optimal spatial resolution. *Remote Sensing of Environment* 49: 105–117.

Markham, B. L., and J. R. G. Townshend. 1981. Land cover classification accuracy as a function of sensor spatial resolution, *Proceedings of the 15th International Symposium on Remote Sensing of Environment*: 1075–1090.

Martin, L. R. G., P. J. Howarth, and G. Holder. 1988. Multispectral classification of land use at the rural-urban fringe using SPOT data, *Can. J. Remote. Sens.* 14(2): 72–79.

Moller-Jensen, L. 1990. Knowledged-based classification of an urban area using texture and context information in Landsat-TM imagery. *Photogrammetric Engineering and Remote Sensing*, 56(6): 899–904.

Mohanty, K. K. 1997. The wavelet transform for local image enhancement, *International Journal of Remote Sensing*, 18(1): 213–219.

O'Nellis, M. D., and J. M. Briggs. 1989. The effect of spatial scale on Kpnza landscape classification using textual analysis. *Landscape Ecology*, 2(2): 93–100.

Richards, J. A. 1986. *Remote Sensing Digital Image Analysis: An Introduction*. Springer-Verlag, p.281.

Ryerson, R. 1989. Image interpretation concerns for the 1990s and lessons from the past. *Photogramm. Eng. Remote Sens.* 55(10): 1427–1430.

Stein, A., W. G. M. Bastiaanssen, S. De Bruin, A. P. Cracknell, P. J. Curran, A. G. Fabbri, B. G. H. Gorte, J. W. van Groenigen, F. D. Van Der Meer, and A. Saldana. 1988. Integrating spatial statistics and remote sensing. *International Journal of Remote Sensing*, 19(9): 1793–1814.

Strahler, A. H., C. E. Woodcock, and J. A. Smith. 1986. On the nature of models in remote sensing. *Remote Sensing of Environment*, 20: 121–139.

Toll, D. L. 1985. Effects of Landsat TM sensor parameters on land cover classification. *Remote Sensing of Environment*, 17(2): 129–140.

Townshend, J. R. G., and C. O. Justice. 1988. Selecting the spatial resolution of satellite sensors for global monitoring of land transformations, *International Journal of Remote Sensing*, 9: 187–236.

Wang, F. 1990. Improving Remote Sensing Image Analysis through Fuzzy Information Representation. *Photogrammetric Engineering and Remote Sensing*, 56(8): 1163–1169.

Wharton, S. W. 1982. A context-based Land-Use Classification Algorithm for high-resolution Remotely Sensed data. *Journal of Applied Photographic Engineering*, 8(1): 46–50.

Wharton, S. W. 1989. Knowledge-based spectral classification of remotely sensed image data. In: *Theory and Application of Optical Remote Sensing*. (Ed: Assor, Shassen) John Wiley & Sons, New York.

Weiler, R. A., and D. A. Stow. 1991. Spatial analysis of land cover patterns and corresponding remotely sensed image brightness. *International Journal of Remote Sensing*. 12(11): 2237–2257.

Woodcock, C. E., and A. H. Strahler. 1987. The factor of scale in remote sensing, *Remote Sensing of Environment*, 21: 311–332.

Woodcock, C. E., A. H. Strahler, and D. L. B. Jupp. 1988a. The use of variograms in remote sensing I: Scence models and simulated images. *Remote Sensing of Environment*, 25: 323–348.

Woodcock, C. E., A. H. Strahler, and D. L. B. Jupp. 1988b. The use of variograms in remote sensing II: Real digital images, *Remote Sensing of Environment*, 25: 349–379.

Woodcock, C. E., and V. J. Harward. 1992. Nested-hierarchical scene models and image segmentation, *International Journal of Remote Sensing*, 13(16): 3167–3187.

Wulder, M., and B. Boots. 1998. Local spatial autocorrelation characteristics of remotely sensed imagery assessed with the Getis statistic, *International Journal of Remote Sensing*, 19(11): 2223–2231.

A new 3D Internet-based approach for the integration of GIS and digital photogrammetric systems

Mahmoud Reza Delavar, Mohammad Abbasi & Ali Azizi
Department of Surveying and Geomatic Engineering, Engineering Faculty, University of Tehran, Tehran, Iran

ABSTRACT: The demand for efficient display and analysis of 3D spatial data in geo-spatial information systems (GISs) has been growing rapidly, and products of digital photogrammetric systems (DPS) are among the main sources of spatial data for GIS. Therefore, the integration of DPS and GIS offers new possibilities for users. In this study, a number of innovative methods for the integration of GIS and DPS are proposed and compared. These methods are: (1) the 'external interface system' where the output of a DPS is connected to the input of a GIS, (2) the 'joint interface system', where the two systems connect together using functions of the two systems, and (3) the 'integrated system' that contains the two systems. Based on the integrated system proposed, an Internet-based integration approach, consisting of DPS, GIS and CAD systems, has been designed and implemented. The efficient parameters in the designed system are: (1) new demands (3D modeling, 3D visualization and 3D analysis), (2) data updating (by geo-referenced images), (3) Internet for accessing and transforming data and (4) covering a range of DPS products (3D vectors data, digital terrain modeling, orthophotos and close range data). Internet programming languages, the importance of Internet in transforming and accessing data and 3D visualization in the integration of GIS and digital photogrammetry are discussed. The acquisition of 3D vector data by digital photogrammetric method, the new integration method and the use of CAD systems are elaborated in this paper. In addition, applications of aerial and close-range images in raster GISs using the developed method are presented.

1 INTRODUCTION

Integration of DPS into GIS offers new possibilities for end-users. This integration allows photogrammetric data to be collected and manipulated in a raster/vector-based GIS environment.

The process in a stereo workstation is more sophisticated and requires specialized hardware and measuring systems. The DPS can provide convergent image data into topologically structured datasets suitable for spatial analysis in a GIS environment. Managing and distributing geo-spatial data across an enterprise is a very complex operation. DPS provides the capability of stereoscopic viewing and the possibility of conducting precise 3D measurements. On the other hand, a GIS environment provides the capability of displaying and editing vector-formatted 3D data as conveniently as possible. Therefore, the integration of DPW and GIS provides users with the new capability to achieve optimum management of geo-spatial data in an integrated environment.

2 REVIEW OF PREVIOUS RESEARCH ON THE INTEGRATION OF GIS AND DPS

With the inception of DPS and its growing capabilities, the importance of photogrammetry in acquiring, editing and updating spatial data has been considered more for geo-spatial information (GI)

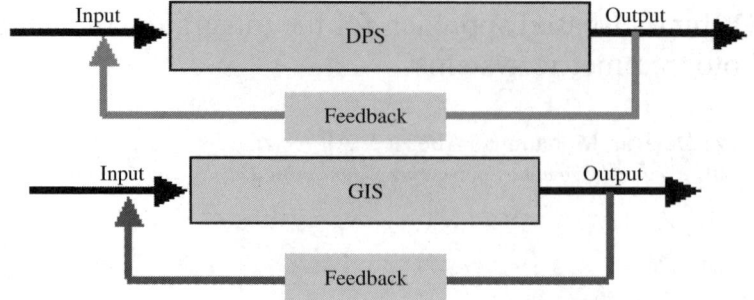

Figure 1. Illustration of digital operation process in GIS and DPS.

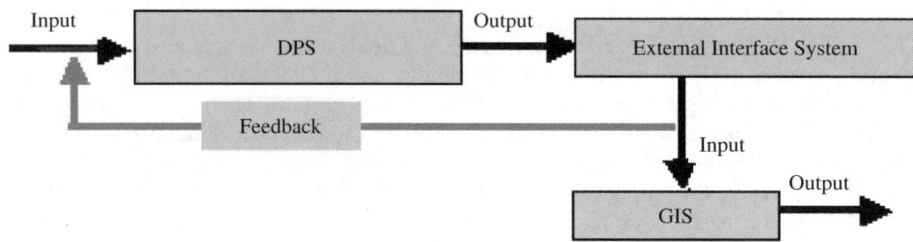

Figure 2. Diagram for using the external interface system.

communities. The integration of DPS, GIS and CAD systems as a basic method to reduce the cost and time required to acquire spatial data for GIS has been in focus in recent years (Albertz and Wiedemann, 1996; Edwards et al., 2000; Madani, 2001; Zlatanova et al., 1999).

Previous research has been focused mainly on the importance of such integration for a specific application, in which the two systems of photogrammetry and GIS have been separately implemented and their products have finally been joined. In this research, some methodologies for the integration of GIS and DPS have been proposed and compared.

3 PROPOSED METHODOLOGIES FOR THE INTEGRATION OF GIS AND DPS

The digital operations in GIS and DPS are illustrated in Figure 1.

The following GIS and DPS integration approaches are proposed in this research:

– Use of an external interface system
– Use of a joint interface system
– Use of an integrated system.

The above-mentioned integration approaches are elaborated below:

3.1 Use of an external interface system

In this approach an interface system is considered that accepts the output of a DPS as its input and, after processing, introduces the output as the input to a GIS system. The approach is shown in Figure 2.

The advantages of this approach are the omission of the feedback of the output of the GIS to the input of the DPS, which reduces the time and cost of editing operations. Furthermore, there is no need to access the internal structure of the DPS and GIS to implement such a system.

126

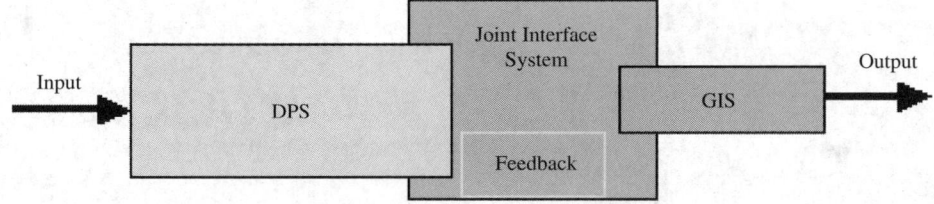

Figure 3.   Diagram of the joint interface system.

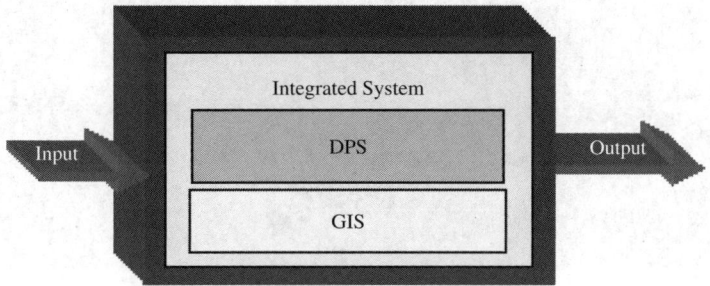

Figure 4.   Diagram of the integrated system.

3.2   *Use of a joint interface system*

In this approach, a system is designed to perform some of the functionalities of the DPS and GIS. This methodology provides a closer relationship between the two systems. Figure 3 illustrates the approach.

The advantage of this approach, in addition to the omission of feedback between the two systems and to the ability to perform the editing operation simultaneously, is to provide closer relationships between the users of both systems. The implementation of this approach is only possible when the two systems enjoy open functions.

3.3   *Use of an integrated system*

The integrated system consists of both the DPS and the GIS. Figure 4 illustrates the integrated system.

The integrated system is an ideal approach enjoying the capabilities of both systems, and has the advantages of the two previously mentioned approaches. It is better to provide users of the integrated system with separate modules to reduce the cost based on the needs of the user. The implementation of such a system is only possible when the functionalities of the two systems can be accessed.

4   THE MAIN IDEA FOR THE INTEGRATION OF THE SYSTEMS

In this research the aim is to develop an ideal GIS and DPS integrated system considering all effective parameters of both systems. A detailed explanation of the developed integrated system and some of the applications of the system for a GI community are described below.

4.1   *Description of the developed integrated system*

Considering the importance of using the Internet in geo-spatial information technology as an efficient storage, access and transfer mechanism (Bishr, 2000), the proposed integrated approach has been developed based on the Internet. Figure 5 illustrates the Internet-based integrated system.

127

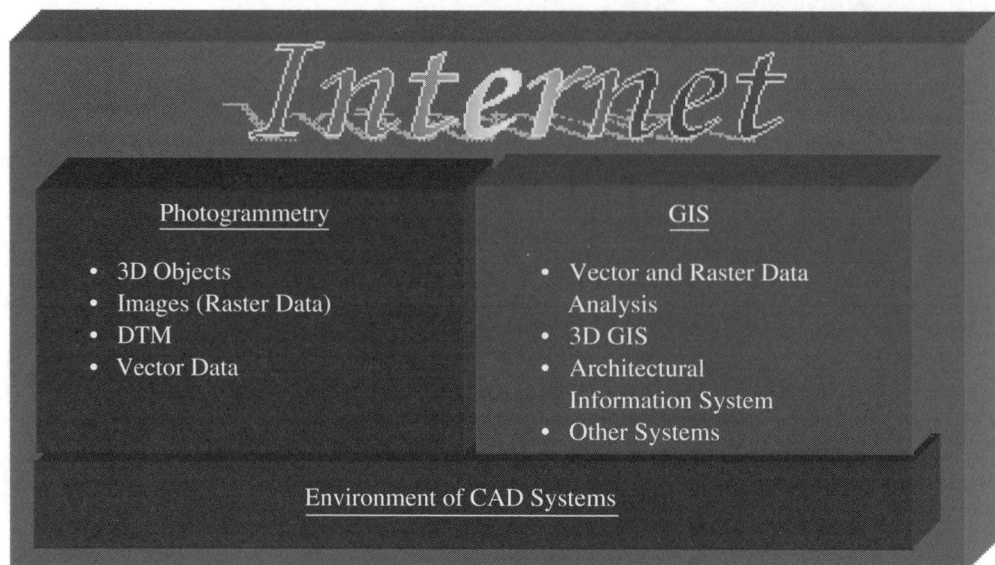

Figure 5.   Illustration of the developed Internet-based integrated system.

With the establishment and development of the World Wide Web (WWW), data of all kinds have been made accessible to anyone with a link to a WWW server. The Internet provides an easy-to-use, flexible and powerful approach to handling and integrating text, images, animation and other information.

The WWW is an information discovery system for browsing and searching the digital information contained in the Internet. If data on landscapes and cities were available on the Web, a great variety of queries would become feasible. It is important to note that to use the Web efficiently, one should be familiar with the offered tools such as HTML, DHTML, VRML, Java and C#, which have been discussed below.

4.1.1   HTML and DHTML

The original WWW format was the hypertext markup language (HTML). Its principle is to have a data structure that is easy to use and which can manage text, hypertext and images.

With the increase in the number of users of the WWW, the demand for more control over data handling grown; thus, HTML has gone through several major updates. In HTML, the basic element is the tag. Labeling each element with a tag informs the browser of how to display it. The major concern about HTML is the compatibility of the format with the browser (Netscape, MS Internet Explore). Dynamic HTML (DHTML) is being widely published for the next generation in the development of the WWW, which is an enhancement of HTML. DHTML is a major change in the data format of the web. It allows photogrammetrists to develop increased user interactions.

4.1.2   VRML

The Virtual Reality Modeling Language (VRML) is an international standard for describing interactive 3D objects. The VRML 97 specification is sufficient for geo-spatial modeling, such as the handling of digital terrain models (DTM), modeling of the earth and database linking. The VRML browsers are different according to processing speed and visualization.

4.1.3   Java

Java is a programming language, which the language compilers have made for a virtual machine instead of for a particulate computer. Therefore, the software produced based on Java is system-independent. Java has been developed by Sun Microsystems Inc. and is used to develop web-based programs.

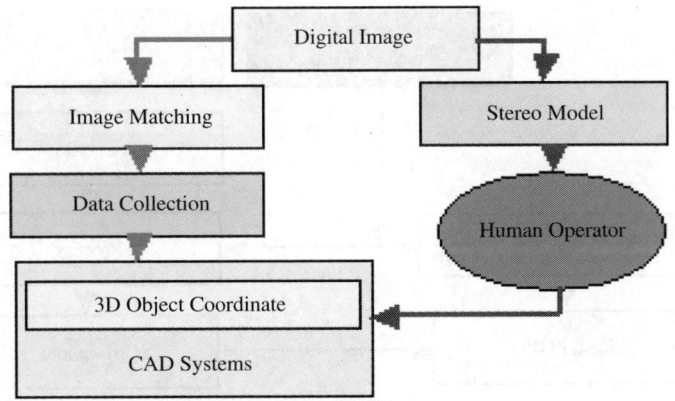

Figure 6. Flow diagram of data between a DPS and CAD system.

4.1.4   C#

C# is a programming language based on extended versions of C and C++. It is similar to Java. C# has been provided as a part of Microsoft Virtual Studio Version 7. C# is an object-oriented language, whose aim is to compile the capabilities of C++ and Visual basic. The Microsoft platform has some types of interface language that connect the class libraries and procedure languages with a common languages subset. Therefore, C# has access to class libraries and is used by Visual C++ or Visual basic (Microsoft, 2002).

4.2   *The position of CAD systems in the developed integrated system*

Although there are now a number of DPSs operating under CAD software, the fact that such systems offer limited image processing options is an impeding factor. Nevertheless, most DPSs run under a specific CAD system. This is because CAD systems have a great ability to handle vector data and cartographic operations. Therefore, a CAD system has been used in the development of the integrated system. Figure 6 shows the flow diagram of data between a DPS and a CAD system in the developed system.

The extraction of spatial data in the integrated system can be done automatically using image matching techniques, or manually in a 3D environment by a human operator. The generated vector data can be input to a CAD system (El-Hakim, 2001). Such an environment can be a specific CAD system with the minimum required facilities. Some examples of using the developed system in handling geo-spatial data are given below.

5   SOME EXAMPLES OF USING THE INTEGRATED SYSTEM

Some examples of using the developed method to handle aerial and close-range images are described below.

5.1   *Aerial images and its applications in the integrated system*

Geo-referenced images have a number of applications in updating spatial data. The images can be easily used in a raster-based GIS. Figure 7 illustrates the use of an aerial image in the integrated system.

As shown in Figure 7, generated orthophoto images, can be imported to a GIS system for raster-based analysis. It is also possible to use such images along with a DTM for 3D visualization analysis (Grussenmeyer, 1999). The representation of 3D visualization can be performed through movement in a 3D model, which can be done under a web system.

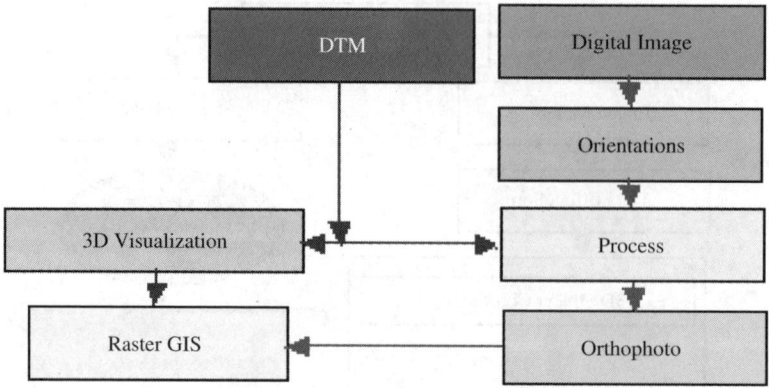

Figure 7.   Diagram of the use of an aerial image in the integrated system.

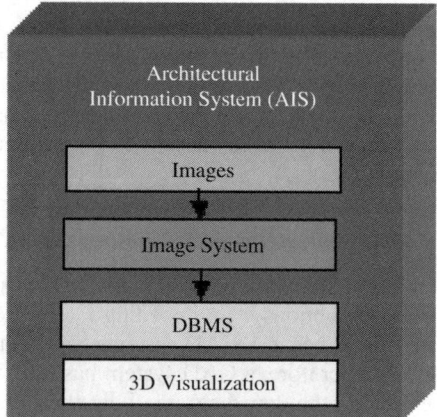

Figure 8.   Diagram of the integrated GIS and close-range photogrammetry.

5.2   *Close-range photogrammetry and its applications in the integrated system*

The distance of the object in close-range photogrammetry (CRP) is usually less than three hundred meters. In CRP, an object is photographed from a number of directions so that a stereoscopic view can be achieved from all parts of the object. One of the applications of CRP is in architectural projects. In this case, the integration of CRP and GIS provides an architectural information system (AIS). In the establishment of an AIS, data generated through CRP operations can be easily transformed to the AIS system, as shown in Figure 8.

The implemented AIS has two components including a DPS, which is shown in Figure 8 as an imaging system; and a database management system (DBMS), which is a part of the GIS. The integrated system is implemented using some programming languages (Visual Basic.Net and C#), GIS software (ArcGIS ESRI), a digital photogrammetric system (ParadEyes developed by Miaad Andisheh Saz Co.) and CAD software (MicroStation 8 Bentley).

The main advantages of the implementation of the integrated system as used in the pilot projects are, as follows:

- Easy access to spatial and non-spatial data
- Use of capabilities of GIS, DPS and CAD software
- Ease of use, flexibility and low cost of geo-spatial handling.

6 CONCLUSIONS

A number of innovative methods for the integration of GIS and DPS are proposed and compared in this study. These methods are: (1) the 'external interface system', by which the output of a DPS is connected to the input of a GIS, (2) the 'joint interface system', in which the two systems are connected together using functions of the two systems, and (3) the 'integrated system' that integrates the two systems in a unified manner.

Based on the proposed integrated system, an Internet-based integration approach consisting of DPS, GIS and CAD systems has been designed and implemented. The result of the experiments performed showed that, in the integrated system, a huge savings can be achieved in the cost and time of running data editing operations.

REFERENCES

Albertz, J. and Wiedemann, A. 1996. Acquisition of CAD data from existing building by photogrammetry, Technical University of Berlin, Germany.

Bishr, Y. A. 2000. Internet based large distributed geospatial databases, International Archives of Photogrammetry and Remote Sensing. Vol. XXXIII, Part B4, Amsterdam 2000, PP. 126–131.

Burrough, P. A. and McDonnell, R. A. 1998. Principles of Geographic Information Systems, Oxford University Press Inc., New York.

Cornford, D. 2001. Geographic Information Systems, Module CS321, Computer Science, Aston University, PP. 1–15.

Edwards, D., Simpson, J. and Woodsford, P. 2000. Integration of Photogrammetric and Spatial Information Systems. IAPRS, Vol. XXXIII, Amsterdam 2000, PP. 603–610.

El-Hakim, S. F. 2001. 3D Modeling of complex environments, SPIE Proceedings Vol. 4309, Video metrics VII, San Jose, January 21–26, 2001, PP. 12–24.

Grussenmeyer, P., Kohel, M. and Noureldin, M. 1999. 3D Geometric and semantic modeling in historic sites, XVII CIPA International Symposium, October 3–6, 1999, Olinda, Brazil, PP. 201–210.

Madani, M. 2001. Importance of digital photogrammetry for a complete GIS, 5th Global Spatial Data Infrastructure Conference, Cartagena, Columbia, May 21–25, 2001, PP. 1–10.

Microsoft, 2002. Microsoft C# language Specifications.

Zlatanova, S., Paintsil, J. and Tempfli, K. 1999. 3D Object Reconstruction from Aerial Stereo Images, International Institute for Aerospace Survey and Earth Sciences (ITC).

www.campus.esri.com/
www.casaucl.dc.uk/gistimeline/
www.CS.aston.ac.uk/
www.dcs.uky.edu/~jaynes/classes/
www.gsd.harvard.edu/brc/brc.html/
www.ncgia.ucsb.edu/Pubs/Core.html/
www.ncrg.aston.ac.uk/~cornfosd/gis/
www.wiley.com/GIS/

Advances in Spatial Analysis and Decision Making, Li, Zhou & Kainz (eds)
© 2004 Swets & Zeitlinger, Lisse, ISBN 90 5809 652 1

Spatial decision support applications based on three-dimensional city models

Chaokui Li[1,2], Qing Zhu[1], Yeting Zhang[1], Duo Huang[1], Jie Zhao[1] & Songlin Chen[1]
[1] State Key Laboratory of Information Engineering in Surveying Mapping and Remote Sensing, Wuhan University, Wuhan, P.R. China
[2] Research Institute of Geomatics, Hunan University of Science and Technology, Xiangtan, P.R. China

ABSTRACT: With the increasing application of three-dimensional city models (hereafter, 3DCM) in urban planning, design and management, it is expected that there will be an urgent need for spatial decision support. In this paper, the basic mathematic models and some typical methods of analysis based on 3DCM are proposed and discussed, such as the statistics model, the time-serials model, the spatial dynamics model, and so on. A few typical spatial decision-making methods integrating the spatial analysis and basic mathematical models are also introduced, for example, visual impact assessment, dispersion of noise emissions, the base station plan for wireless communication, etc. Also, ideas on further applications and value-added services of 3DCM are discussed. For example, sunshine analysis is studied and some helpful conclusions are drawn.

1 INTRODUCTION

The Urban Geographic Information System (UGIS) is a kind of technology service system that automatically collects information about urban infrastructure, functions and mechanisms, and provides dynamic supervision and spatial decision support (SDS) by means of the comprehensive use of high-technology, such as GIS, RS, remote surveying, networks, multimedia and virtual reality (Zhu et al., 2001). Many scholars have done their best research on the spatial analysis and spatial planning of UGIS (Edmond and Shang, 1997; Monika and Gunther, 1997; Isabelle and Gaetan, 1999; Theresa and Martin, 2001; Sivacoumar and Thanasekaran, 1999; Peter, 1999; Tsuyoshi and Takehito, 1998; Klungboonkrong and Taylor, 1998; Yilmaz et al., 2002; Liu et al., 2001; Xu, 2000; Zhang and Gu, 2000). However, many urban problems such as electro-magnet radiation, air pollution, noise pollution, visibility and sunlight have obvious temporal-spatial features, which have been ignored in past studies. The present two-dimensional (2D) UGIS technology cannot satisfy the requirements of spatial analysis for the 3D building models needed by many disciplines. On the other hand, for SDS applications based on 3DCM (i.e., the union of the digital elevation model (DEM) digital orthoimage and 3D building models with photorealistic attributes), the proper mathematical and physical models are also not perfect.

With more and more 3DCMs available as the fundamental database content of UGIS, the value-added applications of 3DCM have limited the further advance of UGIS. Since much more cost and time needed to produce 3DCM than traditional 2D GIS data, inadequate utilization will result in serious waste and in a crisis for 3D GIS. Therefore, the in-depth development and efficient utilization of 3DCM is of great importance. In the mean time, it is hoped that 3DCM applications can provide solutions to the problem of planning for the sustainable development of regional and urban areas, by such means as urban temporal-spatial modelling and prediction analysis.

This paper discusses the models related to the SDS based on the 3DCM. In section 2, some useful basic mathematical models are introduced, as well as the typical methods of SDS. Section 3 provides several important applications of SDS in urban planning, and section 4 presents a detailed analysis of sunlight based on the 3DCM. The paper concludes with some observations and a discussion of possible future work.

2   MODELS AND METHODS OF 3DCM-AIDED SPATIAL DECISION SUPPORT

2.1   *Mathematical models of spatial decision support*

The mathematical models of spatial decision-making support rely on the problems to be solved. Not only are spatial data and SDS necessary, but related special physical, geographical and biological models also need to be integrated. The 3DCM includes DEM, 3D buildings and their attributes. The essence of 3DCM-aided SDS is to provide information, such as words, data, diagrammes, tables, images and knowledge for a visible decision under a virtual urban geographic environment. To obtain so much information, the influences of 3DCM must be considered, and we should dig through vast spatial data such as on geometry and texture to discover the necessary knowledge for decision-making according to some mathematical rules. From the view of cognition, After adopting the exploratory and empirical approach to analysis of 'investigating data – simulating – forecasting', we can obtain a theoretical understanding of 3DCM by applying the statistical model, time-serials model, and systematic dynamic model to vast, messy-looking data such as noise data, pollution data, electro-magnet distribution data and data on water erosion on hill-slopes, etc. It is also feasible and convenient to conduct temporal-spatial simulations for dynamic variations in the city and to extrapolate the developments and changes in the future. Among the 3DCM-aided SDS applications for urban phenomena, selected mathematical models that are applied frequently are enumerated below.

2.1.1   *Optimization and planning models*

They are consist of linear planning, non-linear planning, multi-purpose planning, dynamic and static planning, cooperative-divided planning, and so on. Such models are adopted for district-selecting issues in urban planning and for optimizing problems such as selecting the shortest path between two places. The planning model is a mathematical support for decision-making on optimization. For example, in selecting the proper sites for some buildings taking into account many influence factors, the role of the optimization model is to search for the optimum quality-price ratio by establishing a multi-factor quantitative affection model.

2.1.2   *Spatial statistics models*

They include regression model, time series models, gray models, etc. The methods for constructing such models are classified into the white-case approach, black-case approach and the gray-case approach. In district planning, prediction support tends to depend on statistical models for the solving of such problems as increases in population, economic growth, noise pollution, and so on. For instance, because of the affection of 3DCM, there is still no good analytic formula to describe the influence of noise or air pollution at present, and we have had to analyse and synthesize the spreading laws by means of statistical models from vast observational data.

2.1.3   *Spatial dynamics models*

The mathematical analysis model constructed from the perspective of systematology after having introduced the non-linear dynamic theory, mutation theory, etc. Spatial dynamics models can more completely describe systematic urban features because spatial structural dynamics and individual spatial behaviour are introduced into systematic dynamics. The spatial dynamics model is fully applied in simulations and forecasts of increased urban land-use, urban air pollution, urban economic growth and the state of the ecological environment. In the urban system, any model of 3DCM is regard as individual behaviour. Its increase, decrease and displacement will affect other factors in the system.

2.2 *Typical methods of spatial decision support*

Decision-making and decision-making support are two different conceptions. Decision-making is a subjective and active course in which decision-makers analyse and compare decision-making information according to effectiveness theory. Decision-making support (or DS) is an interactive, computer-based system designed to help a user or a group of users achieve greater effectiveness in decision-making while solving a semi-structured spatial decision problem (Jacek, 1995). Abundant decision-making information such as words, diagrammes, images and data are usually necessary to solve a semi-structured spatial decision problem. Because space analysis problems are concerned with the geometric scale and form of a piece of terrain, it is important to study the affection rules of the terrain model, architecture model, etc. In the past, it was common to use statistic regression analysis to study SDS under the frame of MIS and 2DGIS. The statistic regression analysis model relies heavily on observation data. In practice, however, it is difficult to obtain a great deal of correlation data on some objects, and some matters are closely related to man-made factors; for example, government policy may rapidly change the situation of air pollution in a city. Therefore, when studying 3DCM-aided SDS, it is valuable combine multi-analysis methods with the operational research method, statistics method and systematic theory method to improve the dependability of decision-making information.

2.2.1 *Time-serial analysis method*

The time-serial analysis belongs to the category of statistical analysis. Compared with regression-analysis, time-serial analysis does not consider the factor of affect, and the method directly find the inherent rule of things changing according to the order and size of result-variable. This rule mainly includes the trend rule and the periodicity rule. Time-serial data have two characteristics: the data are arranged by time-serials and between border upon data there is a certain relativity. Because of complicated factors that influence the data, the time-serials contain some non-linear features. We can deal with the problem of non-linear features in the following two ways:

– Building models with non-linear time-serial data;
– Decomposing the structure of non-linear time serials.

The methods used to build a non-linear time-serial model include the black box, grey box and the white box three ways. In decomposing the structure of non-linear time serials the parallel connection method and serial connection method may be adopted. Information such as noise information related to position and to changes in the environment, can be described in the time-serials model.

2.2.2 *Spatial dynamic analysis method*

Spatial dynamics is used to study the flow and feedback of spatial information in a complex system. Thus, the object being analysed by spatial dynamics is an information feedback system. Compared with statistical analysis, the spatial dynamics analysis method has many advantages, as follows (Cheng, 2001):

– Expressing the factors external or internal to the system and their relations;
– Forecasting the dynamic advancing trends of the system;
– Setting control factors for the system for decision analysis;
– Combining quantitative analysis with qualitative analysis;
– Dynamic emulation and analysis of the system.

Therefore, spatial dynamics analysis can be used for 3DCM-aided SDS to analyse dynamic changes in the extension of the urban environment.

3 TYPICAL APPLICATIONS OF 3DCM-AIDED SPATIAL DECISION SUPPORT

3.1 Intervisibility analysis based on 3DCM

Intervisibility is widely used in urban planning; for example, in determining the location of a telecommunications launch tower, in analysing sunlight in shadows with regard to buildings, and so on. Figure 1, shows a launch tower A with an elevation of HA. C is the receiver station with an elevation of HC, B is the top of the highest building (an obstacle) between A and C with an elevation of H_B. The heights above the terrain surfaces at A, B, C are h_A, h_B, h_C, respectively. It is obvious that the data on elevation and height can be obtained directly from 3DCM. For a certain height of h_C, according to the highest elevation between A and C, the height of h_A can be calculated to ensure that A is visible from C.

3.2 Atmospheric pollution analysis based on 3DCM

Atmospheric pollution may be affected by the following factors:

- *Sources of pollution*
  The main sources of air pollution in cities are:
  1. Pollution from industrial production, mostly of which is point pollution;
  2. Pollution from traffic and transportation, most of which is linear pollution;
  3. Living stove pollution, most of which is plane pollution.
- *Gravity*
  Because of the influence of gravity, dust may easily settle on the inter-surface between the atmosphere and the ground.
- *Special climate*
  Because of the density of population and industries, urban pollution is concentrated under a layer of gas, under which people live.

If there is urban area S, the height of the gas-layer above S is H. The total amount of drainage from living stoves and automobile tail gas is M, and industrial drainage amounts to Q. Neglecting the influence of wind, if the polluted gases distribute evenly in space V, then V is as follows:

$$V = S \cdot H \tag{1}$$

When the density of the polluted gases is ρ, according to the gaseous equation, we should have the following equation:

$$V \cdot \rho = S \cdot H \cdot \rho = M + Q + P \tag{2}$$

Here, P is the quantity of air within V. Other parameters in formulas (1) and (2) may be obtained from the environmental protection branch. If there are terrace and terrestrial objects in area S,

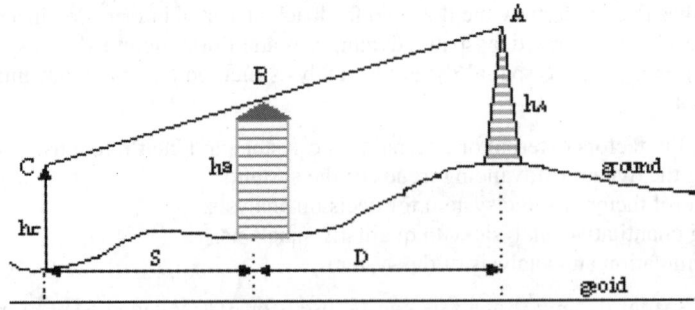

Figure 1.   Intervisibility analysis based on 3DCM.

occupying space V_1 (coming from 3DCM data), we should have:

$$(V-V_1) \cdot \rho_1 = (S \cdot H - V_1) \cdot \rho_1 = M+Q+P \tag{3}$$

Here, ρ_1 is the quantity of gas in the unit volume, which also is called the pollution index.

3.3 *Noise pollution analysis based on 3DCM*

With the development of industry and traffic, noise pollution is becoming serious. In modern cities, noise comes mainly from three sources: (1) transportation; (2) industry; (3) public actions. Noise spreading in the form of waves has features of direction and power decreasing. In Figure 2, the spatial distance between the centre of S and the sound source A, B, C and D is d_1, d_2, d_3 and d_4, respectively.

If the sound intensity from sound source A, B, C and D are P_A, P_B, P_C and P_D, respectively, then the sound intensity in centre S may be obtained according to the overlapping principle of wave

$$P_s = \frac{P_A}{k \cdot d_1} + \frac{P_B}{k \cdot d_2} + \frac{P_C}{k \cdot d_3} + \frac{P_D}{k \cdot d_4} \tag{4}$$

Here, k is a constant factor. If there is an obstacle V between A and S (see Figure 3) and the sound absorbing proportion of V is m, then (4) may be written as:

$$P_s = \frac{(1-m)P_A}{k \cdot d_1} + \frac{P_B}{k \cdot d_2} + \frac{P_C}{k \cdot d_3} + \frac{P_D}{k \cdot d_4} \tag{5}$$

Then judge the degree of pollution in this area by comparing P_S with a given index. If the P_S is beyond a given index, we should take measures to prevent S from noise pollution. Here, the noise pollution index P_S may be affected by many factors, such as DEM, an insulation sound wall, buildings, bridges, streets, the height of a railway, the height of vegetation, the surface structure of walls, etc. Figure 4 shows the power and direction of the change in noise when the noise comes up against obstacles.

3.4 *Analysis of sunlight based 3DCM*

Sunlight has a great influence on human health and human compatibility with the environment. Sunlight analysis and shadow analysis are important means by which we evaluate planning and

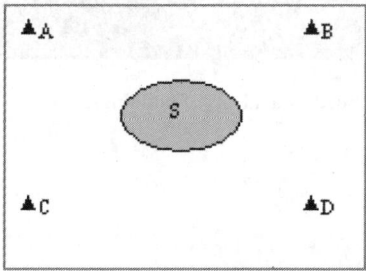

Figure 2.   Point noise pollution distribution.   Figure 3.   Influence from 3DCM on noise spread.

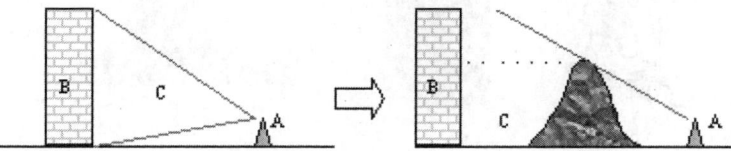

Figure 4.   Left: C as the intensive sound area. Right: C as the weak sound area.

design. Not only is the analysis of sunlight concerned with the speed of movement of the earth and the sun, but also with seasons and the geographical position of objects. Here, we discuss sunlight analysis generally. If there are three blocks A, B and C as shown in Figure 5, simulating the movement of sun we can attain the period of sunlight per day when block B is sun-shined by A and C.

According to Figure 5, the time T when sun is shining on building B is expressed as

$$T = \theta/\omega \qquad (6)$$

Here, ω is the revolution angle speed of the earth. Sunlight time of some concrete floors will be discussed in specialized papers. The analysis of sunlight will be of great help in planning the distance between two buildings, designing the direction in which the buildings should face, the layout and width of the streets, and planning the distribution of the blocks. Figure 6 is the shadow of a building.

3.5 Electro-magnet coverage analysis based on 3DCM

Buildings in cities are often very closely located to each other. If electro-magnet signal are to be received in each corner of a building, many measures should be taken, such as increasing the emitting power of the base station and directing waves to enlarge the coverage of electro-magnet signals. We know that electro-magnets spread in the form of waves. The distance at which the wave spreads is related to the power of the emission; when the electro-magnet wave encounters a building or another obstacle, it will be reflected, absorbed, refracted or will penetrate the object. Therefore, if we want to establish location-based services (LBS) in cities, we must analyse the spatial distribution of electro-magnets in cities. Figure 7 shows A as the emitting base station of the electro-magnet, B as an obstacle, and a man C on the back of B. If the emitting power of the electro-magnet is E, and the

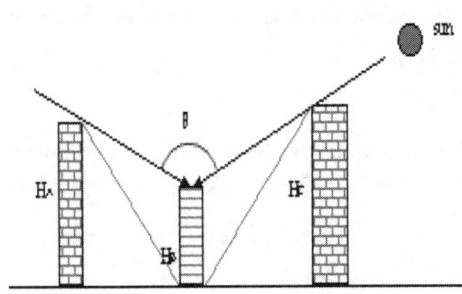

Figure 5.   Sunlight analysis.

Figure 6.   Shadow analysis of buildings.

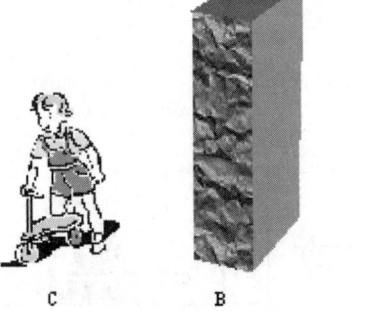

Figure 7.   The spread of an electro-magnetic wave.

distance between A and B is Z, then the power E′ of the electro-magnet at B is calculated according to following equation:

$$E' = E - (1 - 2kZ) \qquad (7)$$

Here, k is a constant. E′ may be divided into four parts after reaching B: (1) the reflecting component; (2) the refracting component; (3) the absorbing component; (4) the penetrating component. The lost power has a relationship with the material of B.

What should be emphasized here is that the electro-magnet emits in three-dimensional space; therefore, the propagation of the electro-magnet was limited by the height of the buildings. When analysing the coverage of the electro-magnet and the distribution of field strength, apart from the power and frequency of the base station, the following factors should be considered:

– Distance between the base station and the moving platform;
– Height of the antenna at the base station;
– Height of moving antenna;
– Typical heights of buildings;
– Distance between one side of the buildings in the street and the moveable station;
– The included angle from the street strike and moving station to the emission direction of the base station.

4 EXAMPLES OF APPLICATION

3DCM-aided spatial decision support is the further application of 3D spatial data and its models. There have been many studies in the past forty years on DSS (Decision Support System) models, but most have focused on the transition from a structural decision to a semi-structural decision, and on connection model-base to data-base. 3DCM-based sunlight analysis of building shadow analysis concern with the Sun's moving rule provides decision proofs for urban spatial planning. There are at least three difficulties in obtaining the decision proofs necessary to optimize and lay out the spatial distribution of buildings; i.e., the display of various shadows, calculation of period of sunlight and intervals of sunlight.

4.1 Display of various shadows

The shadows of a building consist of two categories: the shadows cast by other building and the shadows cast by one's own building. To calculate the influence of the shadows from other buildings is called shadow superimposition (Zhang, 2002), to compute the shadow of one's own building is simply to carry out projection transformation to transform 3D building surfaces to an arbitrary 2D plane along the direction of the rays of sunlight, as shown in Figure 6.

4.2 Calculation of period of sunlight

The period during which sun shines on a building depends greatly on the features of the objects around it. When we investigate the period of sunlight of a building, we have to consider the rule of the movement of the sun, and the time and coverage of shadows resulting from the terrain and objects around the building. The theoretical period of sunlight minus the period of shadow equals the actual period of sunlight. The period of sunlight depends on the season, geographic location (longitude and latitude) and the geometrical shapes of buildings. With a knowledge of the geometric data of geo-objects, we can work out the shadow polygon on the definite height plane in accordance with the shadow project matrix. By judging whether the bottom polygon (2D) of a building inter-crosses with the shadow polygon (2D) of other objects around the building, we will be able to determine whether the building is shaded by other geo-objects around it. Starting from sunrise and using the above measure every moment until sunset, the total amount of time accumulated is the real amount

Figure 8.   Results of the analysis of period of sunlight.

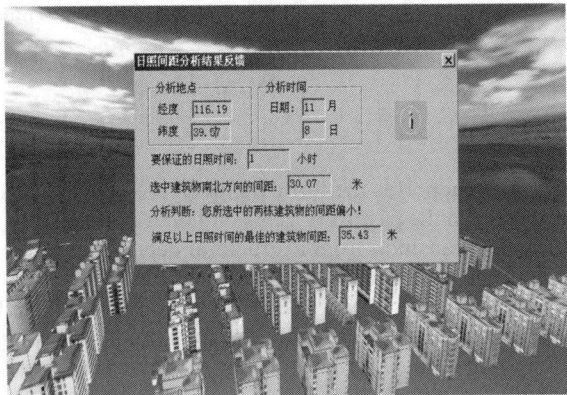

Figure 9.   Results of the analysis of intervals of sunlight.

of sunlight during a certain day. The moment the sun rises and the moment it sets may be supposed to be the earliest and the latest values, respectively. Figure 8 shows the results of the analysis of the experiment on the period of sunlight, using the above parameters.

4.3   *Calculation of intervals of sunlight*

In calculating the intervals of sunlight, the hours for which people demand sunlight should be considered. First, the beginning of the period of shade is calculated according to the height of the angle of the sun at this time, and the angle of the direction of two buildings, then the factor of interval of sunlight is calculated. The sunlight interval factor times the height of one building lying to the east of another building, will yield the reasonable sunlight interval between two buildings, as shown in Figure 9, which may satisfy the special demands.

5   CONCLUSION

To employ 3DCM-based SDS, it is necessary to mine the relevant decision support proofs by way of 3DGIS spatial analysis. In this paper, the basic mathematical models of 3DCM-based SDS such as the spatial statistics model, dynamic analysis model, and time-serials model have been investigated, and their applications studied. These models have been compared with the functions of the current spatial analysis in giving full consideration to the main contents of a cyber-city, named 3DCM. The following are the concluding remarks:

- 3DCM-based SDS is significant to the value-added applications of UGIS;
- 3DCM-based SDS emphasizes 'decision support'; with the support proofs as the critical content. Obtaining decision support proofs depends on the integration of GIS, mathematical models and professional physical models;
- 3DCM-based SDS relies on some basic methods of spatial mathematical analysis, such as spatial statistics, spatial dynamics and spatial optimization.

ACKNOWLEDGEMENTS

This paper is one of the projects supported by the NSFC (No. 40001017), the LIESMARS FOUNDATION (WKL(02)0301). The authors also thank Dr Zhilin Li very much for his critical comments and helpful suggestions on this paper.

REFERENCES

An Zhihong, Chenmin, 1998, Non-linear Time-serial Analysis, Shanghai: Shanghai Science and Technology Press.

Cheng Jianquan, 2001, Urban Systematic Engineering, Wuhan: Wuhan University Press.

Edmond D.H. Cheng, Jie Shang, 1997, Complex terrain surface wind field modelling, Journal of wind engineering and industrial aerodynamics, 67 & 68: 941–962.

Isabelle Lariviere, Gaetan Lafrance, 1999, modelling the electricity consumption of Cities, effect of urban density, Energy Economics, 21(53): 66–82.

Jacek Malczewski, Spatial Decision Support Systems. http://www.ncgia.ucsb.edu/giscc/units/u127/

Klungboonkrong P., M.A.P. Taylor 1, 1998, a microcomputer-based system for multi-criteria environmental impacts evaluation of urban road network computation, Environ. and Urban Systems, 22(5): 425–446.

Liu Yaolin, Liu Yanfang, Liang Qinou, 2001, Urban Environment Analysis, Wuhan: Wuhan University Press.

Monika Ranzinger, Gunther Gleixner, 1997, GIS datasets for 3D urban planning, Computation Environment and urban system, 21(2):159–173.

Peter W.G. Newman, Sustainability and cities, 1999, extending the metabolism model, Landscape and Urban Planning, 44: 219–226.

Sivacoumar R., Thanasekaran K., 1999, Line source model for vehicular pollution prediction near roadways and model evaluation through statistical analysis, Environmental Pollution, 104: 389–395.

Theresa M. Heneker, Martin F. Lambert, George kuKzera, 2001, Journal of hydrology, 247: 54–71. Tsuyoshi Horiguchi, Takehito Sakakibara, 1998, Numerical simulations for traffic-flow models on a decorated square lattice, Physica A, 252: 388–404.

TongJi University and Chongqing Architecture Engineering College. Theory of City Planning. Beijing: Chinese Architectural Industry Press, 1987.

Wang Dan, 2001, Situation and Foreground of Urban Spatial Data and GIS in China, Investigation of Engineering, 28(1): 34–38.

Xu Zhaozhong, 2000, Urban Environment Planning. Wuhan: Wuhan Technical University of Surveying and Mapping Press.

Yilmaz Yildirim, Nuhi Demircioglu, Mehmet Kobya, Mahmut Bayramoglu, 2002, A mathematical modelling of sulphur dioxide pollution in Erzurum City. Environmental Pollution, 118: 411–417.

Zhang Wei, Gu Chaolin, 2000, Urban and Regional Planning Model System, Nanjing: South East University Press.

Zhu Qing, Li Deren, Gong Jianya, Xiong Hanjiang, 2001, The Design and Implementation of Cyber-City GIS (CCGIS), Journal of Wuhan University (Information Version), 26(1): 8–11.

Zhang Ziping. Some Problems on Algorithm of 3D-GIS Spatial Analysis. Wuhan University: Doctoral Paper, 2002, 10s.

Advances in Spatial Analysis and Decision Making, Li, Zhou & Kainz (eds)
© 2004 Swets & Zeitlinger, Lisse, ISBN 90 5809 652 1

A noise model based on a wavelet multi-scale expression and its application to the visualization of spatial activity anomalies of earthquakes

Tao Pei, Cheng-hu Zhou & Jiancheng Luo
State Key Laboratory of Resources and Environmental Information System, Institute of Geographic Sciences and Natural Resources Research, CAS, Beijing

Jiang-she Zhang
Faculty of Science, Xi'an Jiaotong University, Xi'an

ABSTRACT: A noise model based on an á trous wavelet algorithm produces a multi-scale expression of images through the combination of wavelet transform and a testing model of statistical significance. This kind of expression not only gives the formation and location of image structures in different scales, but also eliminates the influence of noise. Since the algorithm does not need any a priori hypotheses, it is suitable for data with complex structures. The aim of this paper to analyse the spatial activities of earthquakes. Specifically, the method of visualizing the multi-scale spatial activities of earthquakes is discussed in this paper. Taking typical sequences in Southwest China as research cases, we systematically study the structural characteristics of the spatial activities of earthquakes on different scales. The results show that, to some extent, multi-scale spatial structures possess an indicative effect on strong epicenters. The foreshock anomalies of the Songpan seismic sequence also reveal interesting patterns during spatial-temporal evolution.

1   INTRODUCTION

How to duly and accurately estimate anomalies in earthquake activity, and how to forecast the three attributes of strong earthquakes in accordance with the temporal-spatial characteristics of the anomaly, are among the most significant issues in studies on earthquake forecasting. In recent years, many advances have been made in the area of seismicity parameters, of which more than 20 such as b, C, D, η, λ, etc. have been proposed (Cheng 1999). These different parameters completely and systematically describe the characteristics of seismicity from different angles. These parameters can be relied upon to serve as important references when forecasting earthquakes. Nonetheless, challenges arise when trying to quantitatively discover the relationship between attributes of strong earthquakes and all kinds of anomalies. These challenges are mainly featured in how to efficiently separate the anomaly from the background and, on this basis, how to determine the spatial-temporal range and its regulation of various kinds of anomalies, and their statistical relationship with the attributes of strong earthquakes.

The concept of background earthquakes has been widely applied, yet there is currently no precise definition of the phenomenon. According to Wyss et al. (2000), background earthquakes are the ones that break out independently and never interact with each other. Other scholars believe that they break out at the equability stage or at the original stage within the active period (Diao et al. 1994). This paper presents the argument that earthquakes within an area consist of background earthquakes and activity anomaly earthquakes. On the one hand, their intensity and range, to a great extent, rely on the spatial-temporal scope concerned, and to observe the anomaly becomes difficult because the background obscures it. On the other hand, background and anomaly are two relative concepts, which can transform into each other on different scales. For instance, within the series of China Mainland-North China – Jingjintang area – Tangshan seismic sequences, each of these areas could have functioned as both the activity anomaly region of the previous neighbouring area and the background of the next neighbouring area. Compared with activity anomaly earthquakes, background earthquakes almost break out at random with weak relativity or even no relativity lying between them. Therefore, the idealistic approach to separating background from anomaly is to set up a multi-scale expression of the earthquake data, and make full use of it to understand the background and anomaly of earthquakes and to differentiate the former from the latter in terms of systematic and dynamic aspects.

The idea of this paper is to set up the multi-scale expression of a spatial of an earthquake using a two-dimensional á trous wavelet algorithm, and to separate the seismic background from the spatial activity anomaly by applying a noise model. It is on such a basis that the relationship between the spatial change characteristics of the anomaly and the main quake's spatial attributes is discussed.

2   NOISE MODEL BASED ON AN Á TROUS WAVELET ALGORITHM

In a number of studies, an á trous wavelet algorithm was applied to the multi-scale decomposition of spatial patterns (Bethoux et al. 1998; Gao et al. 2000; Yang et al. 2001; Pei 1998) and, in practice, noises are usually eliminated by setting soft or hard threshold values. This practice is to some degree subjective when determining noise levels and it neglects the spatial structure of data when eliminating noises. In this paper, the algorithm model has been greatly improved. The following is a brief introduction.

A structure of data or an image usually appears on spatial scales of a certain range, and different structures have different spatial-temporal scales. A wavelet transform provides a multi-scale expression of data and image. In a practical application, strategies of wavelet transform can be classified into a Mallat, namely a pyramid algorithm (Mallat 1989) and into a redundant transform originating from an á trous algorithm (Shensa 1992). A pyramid algorithm is a type of data that transforms without redundancy; however, as the transformation continues, the image will constantly shrink because of the interval sampling. Therefore, we can not make use of this transform to analyse the character of every pixel on different scales, as the wavelet coefficients between different scales can not match respectively. Such a disadvantage can be avoided with the á trous algorithm (Shensa 1992; Starck et al. 1998). Wavelet coefficients originating from this algorithm are redundant, yet the size of the image after each transformation is identical to the size of the original. (Murtagh et al. 1998; Qing et al. 1998; Peng et al. 2000; Yang et al. 2000). Accordingly, we can use this transform to analyse the character of every pixel on different scales.

As both image and data usually include noise, wavelet coefficients produced by the transform also have 'noise'. Thus, the key step to eliminating the noise from the original data is to judge whether the coefficients are from a signal or from a noise. Provided that the noise is a known type, this paper adopts a special statistical testing model as the inner core of the noise model. The exact distribution of the wavelet coefficients when the images are all composed by noises is first computed. The statistical significance testing model is then set up in accordance with this distribution. Explicitly, when the probability that the coefficient is derived from images is small, this coefficient is considered significant; that is, the coefficient is not merely caused by noise, otherwise it would not be significant. In this paper, background earthquakes are considered to break out at random and to be weak, with possibly no relativity lying between them. Thus, we can suppose the Poisson noise model applies to the occurrence of background earthquakes. With such preconditions, we are able to infer the statistically significant testing formulas of the Poisson noise model.

On the kth scale, the wavelet coefficients at (x, y) are:

$$w_k(x, y) = \sum_{i \in K} n_i \Psi(\frac{x_i - x}{2^k}, \frac{y_i - y}{2^k}) \qquad (1)$$

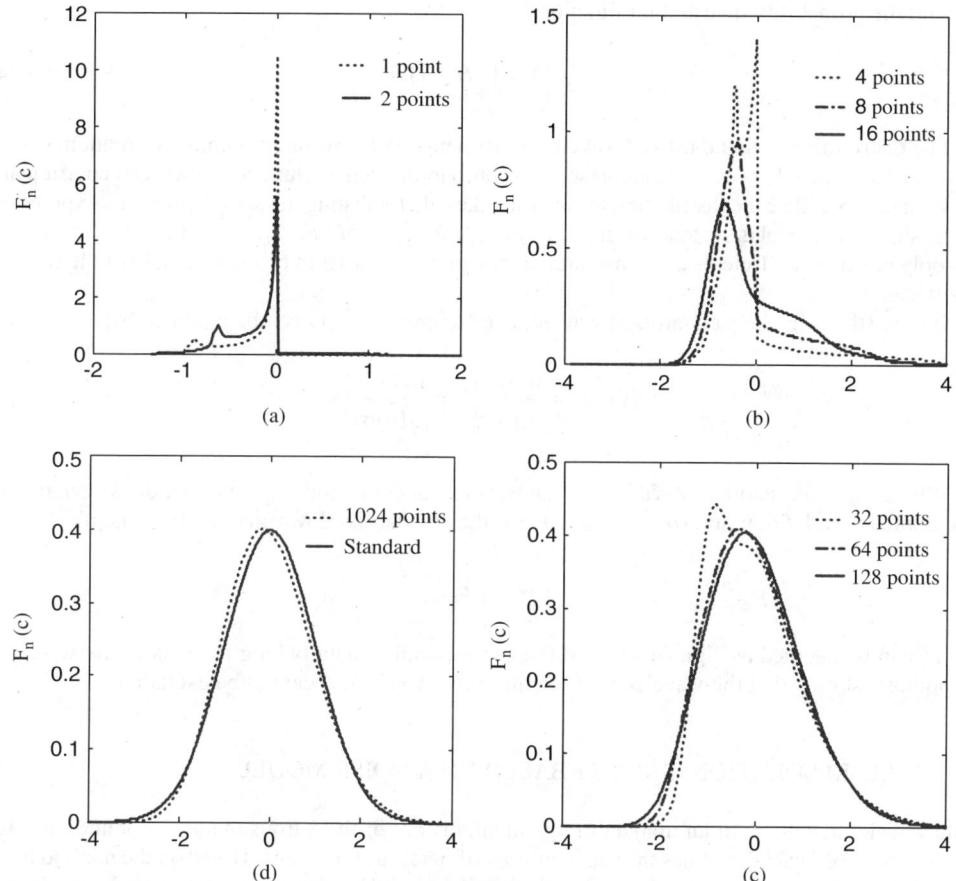

Figure 1. Distribution of wavelet coefficient values for different events (a) 1, 2 events; (b) 4, 8, 16 events; (c) 32, 64, 128 events; (d) 1024 events and standard normal distribution.

where K is the support of wavelet function Ψ, n_i is the number (it is referred to the amount of the events located in the pixel in this paper) of events that contributes to w_k; i.e., the number of events included in the support of the dilated wavelet centred at (x, y). If a wavelet coefficient $w_k(x, y)$ originates from noise, it can be considered a realization of the sum $\sum_{i=k} n_i$ of independent random variables with the same distribution as that of the wavelet function. The distribution of one event in wavelet space can be directly given by the histogram H_1 of the wavelet function Ψ. The distribution of a coefficient w_n related to $n = \sum_{k \in K} n_k$ events is given by an n auto-convolution of H_1:

$$H_n = H_1 \otimes H_1 \otimes ... \otimes H_1 \qquad (2)$$

For details, see Slezak et al. (1990).

The next step is to determine the threshold of statistical testing according to the distribution function mentioned above. In order to facilitate the comparisons, the variable w_n of distribution H_n is standardized by:

$$c = \frac{w_n - E(w_n)}{\sigma(w_n)} \qquad (3)$$

And the probability distribution function is:

$$F_n(c) = \int_{-\infty}^{c} H_n(u)du \tag{4}$$

The distribution of standardized wavelet coefficients with a different number of random events is viewed as Figure 1, where $F_n(c)$ represents the probability density function of wavelet coefficients.

We can now utilize the distribution to construct the rule for testing the significance of a hypothesis. If the significant level is set to ε, we can derive $c_{\min}$ and $c_{\max}$ from $F(c_{\min}) = \varepsilon$ and $F(c_{\min}) = 1 - \varepsilon$. As only positive coefficients are considered in practice, $c_{\max}$ is set to be the threshold of the wavelet coefficients.

On the kth scale, the standardized wavelet coefficients $w_k^r(x, y)$ can be obtained by:

$$w_k^r(x, y) = \frac{w_k(x, y)}{\sqrt{(n)}\sigma_{\Psi_k}} = \frac{w_k(x, y)}{\sqrt{(n)}\sigma_k} 4^k \tag{5}$$

Where σ_Ψ is the standard deviation of the wavelet function, and σ_{ψ_k} is the standard deviation of the dilated wavelet function ($\sigma_{\Psi_k} = \sigma_\Psi/4^k$). For the standardized wavelet coefficients, if

$$w_k^r > c_{\max} \tag{6}$$

w^r can be deemed as significant, and the corresponding data belong to signal; otherwise, the hypothesis stating that the wavelet coefficients come from noise cannot be excluded.

3 MULTI-RESOLUTION SUPPORT BASED ON A NOISE MODEL

The wavelet transform of an image through an algorithm of the á trous method produces, at each scale k, a set of $\{w_k(i, j)\}$. It has the same number of pixels as the image. Based on the method mentioned above, the multi-resolution support of original data will be obtained by detecting the significant coefficients on each scale. The multi-resolution support is defined by:

$$M(k,i,j) = \begin{cases} 1 & \text{if } w_k(i, j) \text{ is significant} \\ 0 & \text{if } w_k(i, j) \text{ is not significant} \end{cases} \tag{7}$$

This shows that a multi-resolution support of an image describes the significance of wavelet coefficients in a Boolean way on a given scale k and at a given position (x, y). The algorithm to create the multi-resolution support is as follows:

(1) Compute the wavelet transform of image by using a á trous algorithm.
(2) Deduce the statistically significant level on each scale by using the noise model.
(3) Compute the multi-resolution support according to (7).
(4) Modify by using a priori knowledge if needed.

In order to visualize the multi-resolution support, the following formula is introduced:

$$S(i, j) = \sum_{k=1}^{p} 2^k M(k,i,j) \tag{8}$$

Where the signification of each parameter is denoted above. Since $S(i, j)$ can generally describe the outline of the structure and location of an interesting signal, we will apply the multi-resolution support originating from the noise model to the analysis of foreshock anomalies of strong earthquakes.

4   EXAMPLES OF 2-D SIMULATING SCATTERING POINTS

Since only discrete point process data (Deng & Liang 1998) are involved in this paper, the gridding method should be employed to convert the point process data into 2-D images.

For a certain dataset A, the symbol function can be defined as: $f(x, y) = I_A(x, y)$ (e.g., when $(x, y) \in$ A, $f(x, y) = 1$; otherwise, $f(x, y) = 0$), and the 3-D point set of $\{(x, y, 1)\}$ is consequently produced. The wavelet transform begins with projecting $\{(x, y, 1)\}$ into a 2-D grid plane by using the interpolating function $\phi(x, y)$, which is employed as the scale function. The result is noted as $\{c_0(i, j)\}$. $\{c_0(i, j)\}$ will be the initial image to carry out the scale transform of wavelets, and the multi-resolution support acquired can be viewed as the multi-resolution expression of data. Next, we will show the efficiency of the algorithm through an example of simulated data.

Simulated data is composed of several patterns, e.g. a point pattern and a linear pattern, and a number of noisy points. The point pattern contains one hundred points obeying the 2-D Gauss distribution with $X \sim (70, 5)$, $Y \sim (180, 5)$; whereas a linear pattern includes 450 2-D Gaussian points distributed along the centre line from (70, 60) to (220, 60). Based on this, 800 Poisson points are added (Hu et al. 2000). The mixed dataset is displayed in Figure 2a.

During the process of the transformation of the wavelet, the B3 wavelet function is employed, following significant threshold values of $\varepsilon = 0.995$. The wavelet coefficients on different scales are shown in Figure 2b–h, with lightness representing the magnitude. The support of wavelet function is displayed in Figure 3. From the wavelet coefficients on different scales, point patterns and linear

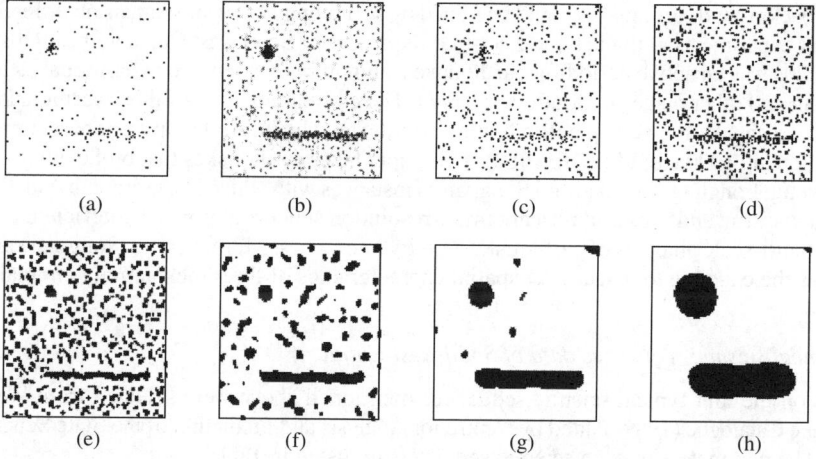

Figure 2.   Simulated data and their wavelet coefficients (á trous method). Simulated data of point pattern with 100 points and line pattern with 450 points and Poisson noise with 800 points; b–e. wavelet coefficients of different scales.

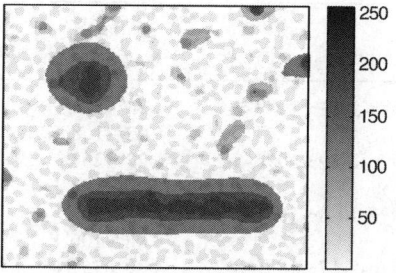

Figure 3.   The support image of the summation of significant wavelet coefficients on different scales.

147

patterns are respectively changed into disks and strips with different brightness in Maps f, g, and h while noises in other places are effectively erased. By weighted averaging, the disk anomaly is centred at (72, 69); and the central points of the strip anomaly are approximately joined into a straight line with the average vertical coordinates approaching 59.

Like the map of wavelet coefficients, point pattern and linear pattern are also very obvious in their support of wavelet function. The difference is that the support can be viewed as the weighted summation of significant degrees of wavelet coefficients on different scales, following the larger weight on the higher scale. Therefore, the deepest colour will be assigned to a pixel whose wavelet coefficient is significant on each scale. Since the support map of wavelet function synthetically expresses the structure of the image in the whole scale sequence, it can be applied to analyse the magnitude and hierarchy of patterns in images.

5 STUDYING SEISMIC SEQUENCES IN SOUTHWEST CHINA

In order to detect an earthquake activity anomaly on different scales, the effects of background earthquakes on different scales must be clarified and then eliminated. The following supposition should be set as the basis when practicing; that is, background earthquakes in areas where no strong earthquakes or seismic swarms break out within a long period could be considered as Poisson Noise. Yet the aftershocks, which break out after a seismic swarm or strong earthquakes, can not be considered as Poisson Noise because of their strong spatial-temporal relativity. Under such a supposition, the anomaly concealed in background earthquakes could be mined by separating noises and signals on different scales. The author experiments with the typical seismic sequences of Southwest China in the 1970s.

The 1970s was an active period for earthquakes, and more than ten strong earthquakes broke out in Songpan and other places in Southwest China. The common feature of these earthquakes is that intensive foreshocks showed up before main earthquake, which formed up an apparent precursory anomaly (Zhang 1990a, b; Ma et al. 1982). This paper takes earthquakes that broke out respectively in Huanglong, Longling, Yanyuan and Songpan as instances with which to experiment with the á trous wavelet transform, and uses a significant multi-resolution support of wavelet coefficient to perform a spatial multi-scale analysis of earthquake activity anomalies. Moreover, the Songpan earthquake is taken as the example to discuss the spatial characteristics of foreshock activity anomalies.

5.1 Typical seismic sequences data of Southwest China

The data on the four typical seismic sequences mentioned above were selected from West China Earthquake Catalogue (1989) edited by Centre for Analysis and Prediction of the State Seismological Bureau. The parameters of each seismic sequence are listed in Table 1.

Setting M = 2 as the floor level, the author experiments with the selected earthquakes in the following application. Because a comparatively complete network of seismic stations was established

Table 1. The statistical data of typical seismic sequences in Southwest China.

Epicentre	Time	Temporal scope	Spatial scope	Fore-shocks	After-shocks	Magnitude
Huanglong	8-11-1973	1973-5-13~1973-9-9	30°–35°N 102°–108°E	51	192	6.5
Longling	5-29-1976	1976-2-29~1976-6-27	22°–26°N 97°–102°E	185	3683	7.4
Yanyuan	11-7-1976	1976-10-8~1976-12-6	22°–30°N 98°–106°E	771	1502	6.7
Songpan	8-16-1976	1976-5-17~1976-8-29	20°–35°N 96°–110°E	7401	3616	7.2

in the area of Sichuan and Yunnan within mainland China (Zhang 1990a, b) in the 1970s, seismic catalogues are largely complete and the parameters of the selected records are exact.

5.2 Multi-scale characters of typical seismic sequences foreshock activity anomaly in Southwest China

The following steps are adopted when applying the above-mentioned algorithm: First, the seismic data is gridded. The grid system is interpolated if a higher resolution is needed and the computational cost is sure to be larger. To obtain a balance between computational cost and precision, the authors gridded the research area of Huanglong and Longling into 150×150, but gridded the research area of Yanyuan and Songpan into 256×256 and added up the number of epicentres within each grid. Of the four typical strong earthquakes, the maximum error of longitude reached 2.82 km, and that of latitude was 2.55 km, which can basically satisfy a wide-range of research requirements.

Limited by the paper length, the authors only compared the multi-scale support of foreshocks (for the Huanglong and Longling sequence, foreshocks within 90 days before main shock were picked up; while earthquakes within 30 days before main shock were selected due to the abundance of foreshocks in the Songpan and Yanyuan sequences) with that of aftershocks within 30 days after main shock. The results are shown in Figure 4 and Figures 5e, f.

These figures synthetically express the structure of the image of the whole scale sequence, and also reflect the multi-scale characteristics of anomalous activity. According to the intensity range of the support, the anomalous activity can be divided equally into five grades from 0 to 4, represented respectively by different colours from shallow to dark. The wavelet coefficients in grade 0 are not entirely significant on each scale and can be deduced from noise; i.e., the background earthquakes. In our paper, the grades of support equal to or greater than 1, respectively, represent anomalous activities on different scales. The higher the grade of the support (with a darker colour), the more significant the wavelet coefficients, indicating more intensive earthquake activity. Based on the assumption discussed above, the grades from 1 to 4 can half-quantificationally describe the anomaly of seismicity.

From the multi-scale support of wavelet coefficients resulting from the foreshocks and aftershocks of four strong earthquakes, we find obvious and intensive anomalies of seismic activity within the region around the main shock. The characteristics of the anomaly show that low-grade anomalies surround the high-grade anomalies, resembling a nesting structure. Compared with aftershocks, the multi-scale anomalies of foreshocks had no prevalent pattern and showed several kinds of shapes, such as strips, disks, and ellipses. The main shock usually occurred in the middle or at the edge of the region of an anomaly of grade 4.

The multi-scale support resulting from the aftershocks possessed a very clear nesting structure, from which we can determine the principal axis of the region formed by aftershocks and the characteristics of the tectonics that will result in strong earthquakes, with high-grade anomalies clearly revealing the spatial distribution of aftershocks.

5.3 The multi-scale expression of anomalies and the spatial-temporal characteristics of change of the foreshocks of the Songpan seismic sequence

In order to study the relationship between anomalies of foreshocks and the temporal-spatial attributes of main quakes, the authors concentrated on examining this kind of relationship within the region 20°–35°N and 96°–110°E by taking the Songpan seismic sequence as an example. The process is implemented as follows: the earthquakes within the temporal-spatial scope are put into a window of 30 days wide according to the order of occurrence, and a á trous wavelet transform is carried out to implement the temporal-spatial scanning by moving the window every 15 days. According to formulas (7) and (8), the multi-scale support S of wavelet coefficients are computed and drawn in Figure 5 by taking the B3 wavelet function and making a significant threshold $\varepsilon = 0.995$.

From the six anomaly maps, epicentres that looked disordered are gathered in several anomaly regions. Two steady anomaly regions appear in each figure (appearing in the first five maps).

149

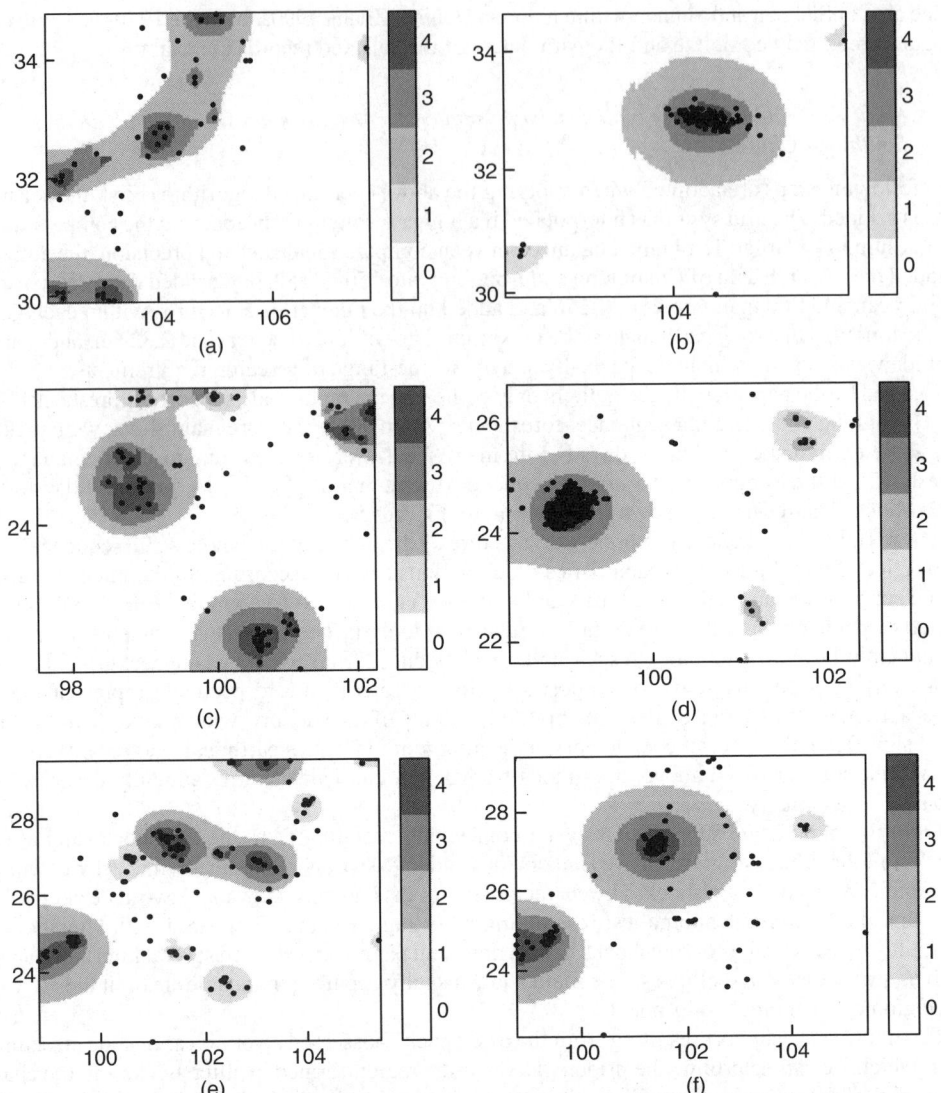

Figure 4.   Multi-scale distribution of epicentres and their anomalies with foreshocks and aftershocks of typical sequences in Southwest China. (a) Distribution of epicentres and anomalies from 13th May to 10th August 1973; (b) Distribution of epicentres and anomalies from 11th August to 10th September 1973; (c) Distribution of epicentres and anomalies from 29th February to 28th May 1976; (d) Distribution of epicenters and anomalies from 29th May to 27th June 1976; (e) Distribution of epicentres and anomalies from 8th October to 6th November 1976; (f) Distribution of epicentres and anomalies from 7th November to 6th December 1976.

They occupy an area of a certain scope and are of a more intensive grade than 2. The larger one, located in the southwest corner of the map with an intensity of between 2 and 4, resulted from the Longling earthquake. The Longling earthquake broke out on May 29th, 1976. It caused many aftershocks and therefore, formed a steady anomaly.

   The other anomaly region is located at the top centre of the map and indicates the scope of the Songpan sequence. In Figure 5a, anomalies of grade 2 and 3 are small at the very beginning, and gradually dilate with the passing of time. In addition, the anomalies are approximately elliptical in shape, with the principal axis extending in the northeast-southwest direction. A comparison of

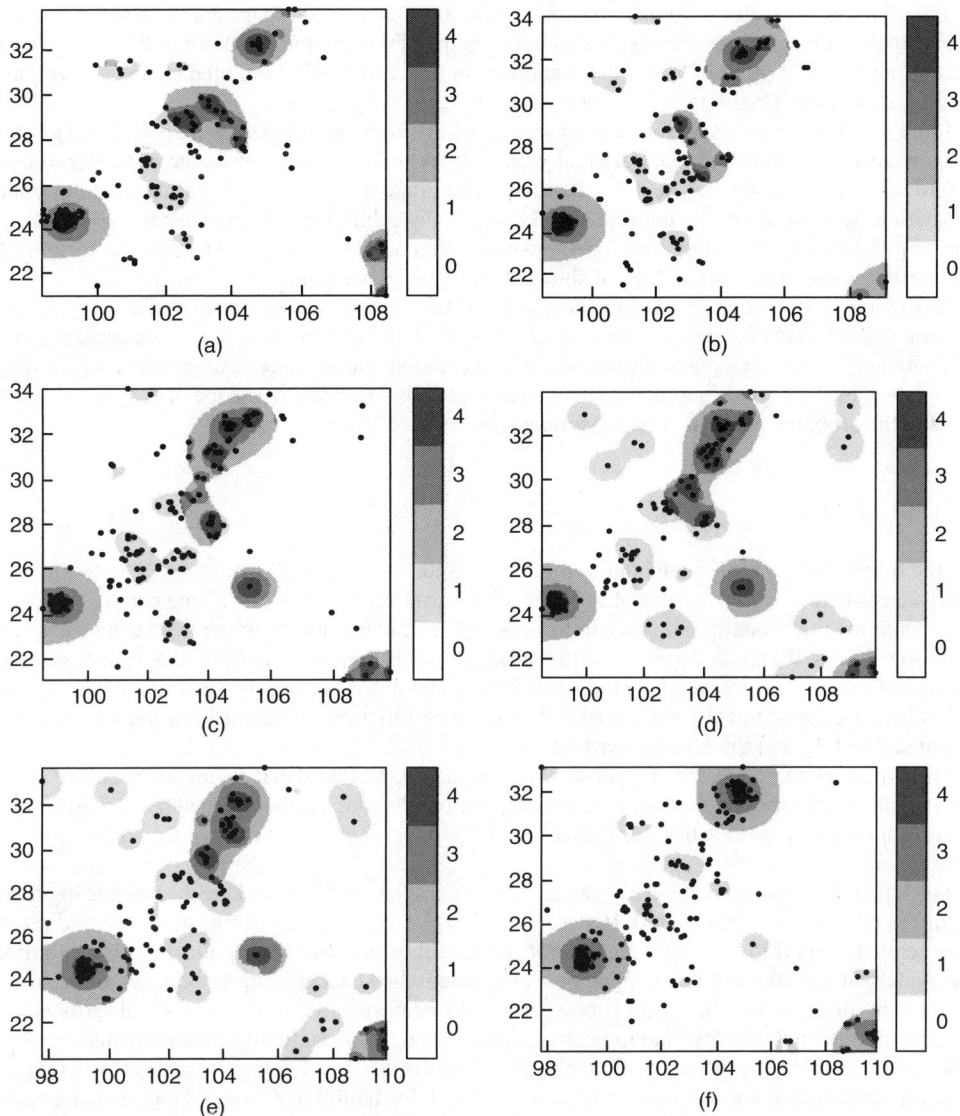

Figure 5.   Distribution of epicentres and their anomalies in Southwest China from 16th May 1976 to 29th August 1976. (a) Distribution of epicentres and anomalies from 17th May to 15th June 1976; (b) Distribution of epicentres and anomalies from 1st June to 30th June 1976; (c) Distribution of epicentres and anomalies from 16th June to 15th July 1976; (d) Distribution of epicentres and anomalies from 1st July to 30th July 1976; (e) Distribution of epicentres and anomalies from 16th July to 15th August 1976; (f) Distribution of epicentres and anomalies from 31st July to 29th August 1976.

Figures 5c and 5d reveals that the area of the anomaly (grade 3) in Figure 5d has been increasing. At the same time, the area and the shape of the anomaly (grade 2) has also changed dramatically. The area of the anomaly (grade 2) is shrinking in Figure 5e; anomalies of grade 3 at the top centre have merged into a larger anomaly, while grade 4 is split into three pieces with equivalent areas. On August 16th, 1976, the Songpan main quake broke out and the epicentre was within the anomalous region of grade 4, on the north side.

The aftershocks resulting from the Songpan main shock caused the shape and area of the anomaly in Figure 5f to change dramatically. In detail, the shape of the anomaly transformed into a disk, the area of the anomaly (grade 2) shrank remarkably, anomaly of grade 3 deflated to some extent, and the scope of grade 4 became more convergent.

From the above analysis, before the occurrence of the Songpan main quake, grade 2 and grade 3 areas were seen to have enlarged gradually until the 60th day before the occurrence, but to have shrunk until the 30th day before the occurrence. The grade 4 area had been increasing throughout this period, being quite similar to the process of mitosis by exhibiting a trend of expanding–splitting–merging. Gradually extending, the foreshocks of Songpan were distributed along the tectonic in the northeast–southwest direction and showed a trend of expansion.

In summary, the Songpan main quake exhibited the obvious characteristics of anomalies, and its anomalies of all grades showed a nesting structure. With the development of the sequence, the areas of all anomalies increased gradually until the 60th day before the occurrence of the main shock, while the shape of the high-grade anomalies changed dramatically. However, the scope of the grade 4 anomaly was the probable place in which the main shock would occur.

6 DISCUSSION

On the limitations of applying the noise model, the authors feel that this model will be feasible for local areas where no strong earthquakes or seismic swarms break out within a long period. With such a precondition this model can be used to forecast (in the mid or short term) strong earthquakes with more foreshocks. We applied this model to the Tangshan earthquake, which was the most destructive disaster of last century but which had few foreshocks, and failed to mine the structure of the anomaly before the occurrence of the main quake. The same results were obtained from the testing of the Bohai earthquake and the Xingtai earthquake.

From many attempts to predict other strong earthquakes, we found that the model discussed in this paper is adept at predicting the spatial-temporal position of strong earthquakes with a certain number of foreshocks, such as the strong earthquakes in Southwest China, Northwest China and a few in North China.

Earthquakes are acknowledged to be an extremely complex problem, and the mechanism to predict them has still not been discovered. Hence, the indicative significance of this model should not be exaggerated, even if there have been several successful examples of its application. For example, the model can sometimes produce fraudulent anomalies when background earthquakes are relatively frequent in some places. That is, no strong earthquake occurred after a swarm of small earthquakes. As a result, the 'authenticity' and indicative significance of the multi-scale anomaly derived from this model should be comprehensively analysed and confirmed by other means, such as by the changing of the chemical composition of groundwater, the deformation of rocks and regional anomalies in gravity.

7 CONCLUSION

Earthquakes are one of the signals released by a gigantic, complicated system: the earth, revealing a complex spatial structure. It can lead to significant values of theory and application to research into earthquake background and activity anomaly. Yet the fact that background and anomaly are related to position and scale allows wavelet transform to reveal a number of advantages in such an analysis.

The algorithm presented in this paper is superior as no supposition of data structure and data form is needed. This superiority is quite applicable to the analysis of seismic data. From the analyses above, it is clear that this model provides an effective means of visualizing anomalous activities. It can be used not only to locate the region of the anomaly, but also to display the characteristics of evolution in the area and the shape of the anomalies by spatial-temporal scanning, and to present references to forecast the spatial-temporal attributes of strong earthquakes as well.

Besides its application in this paper, a noise model based on a á trous wavelet algorithm can also be used to experiment with other parameters of seismicity, such as b, C, D, η, and λ, to mine the multi-scale information of spatial anomalies after gridding and to study how they relate to the spatial-temporal attributes of strong earthquakes.

ACKNOWLEDGEMENTS

This research was supported by the National Natural Science Foundation of China (40101021) and by the 863 program of China 2002AA135230. In addition, the authors would like to thank to the reviewers for their constructive suggestions, which significantly improved this paper.

REFERENCES

Bethoux N, Ouillon G, Nicolas M. 1998. The instrumental seismicity of the western Alps: Spatio-temporal pattern analysed with the wavelet transform [J]. *Geophys J Int*, 135: 177–194

Center for Analysis and Prediction of the State Seismological Bureau eds. 1989. *A Catalogue of earthquakes in western China (1970–1975, M $\geqslant$ 1)* [M]. Beijing: Seismological Press: 1–292

Center for Analysis and Prediction of the State Seismological Bureau eds. 1989. *A Catalogue of earthquakes in western China (1976–1979, M $\geqslant$ 1)* [M]. Beijing: Seismological Press: 1–361

Cheng Wanzheng. 1999. Mathematical basis, correlativity and selection of seismicity parameters [J]. *Acta Seismologica Sinica*, 12(2): 183–192

Deng Yonglu, Liang Zhishun. 1998. *Random point process and its application* [M]. Beijing: Science Press: 1–16

Diao Shouzhong, Guo Aixiang, Wang Hongwei. 1994. Characteristics of early background seismicity predictive of a strong earthquake risk region with M $\geqslant$ 7 [J]. *Earthquake research in plateau*, 6(2): 40–46

Gao Dezhang, Hou Zunze, Tang Jian. 2000. Multiscale analysis of gravity anomalies in the East China sea and adjacent regions [J]. *Chinese journal of geophysics*, 43(6): 842–849

Hu Changhua, Zhang Junbo, Xia Jun, Zhang Wei. 2000. *Matlab Based System Analysis and Design-Wavelet analysis* [M]. Xi'an: Xidian university press: 1–271

Ma Zongjin, Fu Zhengxiang, Zhang Yingzhen, WANG Cheng-min, ZHANG Guo-min, LIU De-fu. 1982. *Nine greatest earthquakes in China from 1966 to 1976* [M]. Beijing: Seismological Press: 1–157

Mallat S. 1989. A theory for multiresolution signal decomposition: the wavelet representation [J]. *IEEE Trans On Pattern Anal Mach Intell*, 11(7): 674–693

Murtagh F, Starck J L. 1998. Pattern clustering based on noise modelling in wavelet space [J]. *Pattern Recognition*, 31(7): 847–855

Pei Tao, Bao Zhengyu. 1998. Research on noise erasing methods with geochemical data [J]. *Geology Geochemistry*, 26(4): 86–90

Peng Yuhua. 2000. *Wavelet transform and engineering application* [M]. Beijing: Science Press: 28–59

Qin Qianqing, Yang Zongkai. 1994. *Applied wavelet analysis* [M]. Xi'an: Xidian University Press: 1–35

Shensa M J. 1992. The discrete wavelet transform: wedding the á trous and Mallat algorithms [J]. *IEEE Trans Signal Processing*, 40: 2464–2482

Slezak E, Bijaoui A, Mars G. 1990. Identification of structures from galaxy counts: use of the wavelet transform [J]. *Astronomy Astrophys*, 227: 301–316

Starck J L, Murtagh F, Bijaoui A. 1998. *Image processing and data analysis* [M]. London: Cambridge University Press: 1–46

Wyss M, Toya Y. 2000. Is background seismicity produced at a stationary poissonian rate [J]. *BSSA*, 90(5): 1174–1187

Yang Fusheng. 2000. *Engineering analysis and application of wavelet transform* [M]. Beijing: Science Press: 1–260

Yang Wencai, Shi Zhiqun, Hou Zunze, CHENG Zhen-yan. 2001. Discrete wavelet transform for multiple decomposition of gravity anomalies[J]. *Chinese journal of geophysics*, 44(4): 534–541

Zhang, Zhaocheng. 1990a. *Earthquake cases in China (1966–1975)* [M]. Beijing: Seismological Press: 117–129

Zhang, Zhaocheng. 1990b. *Earthquake cases in China (1976–1980)* [M]. Beijing: Seismological Press: 29–164

Advances in Spatial Analysis and Decision Making, Li, Zhou & Kainz (eds)
© 2004 Swets & Zeitlinger, Lisse, ISBN 90 5809 652 1

Regional economic disparities in China and their evolution from 1952 to 2000: evidence by Theil coefficient based on comparable prices

Jianhua Xu & Yan Lu
Department of Geography, East China Normal University, Shanghai, China

Nanshan Ai & Yong Chen
Architecture and Environment College, Sichuan University, China

ABSTRACT: Since the late 1970s scholars have conducted many studies on the issue of regional economic disparities, but different scholars have reached different conclusions on the subject. This is mainly because the studies adopted different analytic approaches, perspectives, spatial units and statistical indicators, and examined different periods. On the basis of previous analyses and findings, this paper calculated and decomposed Theil coefficients based on comparable prices, and revealed inter-regional disparities and intra-regional disparities in economic development in China and trends in their evolution from 1952 to 2000.

1 BACKGROUND

In the late 1970s, an increasing number of foreign scholars began studying the issue of disparities in regional development in China. Lardy (1978, 1980) studied the disparities in output and income between rural and urban areas, between agriculture and industry, and between inland and coastal regions before China began to reform and open up to the outside world. He found that due to limited data there was no solid evidence of an expansion in income disparities in different regions of China. Studies by Car (1978) and Lippit (1987) showed that disparities in income between provinces had drastically prior to reduced reform and opening up. Compared with other developing countries, China had achieved remarkable progress, especially in social security. By contrast, Friedman (1987) and Selden (1988) et al. argued that prior to the introduction of reforms, disparities in regional development in China had been growing. Apparently, there are great differences in the way these scholars view the issue of disparities in regional development in China. Moreover, because of the low reliability of the data they had used, their conclusions lack credibility.

In the 1980s, the release of a substantial amount official data in China and the opening up of the country allowed new methodologies to be employed in research. Some scholars have turned their attention to the evolution of regional disparities in development in China since the period of reform and opening up (Keidel, 1995; Jian et al., 1996). Among them are Aguighier (1988) and Yang (1990), who studied regional disparities since the late 1970s and analysed the strategic modes of regional development in China and their evolution. They argued that a policy of bias in development enlarged the disparities in development between the coastal areas of the country and the western region. Tsui (1991, 1998) analysed average per capita national income (NI), and found that disparities in regional development in China showed little change during the period 1952–1970, but grew in the period 1970–1985.

Lyons (1991) examined the per capita net output value in each region of China from 1952 to 1987 using the data released by China's State Statistical Bureau (SSB). He discovered that relative disparities grew in the periods of the 'Great Leap Forward' (1958–1960) and the 'Great Cultural Revolution' (1966–1976), but diminished in the period 1978–1987.

In the 1990s, a series of new analytical methods was adopted to further analyse the composition and sources of regional disparities in China. Rozelle (1994) argued that regional disparity greatly expanded in the coastal provinces from 1984 to 1989. When the Gini coefficient was decomposed, he discovered that the development industrialization in rural areas was the main reason for this. Ying (1999) decomposed the Theil coefficient with the figures for per capita GDP. The result was a 'U-shaped' pattern in regional disparities from 1978 to 1994: before 1990 the disparities between the coastal and interior provinces declined; after 1990 they began to expand. When decomposing the Gini coefficient and the Theil coefficient, Kanbur & Zhang (1999) found that from 1983 to 1995 the disparities between rural and urban areas were greater than those between the coastal and interior provinces. Kim & Knaap (2001) studied regional disparities in agriculture, industry, construction industry and transportation using the Theil coefficient. The results indicated that from 1952 to 1985 disparity of coastal provinces in agriculture industry, construction industry and transportation contributed more than that of interior region in the same sectors to the overall disparity. Furthermore, they argued this mode was obvious in the late 1970s, but that it had not been promoted by any strategy for economic development. Fujita & Hu (2001) decomposed the Theil coefficient with GDP and gross industrial output value, and concluded that disparities between coastal and interior provinces had been growing. Although development disparity of coastal provinces was reducing, industrial development in coastal regions still developed fast. Moreover, they discussed the reasons behind the evolution of regional disparities from the perspective of policies in regional development, economic globalization and liberalization. Lyons (1998) focused his study on a smaller area. He analysed the evolution of regional disparities among counties Fujian Province from 1978 to 1995. He discovered that the interior disparity of Fujian Province was expanding in terms of both absolute disparity and comparative disparity.

Later, Long & Ng (2001) also studied disparities in economic, social and cultural development among the counties of Jiangsu Province. They found that the disparities had grown since 1978. They also analysed the political, economic and social factors behind this growth in disparities. By using Solow's growth model, Chen & Fleisher (1996) found that there was a conditional convergence in the growth of per capita GDP among the various provinces of China in the period 1978–1993. They argued that regional disparities in China have diminished since implementation of the policy of reform and opening up.

In the 1990s many Chinese scholars began studying disparities in regional development in China. Yang (1992) calculated the Gini coefficient with per capita GNP, and analysed the evolution of income disparities between the coastal, middle and western regions of China in the 1980s. He concluded that China's biased strategy of giving priority in development to the coastal regions had led neither to the expansion of income disparities all over China, nor to the expansion of income disparities between the coastal areas and the middle and western regions of the country. On the contrary, there was an overall decline in income disparities in China. Yang (1994) worked out the coefficient of weighed variation by per capita GDP among provinces. He concluded that the evolution of economic disparities among provinces in China was in the approximate shape of an inverted U-curve, with an inflexion in the year 1978. Prior to 1978, the disparities had grown; after 1978 they began to decline. Wei (1992, 1996, 1998) analysed the evolution of disparities among the three supra-provincial regions of coastal, middle and western China in the period 1978–1992. The results indicated that the economies of all of the three supra-provincial regions had grown. The middle and western region, however, obviously lagged behind in terms of pace of development. The disparities between the coastal region and the middle and western regions were still growing; Wei & Liu (1994) also forecasted trends in the economic development of the three supra-provincial regions. They felt that from 1993 to 2010, economic growth would remain unbalanced. Absolute disparities between the coastal and middle-western regions would not diminish, and comparative disparities might grow in the near future. Lu & Xue et al. (1998) and Hu & Zou (1999) argued that overall disparities in regional development in China had expanded before 1978 and then began to diminish until the 1990s, when they again began to increase. Yuan (1996) believed a remarkable trend in regional development in China since the beginning of the reforms was the growing economic differences among the three supra-provincial regions. Lin (1998) and Cai et al. (2000, 2001) studied the evolution of regional disparities during the period of economic transition (1978–1995) using per capita GDP and per capita income as measures. They found that the disparities among the three supra-provincial regions were greater than those within the regions. By studying regional disparities based on income, Song (1998) found national regional disparity was in the shape of an inverted 'U'. Prior to 1990, regional disparities had diminished, but after that year they gradually expanded. Cai & Du (2000) examined inter-regional disparities and intra-regional disparities and found that in the period 1978–1999, intra-regional disparities in the coastal region contributed greatly to overall disparity but in a declining trend. Intra-regional disparities in the middle region contributed a little to the overall disparity, and in a declining trend as well. Disparities within the western region contributed very little to overall disparity, and what contribution there was also showed a trend of decline. Meanwhile, the disparities between the three supra-provincial regions contributed substantially to overall disparity in the nation, and in a marked trend of increase. They argued that there was a conditional similarity in economic growth in different regions in China. Li & Qiao (2001) analysed, for the first time at the county level, the spatial evolution of regional economic disparity in China in the 1990s. Their results demonstrated that economic disparities between counties had declined, but that the disparities between coastal and inland regions had widened; the counties with faster economic growth than the national average were chiefly distributed in three growth belts; namely, the coastal belt, the Beijing-Guangzhou railway belt and the Yangtze River belt (from Chongqing to Shanghai). The less-developed counties were mainly located in the western part of China. Liu (2001) argued convergence in regional economic growth in China appeared at different times and in different regions, and that disparities in output between regions was positively correlated to overall economic instability in China.

From the above analysis, it is clear that while domestic and foreign scholars have both conducted a great deal of research on regional disparities and their evolution in China, their conclusions differ greatly. We think that the reason for this lies large in the use of different analytical approaches, perspectives, spatial units, statistical indicators and in the different periods examined in the studies. On the basis of previous analyses and findings, we have conducted further quantitative computations and empirical studies and here reveal our findings on inter-provincial and regional disparities in economic development and trends in their evolution from 1952 to 2000.

2 METHODOLOGY AND DATA

2.1 *The divisions of the spatial unit*

There is usually a spatial criterion for studies of regional disparities. The spatial criterion chosen depends on the specific objectives of the study. The purpose of this paper is to reveal disparities among provinces and regions in economic development from 1952 to 2000, and the evolution of these disparities. A provincial administrative unit is a political and economic region with an integrated function, and each comes with a complete system of statistical data, which is readily available. Thus, we chose the province (which includes provinces, provincial-level municipalities and autonomous regions) as the basic spatial unit for our analysis, and also chose the three supraprovincial regions: coastal, middle and western as the overall spatial units. The coastal provinces are Beijing, Tianjin, Hebei, Liaoning, Shanghai, Jiangsu, Zhejiang, Fujian, Shandong, Guangdong, Guangxi and Hainan. The middle provinces consist of Shanxi, Inner Mongolia, Jilin, Heilongjiang, Anhui, Jiangxi, Henan, Hubei and Hunan. The western provinces are Yunnan, Guizhou, Sichuan, Chongqing, Tibet, Shaanxi, Gansu, Qinghai, Ningxia and Xinjiang (Figure 1).

2.2 *Selection of statistical indicators and sample data*

As for the study of the dynamic evolution of regional disparities in China, the use of the per capita GDP of each province may be appropriate. Since per capita GDP is the best approximation and can

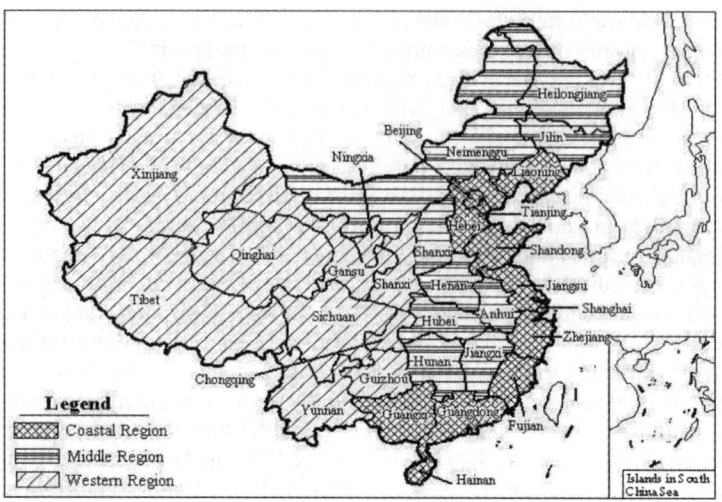

Figure 1.   The divisions of the spatial unit in China.

well reflect the overall level of development and well-being, it is widely used. Moreover, the time series data for per capita GDP in each province is complete, and can be used for temporal and spatial comparisons. Therefore, we chose 31 provinces (municipalities, autonomous region) in China as spatial samples, and the period 1952–2000 as the temporal sample.

The primary data from the National Bureau of Statistics of China, are mainly taken from the following published sources: (1) Comprehensive Statistical Data and Materials on 50 years of New China. Beijing: China Statistics Press, 1999. (2) China Statistical Yearbook 2001. Beijing: China Statistics Press, 2001. (3) Urban Statistical Yearbook of China 2001. Beijing: China Statistics Press, 2001. (4) Historical Data for China Gross Domestic Product (1952–1995). Dalian: Northeast China University of Finance and Economics Press, 1997. In a general way, the data are reliable and authoritative. However, several years of data such as those during the period of the 'Great Leap Forward' are distorted due to non-economic factors, which is proven by the Theil coefficient in section 3.1 of this paper. Consequently, these data are not as credible as those from the other years, and the results calculated by using them are not as precise. However, as we study regional economic disparity in China in as long period, the primary data can only from National Bureau of Statistics of China.

2.3   Data processing method for eliminating the influence of prices

If the price of products and services does not change at all, we can ignore the influence of changes in price. China, however, is a large country with a big spatial difference. During the past 50 years of economic development, China has undergone several different stages during which there were constant changes in the comprehensive price index and inflation rate at different periods and in different places. Therefore, if we discuss disparities in regional development using present-day prices, we may come to erroneous conclusions. In order to accurately reflect the disparities and their evolution, we must consider the influence of price. Thus, we have converted the GDP data for each region into present-day values by using price index (1) as follows (Hu & Zou, 1999):

$$X_i(t) = X_i(t_0) \times \beta_i(t) \tag{1}$$

where $X_i(t)$ is the real GDP data of the ith region at the tth year, $X_i(t_0)$ is the real GDP data of the ith region at the base year (t_0th year), $\beta_i(t)$ is the GDP growth exponent of the ith region from the base year to the tth year.

In this article, 1978 is the base year, and the GDP data of each year is the real data converted in terms of comparable prices.

2.4 *Quantitative exponents*

Many quantitative exponents are used to describe regional disparities (Borts et al., 1964; Friedman, J., 1963; Dunford, 1993; Dowrick & Nguyen, 1989), such as extreme deviation, standard deviation, coefficient of variation, Engel coefficient, location entropy, and so on. After comparing all of these indicators, we chose the Theil coefficient as the quantitative indicator for our analysis of regional disparities in economic development in China. Not only can the Theil coefficient show regional disparities, it also can be decomposed into inter-regional disparities and intra-regional disparities.

The Theil coefficient, also called Theil entropy, was proposed by Theil (Theil & Henri, 1967) in 1967. The Theil coefficient is defined in the following way:

$$T = \sum_{i=1}^{N} Y_i \log \frac{Y_i}{P_i} \tag{2}$$

where N is the number of areas, Y_i is the share of the ith region of the total national GDP, P_i is the share of the ith region of the total national population.

If the Theil coefficient is bigger, the disparities in economic development between various areas will be greater. Otherwise, the disparities will be smaller.

Another characteristic of the Theil coefficient is that it may be decomposed into two parts: inter-group disparity and intra-group disparity, which makes clearer the evolution in both inter-group and intra-group disparities and their respective importance in overall disparity. In China, for instance, the Theil coefficient can be decomposed into inter-regional disparities and intra-regional disparities (Zhou, 1999) as follows:

$$T = T_{\text{inter}} + T_{\text{intra}} = \sum_{i=1}^{3} Y_i \log \frac{Y_i}{P_i} + \sum_{i=1}^{3} Y_i [\sum_j Y_{ij} \log \frac{Y_{ij}}{P_{ij}}] \tag{3}$$

where i ($i = 1, 2, 3$) is one of three supra-provincial regions (the coastal, middle or western region), when $i = 1$, $j = 1, 2, …,12$, which, for instance, respectively corresponds to the 12 provinces of coastal region. Y_i is the share of the supra-provincial region i in national GDP, P_i is the share of the supra-provincial region i in national population; Y_{ij} is the share of province j in the overall GDP of the supra-provincial region, P_{ij} is the share of the province j in the total population of the supra-provincial region i.

3 RESULTS AND DISCUSSION

3.1 *The inter-provincial disparities shown by the Theil coefficient*

As 1978 is the base year, and the GDP data for each year was the real data converted in terms of comparable prices, the Theil coefficients calculated in the period 1952–2000 reveal a dynamic trend in comparative disparities in economic development between provinces in China (Figure 2).

With reference to Figure 2, we can see that the Theil coefficient revealed the trend in the evolution of comparative inter-provincial disparities. From 1952 to 1978, except for several unusual years of data during the period of the 'Great Leap Forward', the disparities display a general upward trend. From 1979 to 1990, the disparities show a slow downward trend. But from 1991 to 2000, the disparities again reveal a slow upward trend. In other words, while the strategy to balance regional development before the period of reform and opening up did not succeed in reducing comparative disparities in regional economic development in China, the lopsided strategy of

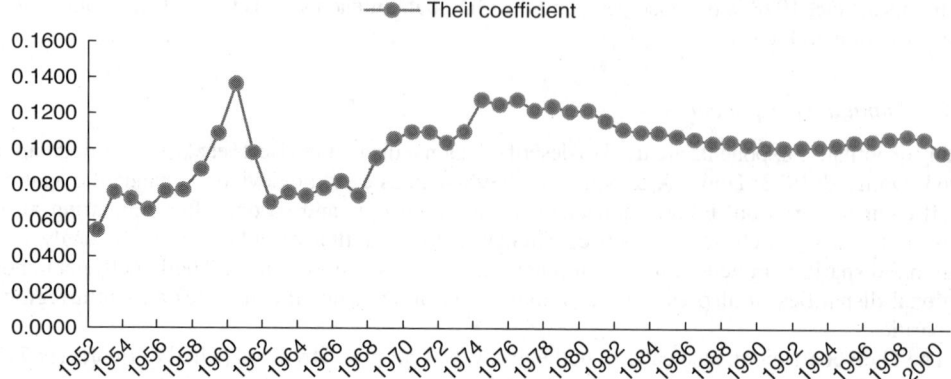

Figure 2.   The Theil coefficient based on comparative prices from 1952 to 2000.

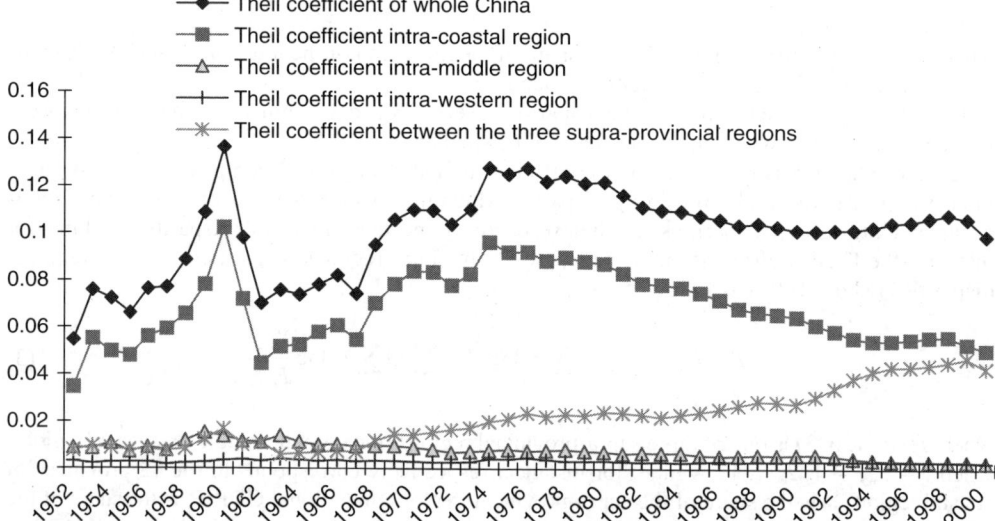

Figure 3.   Decomposition of the Theil coefficient from 1952 to 2000.

development after 1978 has also not enlarged disparities. In other words, the strategy of balanced regional development prior to the period of reforms did not lead to a reduction in comparative disparities in economic development among regions, nor did the lopsided development strategy implemented since then lead to an expansion in comparative disparities in economic development among regions in China. This conclusion is interesting and exciting, and it seems to give rise to another complicated problem of explaining the reasons for this finding. This problem will be studied and explained in another paper.

3.2   *Decomposition of the Theil coefficient: inter-regional disparities and intra-regional disparities*

Decomposing the Theil coefficient can further reveal reasons for the evolution of disparities from a regional perspective, and evolution of intra-regional disparities. Figure 3 reveals the evolution of regional disparities and the evolution of intra-regional disparities. The dynamic trend in comparative inter-provincial disparities in the coastal region is consistent with the dynamic trend in

Table 1. Decomposition of the Theil coefficients: contribution of inter-regional and intra-regional disparities to national disparities from 1952 to 2000.

Years	Whole China	Intra-coastal region	Intra-middle region	Intra-western region	Three supra-provincial regions	Coastal region (%)	Middle region (%)	Western region (%)	Among three supra-provincial regions (%)
1952	0.05458	0.03462	0.00891	0.00325	0.00781	63.42	16.32	5.95	14.31
1953	0.07588	0.05511	0.00867	0.00231	0.00979	72.63	11.42	3.05	12.90
1954	0.07228	0.04975	0.01084	0.00285	0.00884	68.84	14.99	3.94	12.23
1955	0.06614	0.04791	0.00731	0.00291	0.00801	72.44	11.05	4.40	12.11
1956	0.07641	0.05600	0.00924	0.00295	0.00823	73.28	12.10	3.86	10.77
1957	0.07702	0.05929	0.00795	0.00194	0.00785	76.97	10.32	2.52	10.19
1958	0.08870	0.06570	0.01203	0.00255	0.00842	74.07	13.57	2.87	9.49
1959	0.10867	0.07815	0.01535	0.00278	0.01239	71.91	14.13	2.56	11.40
1960	0.13646	0.10221	0.01385	0.00354	0.01687	74.90	10.15	2.59	12.36
1961	0.09798	0.07195	0.01177	0.00400	0.01025	73.44	12.02	4.08	10.47
1962	0.07026	0.04461	0.01128	0.00334	0.01103	63.50	16.05	4.76	15.70
1963	0.07570	0.05164	0.01397	0.00407	0.00601	68.23	18.45	5.38	7.94
1964	0.07381	0.05234	0.01124	0.00371	0.00653	70.90	15.22	5.02	8.85
1965	0.07807	0.05794	0.01002	0.00323	0.00688	74.21	12.84	4.14	8.81
1966	0.08203	0.06076	0.01026	0.00362	0.00739	74.07	12.50	4.42	9.01
1967	0.07411	0.05452	0.00949	0.00305	0.00706	73.56	12.80	4.11	9.52
1968	0.09492	0.07015	0.00960	0.00339	0.01178	73.90	10.12	3.57	12.41
1969	0.10557	0.07828	0.00959	0.00322	0.01448	74.15	9.09	3.05	13.71
1970	0.10978	0.08373	0.00886	0.00292	0.01426	76.27	8.07	2.66	12.99
1971	0.10954	0.08334	0.00805	0.00272	0.01543	76.08	7.35	2.48	14.09
1972	0.10369	0.07776	0.00673	0.00274	0.01646	75.00	6.49	2.64	15.87
1973	0.10989	0.08265	0.00720	0.00285	0.01719	75.21	6.56	2.59	15.64
1974	0.12766	0.09605	0.00771	0.00404	0.01985	75.24	6.04	3.17	15.55
1975	0.12501	0.09184	0.00809	0.00410	0.02098	73.47	6.47	3.28	16.78
1976	0.12767	0.09204	0.00742	0.00465	0.02356	72.09	5.81	3.64	18.45
1977	0.12189	0.08832	0.00767	0.00384	0.02206	72.46	6.29	3.15	18.10
1978	0.12420	0.08976	0.00811	0.00322	0.02311	72.27	6.53	2.59	18.61
1979	0.12125	0.08804	0.00742	0.00321	0.02258	72.61	6.12	2.64	18.62
1980	0.12181	0.08712	0.00721	0.00335	0.02414	71.52	5.92	2.75	19.82
1981	0.11620	0.08303	0.00638	0.00303	0.02376	71.45	5.49	2.61	20.45
1982	0.11110	0.07862	0.00674	0.00280	0.02294	70.77	6.06	2.52	20.65
1983	0.10976	0.07825	0.00663	0.00292	0.02195	71.30	6.04	2.66	20.00
1984	0.10928	0.07699	0.00662	0.00268	0.02299	70.45	6.06	2.45	21.04
1985	0.10740	0.07477	0.00572	0.00296	0.02394	69.62	5.32	2.76	22.29
1986	0.10575	0.07189	0.00539	0.00317	0.02530	67.98	5.10	2.99	23.93
1987	0.10339	0.06809	0.00551	0.00299	0.02679	65.86	5.33	2.89	25.91
1988	0.10417	0.06666	0.00586	0.00305	0.02859	63.99	5.63	2.93	27.45
1989	0.10274	0.06573	0.00561	0.00311	0.02829	63.97	5.46	3.03	27.54
1990	0.10122	0.06452	0.00594	0.00318	0.02759	63.74	5.87	3.14	27.25
1991	0.10095	0.06117	0.00608	0.00321	0.03049	60.59	6.02	3.18	30.20
1992	0.10136	0.05859	0.00539	0.00325	0.03413	57.81	5.32	3.21	33.67
1993	0.10149	0.05555	0.00450	0.00320	0.03824	54.73	4.44	3.15	37.68
1994	0.10254	0.05427	0.00390	0.00314	0.04123	52.93	3.81	3.06	40.20
1995	0.10430	0.05448	0.00349	0.00311	0.04321	52.24	3.35	2.98	41.43
1996	0.10482	0.05505	0.00336	0.00306	0.04335	52.52	3.20	2.92	41.36
1997	0.10637	0.05598	0.00321	0.00308	0.04410	52.62	3.02	2.90	41.46
1998	0.10782	0.05622	0.00326	0.00309	0.04525	52.15	3.02	2.87	41.97
1999	0.10605	0.05305	0.00316	0.00295	0.04689	50.03	2.98	2.78	44.22
2000	0.09871	0.05049	0.00279	0.00271	0.04271	51.15	2.83	2.74	43.27

national comparative disparities. The comparative disparity between the provinces in the middle region and the provinces in the western region remains small, and the evolution was rather slow. Disparities in development between the three supra-provincial regions have been continuously on the increase and have grown even greater since the 1990s.

It is evident from Table 1 that intra-regional disparities among the three supra-provincial regions are increasingly contributing to overall disparity in China. From Table 1 we can see that over half of the disparity results from inter-provincial disparity in the coastal region. The contribution of the coastal region diminished from 63.42% (1952) to 51.15% (2000) and the contribution of the middle and western regions, respectively, diminished from 16.32% and 2.83% (1952) to 5.95% and 2.74% (2000). The contribution of the disparities among the three supra-provincial regions, however, increased from 14.31% (1952) to 27.25% (1990), and to 43.27% (2000). In another words, inter-provincial disparities in the whole of China mainly stemmed from inter-provincial disparities in the coastal region and from the disparities among the three supra-provincial regions (the coastal, middle and western regions).

Figure 3 and Table 1 also imply that the evolutionary process of inter-provincial disparities for the whole of China was mainly controlled by inter-provincial disparities in the coastal region and among the three supra-provincial regions (the coastal, middle and western regions).

3.3   A comparison between Shanghai and Guizhou: the disparity between the richest and poorest provincial-level units

The Theil coefficient and its decomposition have reflected comparative inter-provincial disparities and comparative inter-regional and intra-regional disparities for three supra-provincial regions, but they have covered up the absolute disparities among provinces. Therefore, it is necessary to choose two provinces for comparison to reveal the evolution of absolute inter-provincial disparities. We have chosen to compare Shanghai, the provincial-level unit with the highest level of economic development, and Guizhou, the province with the lowest level of economic development.

The changes in the ratios of per capita GDP in Shanghai and Guizhou not only reflect the disparities between the two places, but also reveals, to some extent, absolute inter-regional disparities. It is clear in Figure 4 that there have been three stages in the evolution of absolute disparities between Shanghai and Guizhou. The disparities in economic development between the two provinces grew until 1978, and diminished in the period of reform and opening up from 1978 to 1989. In 1990 the disparities again began to grow, with a slight drop in 1998. Before 1978, the disparities between Shanghai and Guizhou grew at a comparatively greater rate, with an increase in the ratio of disparities from 12.355 (1952) to 28.076 (1978). After 1978, the disparities between two provinces diminished, with a ratio of 24.026 in 1990. In 1990, the disparities began to grow again, before diminishing in 1998.

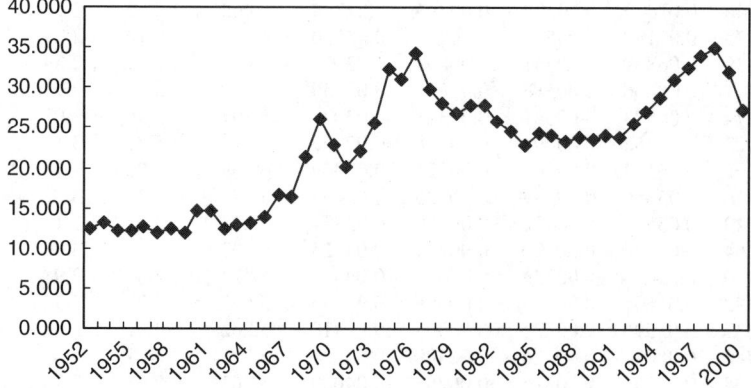

Figure 4.   The ratios of per capita GDP based on comparable prices in Shanghai to that in Guizhou.

Compared with the national average per capita GDP in the period 1952–2000, the ratio of Shanghai to China as a whole was above 4 for every year, while the ratio of Guizhou to the whole China was less than 1, and even under ½. The comparison in the development between Shanghai and China as a whole is approximately consistent with the comparison between Shanghai and Guizhou in the following way: the disparities had increased prior to 1978 and diminished slightly from 1978 to 1990. After 1990, they increased again until 1998, when there was a slight reduction. Before 1976, the disparities between Guizhou and China as a whole increased; and from 1978 to 1990 they diminished slightly. Since 1990 the disparities have grown again.

4 SUMMARY AND CONCLUSIONS

From above analysis, we elicit the following conclusions:

1. Regional disparities in economic development in China, including inter-provincial disparities, inter-regional disparities and intra-regional disparities, have existed for years.
2. The Theil coefficients have revealed a dynamic trend in comparative disparities in economic development between provinces in China. From 1952 to 1978, with the exception of the period of the 'Great Leap Forward', comparative disparities essentially assume an upward trend, and from 1979 to 1990, they assume a slow downward trend. Afterwards, from 1991 to 2000, the disparities again show a slow upward trend. This evolutionary process implies that the strategy of regional balanced development before the period of reform and opening up did not lead to a reduction in comparative disparities in regional economic development, nor did the lopsided development strategy implemented since then bring lead to an increase in comparative disparities in regional economic development in China. This conclusion is interesting and exciting, and will be explained in another paper.
3. The decomposition of the Theil coefficient indicates that the dynamic trend in comparative inter-provincial disparities in the coastal region is in line with the dynamic trend in inter-provincial disparities for the whole China. In other words, inter-provincial disparities gradually grew from 1952 to 1978, and then began to diminish until 1990. The inter-provincial disparities in the middle and western regions diminished during the whole period of 1952–2000, but at a slow pace. Meanwhile, the disparities in economic development among the three supra-provincial regions grew continuous, especially in the 1990s, when the growth was particularly rapid. Inter-provincial disparities for the whole of China and their process of evolution mainly stem from inter-provincial disparities in the coastal region and among the three supra-provincial regions (the coastal, middle and western regions).
4. A comparison between Shanghai and Guizhou shows that absolute inter-provincial disparities have been quite large for many years. The disparities in economic development between the two provinces grew until 1978 and diminished for approximately a decade after the period of reform and opening up. Since 1990, however, the disparities have again grown, with a slight drop in 1998.

REFERENCES

Aguignier, P. 1988. Regional Disparity Since 1978, in Feuchtwang, Sephen, Hussain, Athar and Pairault, Thierry, (Eds), *Transforming China's Economy in the Eighties: the Urban Sector*, 2, 93–106. London: Zed Books Ltd.
Borts, G. H. and Stein, J. L. 1964. *Economic Growth in a Free Market*. New York: Columbia University Press.
Cai, F. and Du, Y. 2000. Convergence and divergence in regional economic growth in China. *Economic Research Journal* (in Chinese), 10: 30–37.
———. 2001. Regional gap, convergence and developing China's western region. *China Industrial Economy* (in Chinese), 2: 48–54.

Car, R. 1978. *China's Political Economy: The Quest for Development since 1949*. Oxford: Oxford University Press.

Carlberg, M. 1981. A neoclassical model of interregional economic growth. *Regional Science and Urban Economics*, 11: 191–203.

Chen, J. and Fleshier, B. M. 1996. Regional income inequality and economic growth in China. *Journal of Comparative Economics*, 22(2): 141–164.

Dowrick, S. and Nguyen, D. 1989. OECD comparative economic growth 1950–1985: catch-up and convergence. *American Economic Review*.

Dunford, M. 1993. Regional disparities in the European community: evidence from the REGIO databank. *Regional Studies*, 27(8): 727–743.

Friedman, E. 1987. Maoism and the liberation of the poor. *World Politics*, 39(3): 408–428.

Friedman, J. 1963. Regional economic policy for developing areas. *Papers of the Regional Science Association*, 11.

Jian, T., Sachs, J. D. and Warner, A. M. 1996. Trends in regional inequality in China. *National Bureau of Economic Researsh Working Paper 5412*. Cambridge, MA.

Hu, A. and Zou, P. 2000. *Society and development: a study of China's regional social development disparity* (in Chinese). Hangzhou: Zhejiang People's Publishing House.

Kanbur, R. and Zhang, X. 1999. Which regional disparity? The evolution of rural-urban and inland-coastal disparity in China from 1983 to 1995. *Journal of Comparative Economics*, 27(4): 686–701.

Kim, T. J. and Knaap, G. J. 2001. The spatial dispersion of economic activities and development trends in China: 1952–1985. *The Annals of Regional Science*, 35(1): 39–57.

Keidel, A. J. 1995. *China: Regional Disparity*. Washington, DC: World Bank.

Lardy, N. R. 1978. *Economic growth and income distribution in the People's Republic of China*. New York: Cambridge University Press.

———. 1980. Regional growth and income distribution in China, In R. F. Denberger (Ed.), *China's Development Experience in Comparative Perspective* (153–190). Cambridge: Harvard University Press.

Li, X. and Qiao, J. 2001. County level economic disparities of China in the 1990s. *Acta Geographica Sinica*, 56(2): 136–145.

Lin, Y. et al. 1998. An analysis of regional gaps in China's economic transition. *Economic Research Journal* (in Chinese), 6: 3–10.

Lippit, V. D. 1987. *The Economic Development of China*. New York: ME Sharpe.

Liu, Q. 2001. An analysis of convergence in China's economic growth. *Economic Research* (in Chinese), 6: 70–77.

Long, G. N. and Ng, M. K. 2001. The political economy of intra-provincial disparity in post-reform China: a case study of Jiangsu province. *Geoforum*, 32: 215–234.

Lu, D. et al. 1998. *Ae Report of China's, Regional Development in 1997* (in Chinese). Beijing: The Commercial Press.

Lyons, T. P. 1991. Inter-provincial disparities in China: output and consumption, 1952–1987. *Economic Development and Cultural Change*, 39(3): 471–506.

———. 1998. Intra-provincial disparity in China: Fujian Province, 1978–1995. *Economic Geography*, 74(3): 201–227.

Masahisa F. and Hu, D. 2001. Regional disparity in China 1985–1994: the effects of globalization and economic liberalization. *The Annals of Regional Science*, 35(1): 3–37.

Roll, C. R., Jr. and Yeh, K. 1975. Balance in Coastal and Inland Industrial Development, in China. A Reassessment of the Economy, U.S. Congress Joint Economic Committee. Washington, D. C.: Government Printing Office.

Rozelle, S. 1994. Rural industrialization and increasing inequality: emerging Patterns in China's reforming economy. *Journal of Comparative Economics*, 19(3): 362–391.

Selden, M. 1988. *The Political Economy of Contemporary China*. New York: Sharpe.

Song, D. 1998. Regional differences in economic development since the reforms. *Quantitative & Technical Economics* (in Chinese), 3: 15–18.

Theil, H. 1967. *Economics and Information Theory*. Amsterdam: North Holland.

Tsui, K. Y. 1991. China's regional inequality, 1952–1985. *Journal of Comparative Economics*, 15(1): 1–21.

———. 1998. Factor decomposition of Chinese rural income inequality: new methodology, empirical findings, and policy implications. *Journal of Comparative Economics*, 26(3): 502–528.

Wei, H. 1992. Patterns in the evolution of regional income disparities in China. *Economic Research* (in Chinese), 4: 51–55.

———. 1996. An Analysis of China's regional income disparities. *Economic Research* (in Chinese), 11: 66–73.

————. 1998. Theories in regional economic science. *Development Research* (in Chinese), 1: 34–38.

Wei, H. and Liu, K. 1994. The evolution of regional disparities in China: analysis and forecast. *Industrial Economics Research* (in Chinese), 4: 28–36.

Williamson, J. G. 1965. Regional inequality and the process of national development: A description of patterns. *Economic Development and Cultural Change*, 13(4): 3–84.

Yang, D. 1990. Patterns of China's regional development strategy. *China Quarterly*, 122: 231–257.

Yang, K. 1994. The evolution of regional economic disparities in China. *Economic Research Journal* (in Chinese), 12: 28–33.

Yang, W. 1992. *Empirical studies. Economic Research* (in Chinese), 1: 70–74.

Yao, S. and Zhang, Z. 2001. On regional disparity and diverging clubs: a case study of contemporary China. *Journal of Comparative Economics*, 29(3): 466–484.

Ying, L. 1999. China's changing regional disparity during the reform period. *Economic Geography*, 75(1): 59–70.

Yuan, G. 1996. Regional economic disparities and macroeconomic fluctuations. *Economic Research* (in Chinese), 10: 49–56.

Zhou, G. 1999. *A comprehensive study of regional discrepancies in China's economic development* (in Chinese). Northwest University of Finance and Economics Press.

Advances in Spatial Analysis and Decision Making, Li, Zhou & Kainz (eds)
© 2004 Swets & Zeitlinger, Lisse, ISBN 90 5809 652 1

A general constrained cellular automaton for urban planning

Anthony G.O. Yeh
*Centre of Urban Planning and Environmental Management, The University of Hong Kong,
Hong Kong, China*

Xia Li
School of Geography and Planning, Sun Yat-sen University, China

ABSTRACT: This paper presents a general constrained cellular automaton (CA) model that can be used as a planning tool for generating different development options for urban planning. The general constrained CA model, like the gravity model, can generate a suite of CA planning models that can explore a large spectrum of urban planning criteria, such as land suitability, urban forms and development densities. It can be used for the generation of different land development options according to different planning objectives. The model can help urban planners to evaluate the potential impacts of various possible development scenarios. Environmental suitability, urban forms and development density are important considerations in sustainable land development. Using Dongguan, a rapidly growing city in southern China as an example, the paper has demonstrated how the parameters for the environmental factors, urban form, development density are defined, and how the combined constrained model derived from the general constrained CA model are applied in urban planning process.

1   INTRODUCTION

Sustainable development has attracted increasing attention in recent years because of the rapid deterioration of environment and depletion of resources in many fast growing regions. Sustainable development can help to mitigate the impacts of economic development on the environment by rational use of resources and proper conservation of the environment. Cities are important to sustainable development because they consume a lot of energy and valuable resources, especially land, and have major impacts on the environment. Urbanization process accompanied by unprecedented urban expansion has caused tremendous changes of the environment. Planning models are needed to help planners and decision makers to generate and evaluate different land development options in the planning for sustainable development.

   Recent studies have indicated that cellular automata (CA) are useful tools for simulating complex urban systems (Couclelis, 1997; Batty, 1997; Batty and Xie, 1994; Batty and Xie, 1997; White and Engelen, 1997). The simulation can help to understand urban dynamics and provide useful implications for urban planning. CA were first introduced in 1948 by von Neumann and Ulam to model complex dynamic systems, such as biological reproduction, chemically self-organizing systems, propagation phenomenon, and human settlements (Goles, 1989). The 'game of life' developed in 1970 by the mathematician Conway can be regarded as an explicit CA game (Gardner, 1971; Portugali, 2000). The merits of CA are that very complex behaviors and global structures can emerge from some simple local actions or rules.

   CA models can be integrated with geographical information systems (GIS) for producing more accurate, realistic and plausible simulation results. Some of early CA models were just focused on testing ideas and assumptions related to urban theories using artificial cities in the simulation.

These models only dealt with very limited hypothetical data. However, recent studies indicate that CA models can be used for simulating real cities by incorporating GIS in the modeling process (Batty, 1997; Li and Yeh, 2000). The simulation can make use of the site-specific data provided by the GIS. Remote sensing can also be integrated with CA in providing detailed land use information which is used as training data to calibrate urban simulation (Li and Yeh, 2001a). They are able to generate cellular cities that have features very similar to those of real cities (White and Engelen, 1997). These attempts can help CA models to produce more realistic and applicable simulation results for urban planning.

CA models have potentials as a planning support system since they can simulate actual cities and generate development options by using GIS data. The simulation provides the procedures for the design of optimal forms (Batty, 1997). Studies demonstrate that CA models can simulate planned development as well as realistic development by incorporating sustainability in the simulation (Ward et al., 2000; Li and Yeh, 2001b; Yeh and Li, 2001; Yeh and Li, 2002). Economic, physical and institutional control factors can be incorporated in CA models to modify, constrain and prohibit urban growth.

This study focuses on the development of a general constrained CA model which can simulate sustainable urban development with various planning considerations. A suite of CA planning models can be developed from this general constrained CA model for exploring a large spectrum of urban planning criteria, such as land suitability, urban forms and development densities. They can be used for the generation of different land development options according to different planning objectives. It has the capabilities to simulate planned development with emphasis on protecting important agricultural land and promoting compact development. Factors related to environmental suitability, urban forms, and development densities, which are defined in GIS, are incorporated in the model to facilitate the simulation. The suite of CA planning models generated from this general constrained CA planning model has been carried out in the rapidly growth corridor between Hong Kong and Guangzhou where a lot of development is taking place.

2 THE GENERAL CONSTRAINED CELLULAR AUTOMATA MODEL

The simulation is based on a dynamic modeling technique, cellular automata (CA). A cellular automaton usually consists of four elements – cells, states, neighbourhood and transition rules. Cells are the smallest units which must manifest some adjacency or proximity. The state of a cell can change according to transition rules which are defined in terms of neighbourhood functions. The notion of neighbourhood is essential to the CA paradigm (Couclelis, 1997), but the definitions of neighbourhood are rather relaxed in urban CA models. There are usually two typical neighbourhoods – the von Neumann neighbourhood which consists of the four cells adjoining the central cell, and the Moore neighbourhood which is composed of the eight adjacent cells. However, circular neighbourhoods with various radiuses have been used to deal with complex spatial information in urban simulation (White and Engelen, 1993; Li and Yeh, 2000).

CA models apply a 'bottom-up' approach in which simple local rules generate complex patterns and behaviors. Studies indicate that complex systems can be more easily and accurately simulated by using CA models rather than by using mathematical equations. Formally, standard cellular automata may be generalized as follows:

$$S^{t+1} = f_N(S^t) \tag{1}$$

where S is a set of all possible states of the cellular automata, N is a neighbourhood of all cells providing input values for the function f, and f is a transition function that defines the change of the state from t to $t + 1$.

Detailed transition rules are required for implementing CA models. A simplified rule-based structure can be defined to deal with neighbourhood effects in simulating urban development

(Batty, 1997):

$$\textbf{IF} \text{ any cell } \{i\pm1, j\pm1\} \text{ is already developed}$$

$$\textbf{THEN} \quad P_{ij} = \sum_{\tau\upsilon\in\Omega} P_{\tau\upsilon} / 8$$

&

$$\textbf{IF } P_{ij} > \text{ some threshold value}$$

$$\textbf{THEN} \text{ cell}\{i,j\} \text{ is developed with some other probability } P\{i,j\}$$

where P_{ij} is urban development probability for cell $\{i,j\}$, cell $\{\tau, \upsilon\}$ are the all cells from the Moore neighbourhood Ω including the cell $\{\tau, \upsilon\}$ itself.

Transition rules decide whether a cell will change its state from one to another. The conversion is in a discrete form, and a binary value is used to represent if there is a conversion (1 for converted and 0 for not). The conversion is usually decided by comparing the probability of conversion with a threshold value or a random number. The cell is selected for conversion when the probability is greater than the threshold value or the random number. However, the binary value has limitations when it is used to define transition rules. It does not allow selecting cells for development based on a cumulative process. Constraints are not easily incorporated in the transitional rules. Practically, a cell will not 'suddenly' mature for development. It will be more appropriate to select a cell for conversion gradually through a couple of iterations in simulation. A 'grey value' can be defined to address the fuzzy state in the continuous selection process.

The fuzzy state is defined by using a continuous value instead of using a binary value to represent the process of development. The value indicates the cumulative degree of development for a candidate cell before it is completely selected for development or conversion. An iteration formula can be defined for the cumulative process:

$$G_{ij}^{t+1} = G_{ij}^{t} + \Delta G_{ij}^{t} \tag{2}$$

where G is the 'grey value' for development which falls within the range of $0\sim1$; and ij is the location of the cell. A cell will not be regarded as a developed cell until the value reaches 1. The value should be assigned to 1 when it is greater than 1 during the simulation. ΔG^{t} is the gain of the 'grey value' at each loop.

The essential part of the model is to calculate the value of ΔG^{t}. ΔG^{t} can be defined using the neighbourhood function which is the basis of CA simulations. According to the neighbourhood function, the conversion probability at a cell depends on the states of its neighbouring cells. There is a higher chance of conversion at a cell if it is surrounded by more converted cells. The increase of 'grey value' should be determined by the amount of developed cells in the neighbourhood.

ΔG^{t} can be simply defined by the following neighbourhood function:

$$\Delta G_{ij}^{t} = f_N(q_{ij}) = \frac{q_{ij}}{\pi \xi^2} \tag{3}$$

where q is the total amount of developed cells in the neighbourhood; ξ is the radius of the circular neighbourhood. A circular neighbourhood is used because it has no bias in all directions (Li and Yeh, 2000).

The transition rules according to equation (3) only address the influences of the states (developed cells) in the neighbourhood. The evolution of real cities is influenced by a series of complicated factors which can be obtained at various local, regional and global levels. The neighbourhood function cannot address the issue of urban structures and environmental problems. Some kinds of constraints should be used to regulate the simulation to improve modelling accuracy. Without constraints, urban simulation will generate patterns as usual based on historical trends. Constraints should be

169

added into urban CA models to reflect environmental and sustainable development considerations. They are the important factors for the formation of idealized patterns.

By taking environmental and other constraints into considerations, equation (3) can be revised into a general constrained CA model as follows:

$$\Delta G_{ij}^{\prime t} = f_N(q_{ij}) \times \prod_{m=1}^{M} \delta_{mij}$$

$$= \frac{q_{ij}}{\pi \xi^2} \times \prod_{m=1}^{M} \delta_{mij} \tag{4}$$

where δ_{mij} is the function to represent various types of constraints in which the values should be normalized within the range from 0 to 1. It can be regarded as a scaling factor to readjust the increase of 'grey value'.

A stochastic disturbance term (γ) is added to the model to represent unknown errors which frequently exhibit in many complex systems. This can allow the generated patterns more similar to realistic development patterns. Incorporating the random variable in equation (4), the final equation for calculating the increase of 'grey value' is then given by:

$$\Delta G_{ij}^{\prime t} = (1 + (-\ln \gamma)^{\alpha}) \times f_N(q_{ij}) \times \prod_{m=1}^{M} \delta_{mij}$$

$$= (1 + (-\ln \gamma)^{\alpha}) \times \frac{q_{ij}}{\pi \xi^2} \times \prod_{m=1}^{M} \delta_{mij} \tag{5}$$

The general constrained CA model developed above does not only take into account the influences of neighbouring states, but also a series of economic and environmental constraints. These constraints may include environmental suitability, urban forms, and development density. The following examples use Dongguan in southern China as an example to show how the general constrained CA model can be used to generate CA planning models that take the environment, urban form and density into consideration respectively and how these factors can be combined into one constrained model. The CA models were programmed using the Arc Macro Language (AML) within a GIS package, ARC/INFO GRID.

The study area is located in Dongguan, a very fast growing city in the Pearl River Delta in southern China (Yeh and Li, 1997; Yeh and Li, 1999). It has a city proper and 29 towns with an area of 2,465 km^2 (Figure 1). Remote sensing and GIS data were used to provide the basic information for the simulation. The 1988 and 1993 TM Landsat images were classified to retrieve land use and land use change information (Li and Yeh, 1998). A GIS database was built to contain the information of land use, transportation, population and administrative boundaries. The database was converted into a raster format for the simulation. The basic unit is cell which has an area of $50 \times 50 \, \text{m}^2$ on the ground. The initial map for the simulation was from the land use classification of the 1988 satellite TM image. The models attempt to generate land development options for 1988–93. The actual urban areas (built-up areas and development sites) in 1993 obtained from the classification of the 1993 satellite TM image were used for as a baseline for evaluating the results of the simulation.

3   MODEL 1: ENVIRONMENTAL CONSTRAINED CA MODEL

The first model is to incorporate environmental constraints in urban simulation. There is growing concern on environmental issues related to urban growth in the world. Urban development should be determined not only by pure economic factors, but also by environmental constraints. Environmental consciousness should be reflected in the model so that idealised urban development patterns could be formulated. It is convenient to obtain environmental constraints and embed them in urban simulation based on the integration of CA and GIS technologies.

Environmental constraints are used to indicate whether a piece of land should be protected from development with regard to environmental considerations. The general constrained CA model in

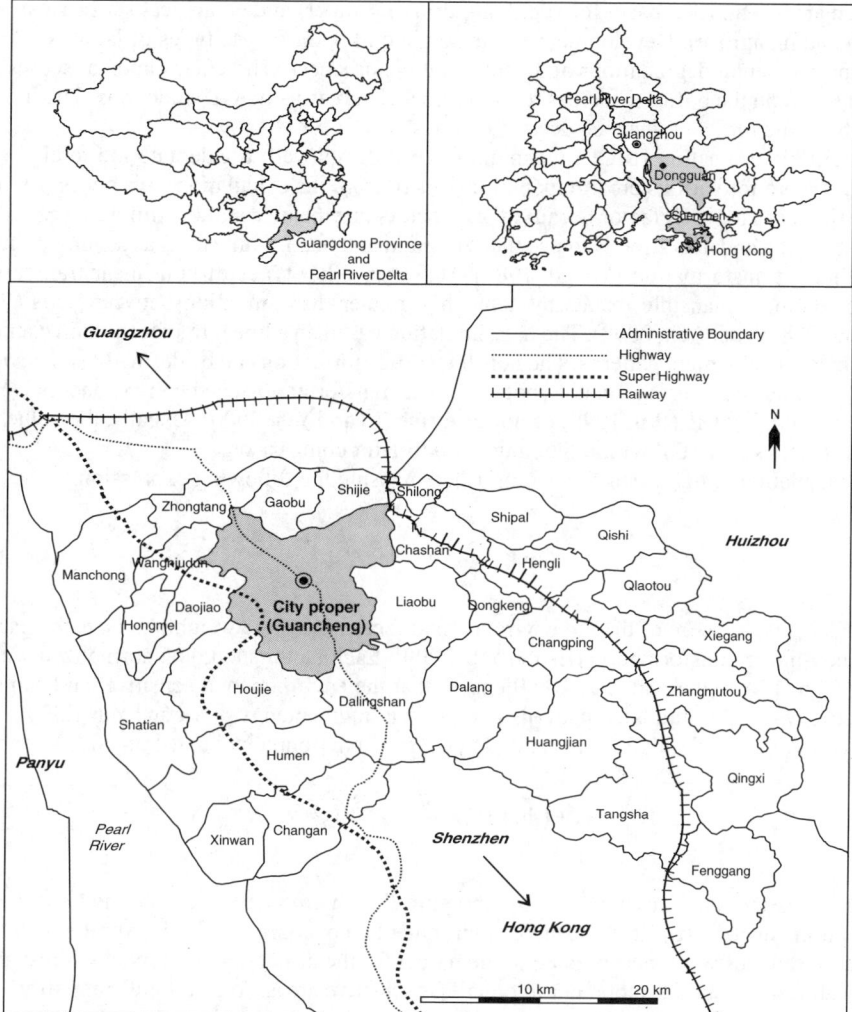

Figure 1.   The location of Dongguan.

equation (5) can be modified into an environmental constrained CA model by incorporating environmental constraints into the general model:

$$\Delta G_{ij}^{'t} = (1 + (-\ln \gamma)^{\alpha}) \times \frac{q_{ij}}{\pi \xi^2} \times \delta_{ENVij}$$ (6)

where δ_{ENVij} is the function of environmental constraints.

The constraint function δ_{ENVij} in equation (6) can be defined by using GIS data. It is related to
a number of environmental factors. This model emphases the protection of strategic agricultural
land and other important ecological areas in urban planning. The score of the constraint function
can be calculated by combining agricultural suitability score and other environmental scores for
protecting resources and environment. Agricultural suitability, which reflects the potential of agricultural production, can be used to address the need to reserve important agricultural land. Higher
costs are associated with the encroachment on the sites of good-quality agricultural land. Other
environmental scores can also be calculated to address the disturbance of development in

171

protected areas (e.g. river basin for supplying drinking water) and ecological sensitive areas (e.g. wetland and mangrove). Development in the neighbourhood of these types of land use can bring about environmental degradations and ecological disturbances. The environmental scores can be defined based on the buffer distances to these sensitive areas using GIS functions. The influences should be in the forms of distance decay functions.

Multicriteria evaluation (MCE) techniques can be employed to calculate the total combined constraint score for various environmental factors. Before the calculation, it is necessary to standardize the score for each factor because these factors may be measured at different scales. A typical method of standardization is to use the minimum and maximum values as scaling points for a simple linear transformation (Voogd, 1983). However, other types of non-linear transformation can provide more plausible results by achieving greater discrimination between cells (Wu and Webster, 1998; Li and Yeh, 2000). The CA simulation based on a linear transformation cannot generate typical development patterns. The non-linear transformation can be defined in an *ad hoc* way because a unique transformation does not exist. The transformation can be in exponential (Wu and Webster, 1998), logistic (Wu, 1998) or power forms (Li and Yeh, 2000). Usually, the values of the adjusted scores should fall within the range of 0 to 1 for comparison.

The calculation of δ_{ENVij} can be accomplished by using the following expression:

$$\delta_{ENVij} = \sum_{\theta=1}^{N} w_\theta (1 - ENV_{\theta ij})^k \qquad (7)$$

where $ENV_{\theta ij}$ is the score of the θth environmental factor, w_m is the weight, and k is the parameter for the non-linear transformation (Li and Yeh, 2000). Each factor should be normalized within the range of 0 to 1. A higher value of k will ensure that the environmental sensitive land can be protected strictly, but the simulated patterns may be in a fragmented way (Li and Yeh, 2000).

The combination of equations (6) and (7) yields the environmental constrained CA model:

$$\Delta G_{ij}^{'t} = (1 + (-\ln \gamma)^\alpha) \times \frac{q_{ij}}{\pi \xi^2} \times \sum_{\theta=1}^{N} w_\theta (1 - ENV_{\theta ij})^k \qquad (8)$$

Figure 2 is the simulation results by applying the environmental constrained CA model as described in equation (8). In this study, environmental constraints of agricultural suitability and forest and wetlands were incorporated in the model for the protection of cropland, forest and wetland by allocating land development in other less sensitive areas. Agricultural suitability which is one of the major environmental considerations is calculated from the slope and soil maps in the GIS. The locations of forest and wetland, which are also obtained from the GIS, are used as environmental constraints of the model (Li and Yeh, 2000). The environmental constraint is essential for the model to find the solution in minimizing the impacts of urban development on agricultural and ecological conservation. Non-linear transformation ($k = 3$) was used for the strict protection. In most situations, environment and resources are in heterogeneous patterns across the space. GIS is essential for providing real data to the modelling process. In this study area, the most fertile agricultural land is concentrated in the alluvial plain in the northwest part near the city proper. This model can preserve productive agricultural land and ecological land by incorporating a series of environmental constraints.

4   MODEL 2: URBAN FORM CONSTRAINED CA MODEL

The planning of urban form will be central to the promotion of sustainable development (Breheny, 1996). There were a lot of debates on how to confine urban sprawl and conserve agricultural land resources (Bryant et al, 1982; Gierman, 1977; Ewing, 1997; Daniels, 1997). An essential part of plan-making is to develop simulation models in generating regional and subregional growth

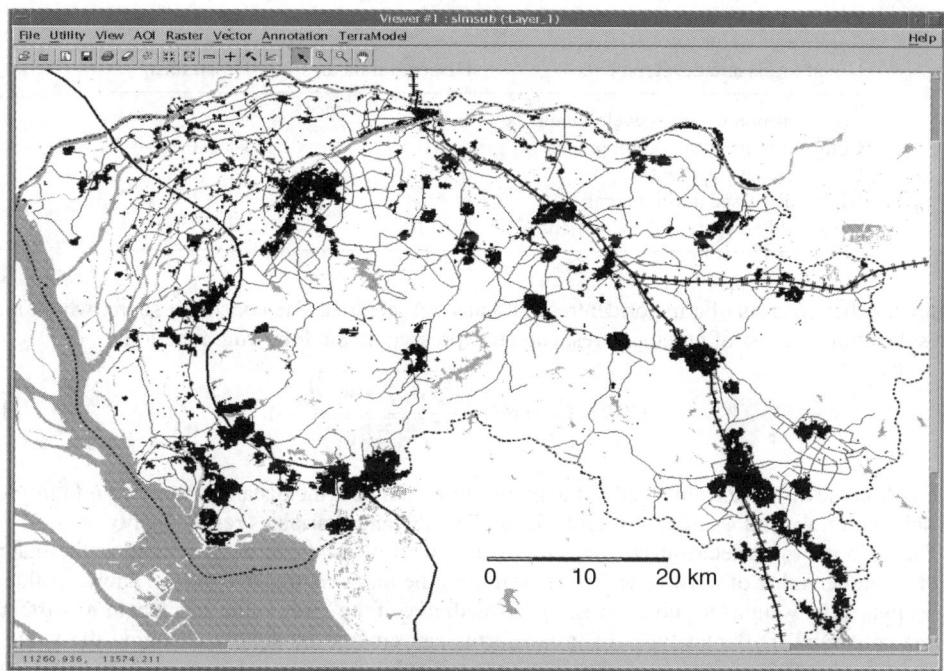

Figure 2. Land development scenario for agricultural and ecological protection in Dongguan in 1988–93 from the constrained CA planning model.

options. The California Urban Future Model (CUF Model) is such model which is valuable for presenting and comparing the details of different development scenarios (Landis, 1995). Clarke et al. (1997) also provide a CA model that can simulate urban dispersion using self-modifying rules. The model has been applied to the simulation of urban growth in San Francisco Bay area.

There are a lot of advantages in using cellular automata and GIS to simulate possible urban forms for the planning of sustainable urban development. Urban CA models can help planners to explore various options and evaluate possible environmental impacts. When urban form is used as the main constraint, the general constrained CA model in equation (5) can then be modified into a urban form constrained CA model:

$$\Delta G_{ij}^{\prime t} = (1 + (-\ln \gamma)^{\alpha}) \times \frac{q_{ij}}{\pi \, \xi^2} \times \delta_{ij} \, (FORM)$$ (9)

The function, $\delta_{ij} \, (FORM)$, is determined by the relationships between urban growth and urban centres. Land development can be concentrated around the main centre for promoting the growth of large cities, or can be shifted to around sub-centres for promoting polycentric growth. There are many possible development patterns which are the main concerns of urban planning. Land development and urban forms is closely related to the efficient use of energy, capital and land resources (Banister et al., 1997; Burchell et al., 1998). The constraint is decided by location factors in terms of the distances to urban centres. Urban centres play an important role in urban growth as they provide the supports for the requirements of energy, materials, capital, and techniques for development. The influence of urban centres can be measured by a distance decay function. The classical measure of urban structure is the density gradient from CBD (Muth, 1969). Although the density gradient is related to the monocentric concept of urban form, nevertheless it gives us an index of the degree of decentralization. Two distances can be defined to capture the hierarchy of urban

173

Table 1. Model parameters for simulating different types of urban forms.

Urban forms and developments	Dispersion factor	Urban form
Compact-monocentric development	$\alpha = 0$	$w_R = 1; w_r = 0$
Compact-polycentric development	$\alpha = 0$	$w_R = 0; w_r = 1$
Dispersed development	$\alpha = 1$	Nil
Highly dispersed development	$\alpha = 5$	Nil
Very highly dispersed development	$\alpha = 10$	Nil

structures that consists of a major centre and many sub-centres. The constraint score, which indicates the attractiveness of urban centres, can be expressed by the following function:

$$\delta_{ij}(FORM) = \exp(-\frac{\sqrt{w_R^2 d_{Rij}^2 + w_r^2 d_{rij}^2}}{\sqrt{w_R^2 + w_r^2}}) \tag{10}$$

where d_R is the distance from a cell to the main centre and d_r is the distance from cell ij to its closest sub-centre. w_R and w_r are the weights for the two distance variables respectively.

The ratio of w_R/w_r determines what kind of urban forms that can be generated at the macro-level. A higher value of w_R/w_r gives more weight to the main centre. In contrast, a lower value of w_R/w_r puts more weights to sub-centres. The growth rate of the 'grey value' at a location is affected by the constraint. Different types of urban forms can emerge by simply changing the value of w_R/w_r. A higher value of w_R/w_r results in monocentric development, whereas a lower value leads to polycentric development. Finally, the combination of equations (9) and (10) yields the urban form-constrained CA model:

$$\Delta G_{ij}^{'t} = (1 + (-\ln \gamma)^\alpha) \times \frac{q_{ij}}{\pi \xi^2} \times \exp(-\frac{\sqrt{w_R^2 d_{Rij}^2 + w_r^2 d_{rij}^2}}{\sqrt{w_R^2 + w_r^2}}) \tag{11}$$

This model simulated five types of urban forms according to equation (11). They include compact-monocentric, compact-polycentric, dispersed, highly dispersed, and very highly dispersed development forms. Table 1 lists the parameter values used in the model. Other mixed urban forms can be easily generated using the same method by changing the parameters.

Figure 3 shows one of the simulated scenarios, the compact-monocentric development, to demonstrate the capability of the CA model in generating development options. The simulation was to emphasize the role of the main centre in supporting urban growth. The growth rate was affected by the distance to the city proper rather than by the distances to sub-centres. Higher growth rates were only allowed to take place around the city proper according to the criterion. There was very limited growth around those sub-centres that were farther away from the city proper. The large part of development only took place within a small area around the city proper (the north-west part). The pattern is much more compact, compared with the actual dispersed development pattern.

5 MODEL 3: DEVELOPMENT DENSITY CONSTRAINED CA MODEL

Development density should be another important factor for compact development. Unfortunately, development density has not been well integrated in the process of general GIS site selection and urban CA simulation. Development density can be represented by the total number of people that can be accommodated by a developed cell. High development density can significantly reduce the expenditures for providing infrastructure and public service and the costs of consuming energy and resources.

Rising development densities has many benefits for sustainable development (Newman and Kenworthy, 1988; Pushkarev and Zupan, 1977). Per capita infrastructure costs almost certainly

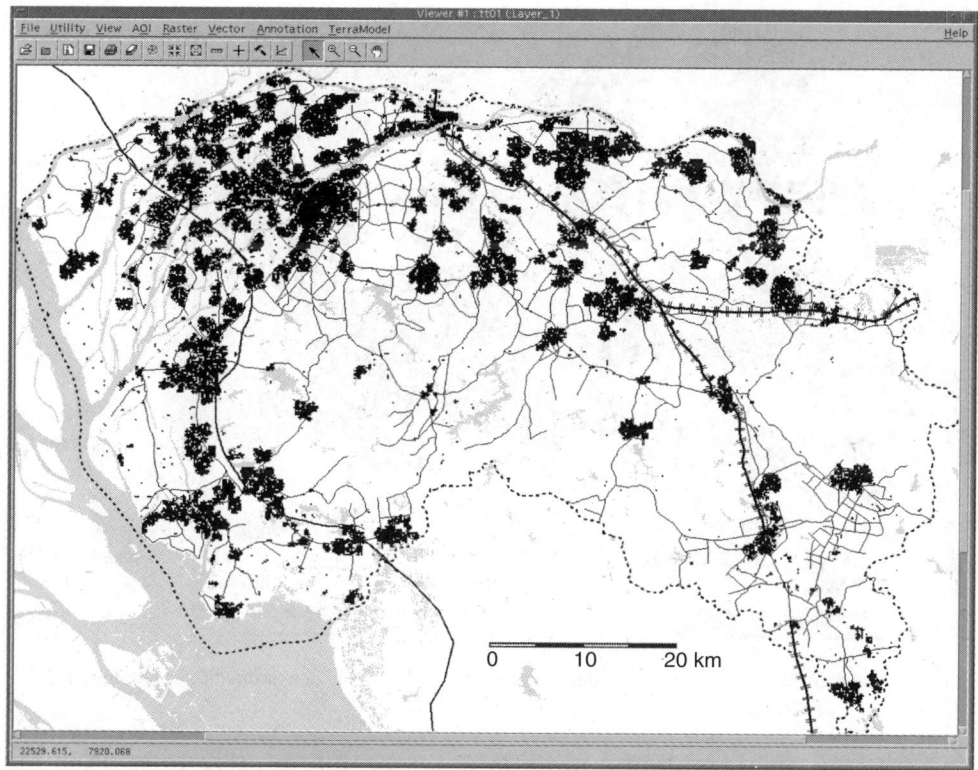

Figure 3. Compact-monocentric development in Dongguan in 1988–93 from the constrained CA planning model.

fall as densities rise although extremely high densities may cause an increase in costs (Ewing, 1997). Urban sprawl will lead to the encroachment on much more land than compact development. For example, an assessment of two development plans for the State of New Jersey indicates that the compact plan can reduce land consumption as much as 60% (CUPR, 1992; Ewing, 1997). An urban-sprawl plan will result in five-time greater loss of environmentally sensitive land and two-thirds greater loss of farmland than compact development. A solution to contain urban sprawl is to increase development density properly. This can reduce a series of costs for land development and increase the benefits for environmental protection.

This model includes the factor of density in urban simulation. The essential part of simulation is to determine the increase of 'grey value' for a cell based on neighbourhood functions. The increase of 'grey value' is proportional to the total population in the neighbourhood. When development density is used as the main constraint, the general constrained CA model in equation (5) can be modified into a development density constrained CA model:

$$\Delta G_{ij}^{'t} = (1 + (-\ln \gamma)^{\alpha}) \times f_N(Den_{ij})$$

$$= (1 + (-\ln \gamma)^{\alpha}) \times \frac{\sum\limits_{ij \in \Omega_N} Den_{ij}}{Den_{max} \pi \, \xi^2} \tag{12}$$

where Den_{ij} is development density, Ω_N is the set of developed cells in the neighbourhood N, ξ is the radius of the circular neighbourhood, and Den_{max} is the maximum value of the development density. A circular neighbourhood is used to calculate the total population.

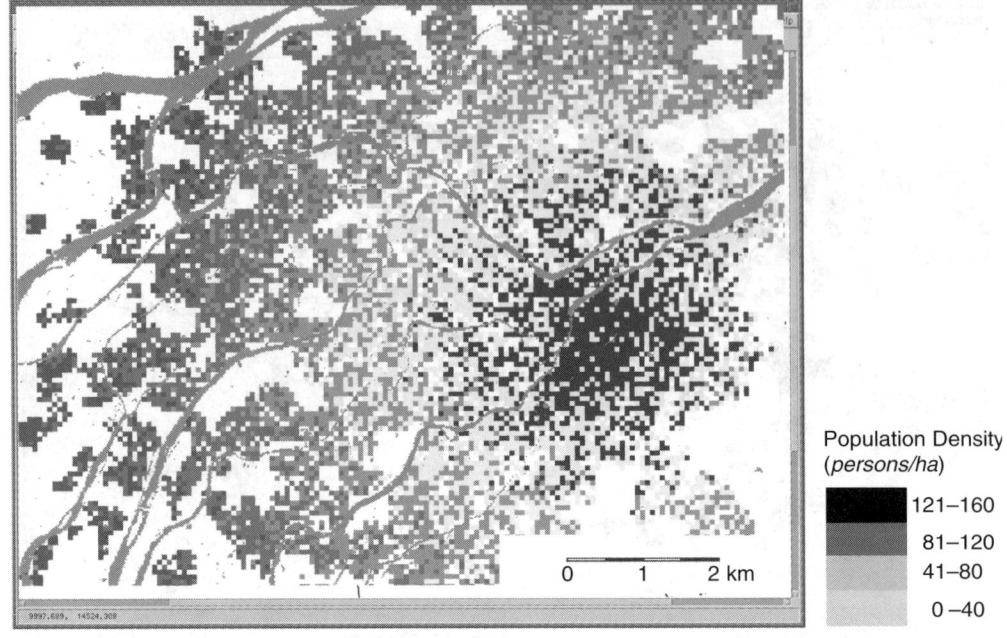

Population Density
(*persons/ha*)

▮ 121–160
▮ 81–120
▮ 41–80
▯ 0–40

Figure 4. Land development scenario of development density from the constrained CA planning model.

When a cell is selected for development, a development density should be assigned to the cell according to local experience or historical data. Development density in terms of population should be dependent on the distance to urban centres. Density decay functions can be used to determine the development density of a developed cell. The density decay function assumes that development density (population density) declines in an inverse exponential way. The notion that population density declines from centres have been well discussed in many studies (Clark, 1951; Thrall, 1988; Papageorgiou, 1971). The function is generally given by (Clark, 1951):

$$Den_{ij}^0 = Ae^{-\beta l_{ij}} \tag{13}$$

where Den_{ij}^0 is the assigned development density, l_{ij} is the distance to a centre, and A and β are the parameters of density decay function. The function had been repeatedly tested for the past 150 years and examined to be statistically significant (Papageorgiou, 1971). Especially, for most large cities, the negative exponential model seems to describe the real-world observations fairly well as a first approximation (King and Golledge, 1978).

This study attempts to generate different scenarios of development densities to provide more detailed information for urban planning. We use the model to generate different combinations of development densities, which are compared with the actual development density. Figure 4 is the example of simulating development density around the city proper of Dongguan by incorporating the density gradient in the CA model according to equation (12). The parameters A and β of the density decay function were set to 80 and 0.005 respectively.

6 CONCLUSION

This paper has generalized different CA models developed by the authors into a general constrained CA planning model (Li and Yeh, 2000; Yeh and Li, 2001, 2002). It has put different CA models under a general framework in which other planning models and considerations can be developed and incorporated in the future. The general constrained CA planning model is developed mainly by the use of 'grey value' in which a cumulative development probability instead of a binary value is used to represent the continuous selection process for land development. The 'grey value' indicates the cumulative degree of development for a candidate cell before it is completely selected for development or conversion. Like the general gravity model in which different forms of gravity models can be generated (Wilson, 1971; Cordey Hayes and Wilson, 1971; Batty, 1976), the general constrained CA planning model can also generate different forms of planning models depending on the planning objectives of the planners and decision makers.

Environmental suitability, urban forms and development density are important considerations in sustainable land development. They are sometimes difficult to be combined into a single planning model. All these factors can now be handled by the general constrained CA planning model developed in this paper. They can be considered individually, in combination, or altogether by the general constrained CA planning model. Using Dongguan, a rapidly growing city in southern China as an example, the paper has demonstrated how the parameters for the environmental, urban form, development density are defined and the combined constrained models is generated and applied. It has demonstrated that the general constrained CA model can meet most modeling requirements in urban simulation by incorporating various constraints in the model. The model can help urban planners and decision makers to examine different development options under different planning objectives.

Integrated with GIS, the general constrained CA model can benefit much from using the social, economic, physical and environmental data in the GIS directly in its simulation. If the general constrained CA model is made available as one of the standard functions in a GIS software, like the gravity and location-allocation models, it can be a very useful planning model for a planning support system.

REFERENCES

Banister D, Watson S, Wood C, 1997, Sustainable Cities: Transport, Energy, and Urban Form, *Environment and Planning B* 24 125–143

Batty M, 1976, *Urban Modelling: Algorithm, Calibrations, Predictions* (Cambridge University Press, Cambridge)

Batty M, 1997, Cellular automata and urban form: a primer *Journal of the American Planning Association* 63(2) 266–274

Batty M, Xie Y, 1994, From cells to cities *Environment and Planning B: Planning and Design* 21 531–548

Batty M, Xie Y, 1997, Possible urban automata *Environment and Planning B: Planning and Design* 24 175–192

Breheny M, 1996, Centrists, decentrists and compromisers: views on the future of urban form, in *The Compact City: A Sustainable Urban Form?* Eds M Jenks, E Burton, K Williams (E&FN SPON, London) pp 13–35

Bryant C R, Russwurm L H, McLellan A G, 1982, *The City's Countryside: Land and Its Management in the Rural-urban Fringe* (Longman Group Ltd, New York)

Burchell, R W, Shad N A, Listokin D, Phillips H, Downs A, Seskin S, Davis J S, Moore T, Helton D, Gall M, 1998, *The Costs of Sprawl – Revisited* (National Academy Press, Washington, D.C.)

Clark C, 1951, Urban population densities *Journal of Royal Statistical Society, Series A* 114 490–496

Clarke K C, Gaydos L, Hoppen S, 1997, A self-modifying cellular automaton model of historical urbanization in the San Francisco Bay area *Environment and Planning B: Planning and Design* 24 247–261

Cordey Hayes M, Wilson A G, 1971, Spatial interaction *Socio-Economic Planning Sciences* 5 73–95

Couclelis H, 1997, From cellular automata to urban models: new principles for model development and implementation *Environment and Planning B: Planning and Design* 24 165–174

CUPR (Center for Urban Policy Research), 1992, *Impact Assessment of the New Jersey Interim State Development Plan* (New Jersey Office of State Planning, Trenton, NJ)

Daniels T L, 1997, Where does cluster zoning fit in farmland protection? *Journal of the American Planning Association* 63(1) 129–137

Ewing R, 1997, Is Los Angeles-style sprawl desirable? *Journal of the American Planning Association* 63(1) 107–126

Gardner M, 1971, On cellular automata, self-reproduction, the Garden of Eden and the Game of life *Scientific American* 224 112–117

Gierman D M, 1977, Rural to urban land conversion Occasional Paper 16, Lands Directorate, Environment Canada, Ottawa

Goles E, 1989, Cellular automata, dynamics and complexity , in *Cellular automata and modeling of complex physical systems* Eds P Manneville, N Boccara, G Y Vichniac, R Bidaux (Springer-Verlag, Berlin) pp 10–20

King L, Golledge R G, 1978, *Cities, Space, and Behavior: the Elements of Urban Geography* (Prentice-Hall, Inc., Englewood Cliffs, New Jersey)

Landis J D, 1995, Imagining land use futures: applying the California urban futures model *Journal of American Planning Association*, 6(4) 438–457

Li X, Yeh A G O, 1998, Principal component analysis of stacked multi-temporal images for monitoring of rapid urban expansion in the Pearl River Delta *International Journal of Remote Sensing* 19(8) 1501–1518

Li X, Yeh A G O, 2000, Modelling sustainable urban development by the integration of constrained cellular automata and GIS *International Journal of Geographical Information Science* 14(2) 131–152

Li X, Yeh A G O, 2001a, Calibration of cellular automata by using neural networks for the simulation of complex urban systems *Environment and Planning A* 33 1445–1462

Li X, Yeh A G O, 2001b, Zoning for agricultural land protection by the integration of remote sensing, GIS and cellular automata *Photogrammetric Engineering & Remote Sensing* 67(4) 471–477.

Muth R F, 1969 *Cities and Housing* (University of Chicago Press, Chicago)

Newman P W G, Kenworthy J R, 1988, The transport energy trade-off: fuel-efficient traffic versus fuel-efficient cities *Transport Research A* 3 163–174

Papageorgiou G J, 1971, A theoretical evaluation of the existing population density gradient function *Economic Geography* 47 21–26

Portugali J, 2000, *Self-Organization and the City* (Springer-Verlag, New York)

Pushkarev B S, Zupan J M, 1977, *Public Transportation and Land Use Policy* (Indiana University Press, Bloomington, IN)

Thrall G I, 1988, Statistical and theoretical issues in verifying the population density function *Urban Geography* 9(5) 518–537

Voogd H, 1983, *Multicriteria Evaluation for Urban and Regional Planning* (Pion, Ltd., London)

Ward D P, Murray A T, Phinn S R, 2000, A stochastically constrained cellular model of urban growth *Computers, Environment and Urban Systems* 24 539–558

White R, Engelen G, 1993, Cellular automata and fractal urban form: a cellular modelling approach to the evolution of urban land-use patterns *Environment and Planning A* 25 1175–1199

White R, Engelen G, 1997, Cellular automata as the basis of integrated dynamic regional modelling *Environment and Planning B: Planning and Design* 24 235–246

Wilson A G, 1971, A family of spatial interaction models and associated developments *Environment and Planning* 31–32

Wu F, 1998, An experiment on the general polycentricity of urban growth in a cellular automatic city *Environment and Planning B: Planning and Design* 25 103–126

Wu F, Webster C J, 1998, Simulation of land development through the integration of cellular automata and multicriteria evaluation *Environment and Planning B: Planning and Design* 25 103–126

Yeh A G O, Li X, 1997, An integrated remote sensing and GIS approach in the monitoring and evaluation of rapid urban growth for sustainable development in the Pearl Rive Delta, China *International Planning Studies* 2(2) 193–210

Yeh A G O, Li X, 1999, Economic development and agricultural land loss in the Pearl River Delta, China *Habitat International* 23(3) 373–390

Yeh A G O, Li X, 2001, A constrained CA model for the simulation and planning of sustainable urban forms using GIS *Environment and Planning B: Planning and Design* 28 733–753

Yeh A G O, Li X, 2002, A cellular automata model to simulate development density for urban planning *Environment and Planning B: Planning and Design* 29 431–450

Advances in Spatial Analysis and Decision Making, Li, Zhou & Kainz (eds)
© 2004 Swets & Zeitlinger, Lisse, ISBN 90 5809 652 1

True ortho-image-assisted decision-making technology for urban planning and management

Guoqing Zhou & Zhihao Qin
Department of Civil Engineering and Technology, Old Dominion University, Norfolk, USA

Jixian Zhao
Department of Surveying Engineering, East China Institute of Technology, Fuzhou, China

ABSTRACT: Geographical Information System (GIS) technology has been an important tool in supporting urban planning and management, bringing efficient accessibility to more, better and more timely information. Ortho-images, especially, urban true ortho-images are important in supporting decision-making in urban planning and management. This paper presents a methodology to automatically ortho-rectify large-scale urban aerial images for decision-making in urban planning and management. The methodology to generate true ortho-photos includes three steps: ortho-rectifying the displacement, detecting the occluded area, and refilling the occluded area. We first compute the coordinates of the input image pixels in the output ortho-image and the distance of the coordinates to the perspective centre. Meanwhile, we use a data matrix to store the coordinate and its distance. We then compare the coordinate with the existing coordinates in a matrix that holds the coordinates and distance. If two image pixels have the same coordinates in the output ortho-image, the one with the shorter distance representing the roof of the building is the true position of the pixel, and the other with longer distance is the bottom of the building. By creating an index image, the occluded area of the buildings can be recorded for next step in processing. Mosaics from neighbouring slave ortho-images can be used to refill the occluded areas in the master ortho-image. The experimental results of the ortho-image of downtown Denver, Colorado, demonstrated that our algorithms can effectively ortho-rectify dense urban building area, with the result that all buildings are placed in their TRUE positions, and all streets are visible.

1 INTRODUCTION

Effective urban planning and management demands efficient accessibility to more, better and more timely information (Kyariga 2001). The Geographical Information System (GIS) has a wide range of applications in urban planning and management operations, and has had an enormous impact on urban planning practices. Decision-makers are using GIS as a decision-support tool to acquire information in an effective and efficient manner that expedites the decision process. Large-scale urban ortho-images are an important source of information in city planning and management for decision-making. This is because they provide valuable information for the GIS model because the model not only represents the actual coordinates of the region but also useful information on the region as seen by photogrammetric cameras. Conventional ortho-rectification only considers the geometric distortion caused by the imaging process and by terrain (Zhou *et al.* 2002, 2003, Chen *et al.* 1993, Baltsavia 1996). This is because conventional ortho-rectification is based on a two-dimensional (2D) or two-and-half dimensional (2.5D) digital terrain model (DTM) that does not consider man-made buildings (Baltsavias & Kaser 2002). In many cases, especially in urban area, conventional ortho-rectification has encountered many difficulties (Rau *et al.* 2002, Zhou & Qin 2003). One of the difficulties is that it cannot solve the problem of occlusions caused by man-made buildings and the shadow caused by buildings due to different solar angles when the images were taken.

Despite the great potential of the large-scale application of urban true ortho-images for urban planning and management, studies on the generation of true ortho-images are still limited. Current studies on the ortho-rectification of aerial images mainly concentrate on landscape images. Vassilopoulou *et al.* (2002) used IKONOS images and DEM (digital elevation model) to generate ortho-images for the monitoring of volcanic hazards on Nisyros Island, Greece. Sharkatos (1999) and Joshua (2001) demonstrated that building occlusions in fact significantly influence not only the quality, but also the accuracy of ortho-images. Amhar *et al.* (1998) and Schicker & Thorpe (1998) considered the hidden effects introduced by abrupt changes in surface height (e.g., buildings and bridges). Schickler and Thorpe (1998) further considered seamless mosaicking around the filled-in areas in order to reduce the discontinuity in grey values, but no study has yet been reported on the radiometric aspect, or shadow enhancement to restore information within the areas of a building that lie in shadow. Rau *et al.* (2002) examined enhancements of image radiometry. A suitable enhancement technique was used to restore the information within the shadowed areas of a building.

This paper presents a methodology for the true ortho-rectification of large-scale urban aerial images using a digital surface model (DSM), which is a combination of a digital terrain model (DTM) and a digital building model (DBM), to describe buildings. We first discuss the principles of true ortho-image generation. We then establish the mathematical models and the procedures for actual computation. The occluded areas are detected by perspective geometry, and the lost information in occluded areas is compensated for from conjugate area in adjacent images. We apply the methodology to Denver, Colorado for verification. Finally, we discuss the issues related to the generation of true ortho-images and future improvements.

2 PRINCPLES AND MATHEMATICAL MODELS FOR TRUE ORTHO-RECTIFICATION

2.1 *Principles for the generating of true ortho-images*

The digital ortho-image should theoretically be a spatially accurate image with ground features represented in their TRUE planimetric positions. Therefore, an ortho-image contains both the image characteristics of a photograph and the geometric qualities of a map. However, it has been demonstrated that the traditionally photogrammetric ortho-rectification algorithm is not appropriate for urban areas because the ortho-rectification of the image is partly geometrically inaccurate and/or incomplete, and the features of the image are distorted from their true location (Zhou & Qin 2003). The problems have significantly limited the usefulness of the digital ortho-image in urban planning and management because the errors in these incompletely rectified large-scale city ortho-image maps can no longer be tolerated when used for updating and planning urban tasks. Thus, the generation of so-called *true ortho-images* has become a matter of some urgency. The generating of true ortho-images from large-scale aerial photographs often involves two important actions: coordination transformation and the geometric correction of the distortions caused by terrains, buildings and the imaging process. True ortho-rectification will rectify the coordinates of pixel into a specified map-based coordinate system, removing all of the geometric distortions.

In urban areas, the occlusion of buildings is another major difficulty in the generation of true ortho-images. Thus, in the true ortho-rectification of large-scale urban aerial images, how to identify the occlusions and how to remove them are critical issues. In fact, geometric distortions in photography are subjected to the collinearity condition. Thus, a method based on a combination of collinearity equations with a comparison of distance will be used to detect occluding and occluded areas to compensate for occluded areas.

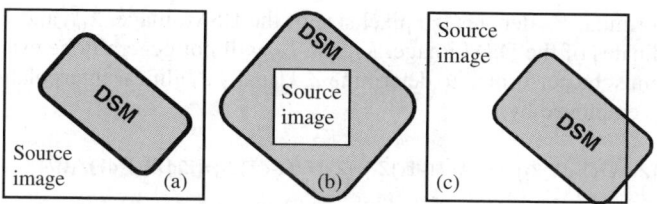

Figure 1.   Three cases of overlap between the source image and the DSM image. (a) DSM image fully inside the source image, (b) source image smaller than the DSM image, and (c) partial overlap between the DSM image and the source image.

2.2   Size of the resulting ortho-image

The resulting ortho-image is generally expressed as a raster image, with pixels arranged in rows and columns. Since the resulting ortho-image is rectified from the inputting of a raster image (called the *source image*) using the DTM data (image), the size of the output ortho-image is a function of three factors: the size of the source image; the size of the DTM image; and the pixel resolution of the output ortho-image. As to the first two factors, there are generally three cases that describe the relationship of the source image to the DTM (see Figure 1). Consequently, the size of the output image can be determined in three ways. In the first case, when the size of the source image is larger than the size of the DTM image (Figure 1a), the size of the output image will be the same as that of the DTM image. In this case, the source image will not be fully presented in the output ortho-image. In the second case, when the source image is smaller than the DTM image (Figure 1b), the size of the output image will be the same as that of the source image. The third case is when the source image partly overlays the DTM image (Figure 1c). The output image then only covers the overlapping portions of the images. The coordinates of the output images for the three cases can be expressed as follows:

Case 1: $X_0 = \min\{X_D\}$ and $Y_0 = \min\{Y_D\}$
$X_1 = \max\{X_D\}$ and $Y_1 = \max\{Y_D\}$
Case 2: $X_0 = \min\{X_I\}$ and $Y_0 = \min\{Y_I\}$
$X_1 = \max\{X_I\}$ and $Y_1 = \max\{Y_I\}$
Case 3: $X_0 = \max\{\min\{X_I\}, \min\{X_D\}\}$ $Y_0 = \max\{\min\{Y_I\}, \min\{Y_D\}\}$
$X_1 = \min\{\max\{X_I\}, \max\{X_D\}\}$ $Y_1 = \min\{\max\{Y_I\}, \max\{Y_D\}\}$

In the above cases, X_0 and Y_0 are the coordinates of the lower-left corner of the output image; X_1 and Y_1 are the coordinates of the upper-right corner of the output image; X_D and Y_D are the X and Y coordinates of the DSM image; X_I and Y_I are the X and Y coordinates of the source image; and max and min denote the maximum and minimum of the elements in the blanket. All of the coordinates here refer to the geodetic coordinate system required in the resulting ortho-image.

2.3   Computing the Z coordinates of the pixel

In order to perform ortho-rectification, we also need to know the Z coordinates of the pixel $P(K, L)$ in an output image. Generally, this is obtained from the DSM image. First, we have to convert back to the pixel's geodetic coordinate into the column and row of the DSM image using the following formula:

$$K_D = (X - X_{0D})/P_D \tag{1}$$

$$L_D = (Y - Y_{0D})/P_D \tag{2}$$

In the above formula, P_D denotes the pixel size of the DSM image. X_{0D} and Y_{0D} are the lower-left corner coordinates of the DSM image. K_D and L_D will not generally be exact integers. Thus, an interpolation must be performed to determine Z. Usually a bilinear interpolation method of the following form is employed by:

$$Z=\{[Z_1 dX+(1-dX)Z_2]+[Z_4 dX+(1-dX)Z_3]+[Z_1 dY+(1-dY)Z_2]+[Z_4 dY+(1-dY)Z_3]\}/4 \qquad (3)$$

where $dX = K_D - K_m$ and $dY = L_D - L_m$, in which K_m is the K_D around to its maximal integer and L_m is L_D around to its maximal integer. After this estimation, we then know the coordinate (X, Y, Z) of the pixel.

2.4 Computing the corresponding coordinates in the source image

In order to ortho-rectify the source image, we have to find out the corresponding coordinates of the source image pixel in the resulting image. This can be done by the back-projection of collinearity; i.e.,

$$x_I = x_{0I} - f\frac{a_1(X - X_s)+b_1(Y - Y_s)+c_1(Z - Z_s)}{a_3(X - X_s)+b_3(Y - Y_s)+c_3(Z - Z_s)} \qquad (4)$$

$$y_I = y_{0I} - f\frac{a_2(X - X_s)+b_2(Y - Y_s)+c_2(Z - Z_s)}{a_3(X - X_s)+b_3(Y - Y_s)+c_3(Z - Z_s)} \qquad (5)$$

where x_I and y_I are the corresponding coordinates of the pixel $P(X,Y)$ in the source image; X_s, Y_s and Z_s are the exposure station; f is the focal length; $a_i = \{a_1, a_2, a_3\}$, $b_i = \{b_1, b_2, b_3\}$ and $c_i = \{c_1, c_2, c_3\}$ are elements of the rotation matrix related to the orientation angles of the three exteriors (ω, ϕ, κ). The elements have to be computed before the equation can be used to transform the coordinates. At least three ground-control points (GCP) that relate the coordinate of the source image and the coordinate of the DSM are required to solve the exterior orientation elements. Generally more than three GCPs are required for a better estimation via least-square estimation.

2.5 Assigning a grey value to the pixel

The grey value in the resulting ortho-image is obtained by interpolating the corresponding pixels in the source image, since the grid of pixels in the source image rarely matches the grid of the output ortho-image. There are several methods of resampling, the most popular of which are the nearest neighbour, bilinear interpolation and cubic convolution. We employed the nearest neighbour method for the assignment of grey values because the nearest neighbour method directly transfers the values of the original data without averaging them. In order to assign a grey value to the output pixel with coordinate $P(X,Y)$, we have to retransform its coordinate back to the source image. Supposed the pixel $P(X,Y)$ in the output image has its corresponding coordinate $P(X_I,Y_I)$ in the source image. Once $P(X_I,Y_I)$ is computed, it is easy to find the nearest pixel to assign an output value for the pixel in the output image. The computational procedure is illustrated in Figure 2.

2.6 Occlusion analysis by DBM

Occlusion removal is a critical step in the generation of large-scale true ortho-images in urban areas where man-made buildings are a common part of the landscape. Here, we propose a mathematical model to automatically detect the occlusions. As illustrated in Figure 3, the roof and bottom of a building have different imaging locations in the image plane. In the real aerial image, the building is visible via roof Points 2 and 4 and a bottom Point 1. The pixels between Points 2 and 4 represent the roof, and the pixels between Points 2 and 1 represent the wall in the aerial image, and

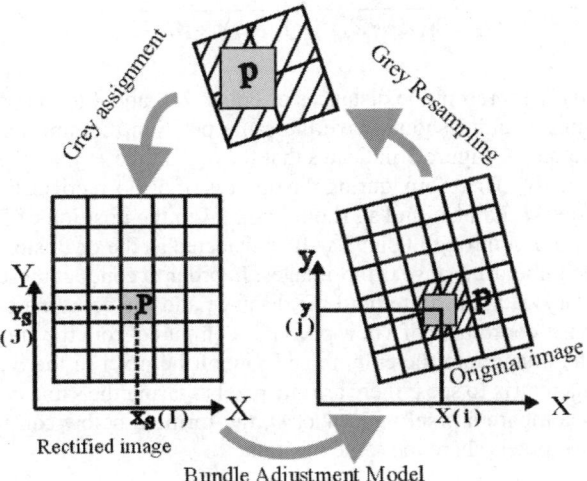

Figure 2.   The procedure of geometric rectification.

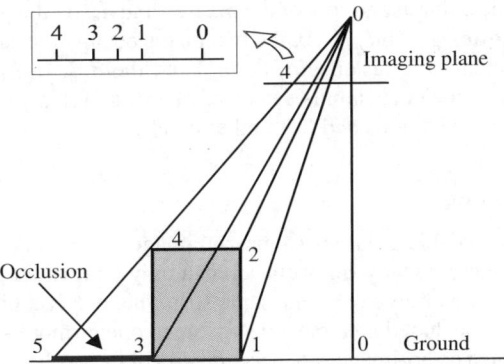

Figure 3.   Occlusion analysis of buildings.

the pixels between 3 and 5 are occluded. In the true ortho-image, what we see is the roof; i.e., the landscape coving Point 2 through 4.

This problem can be solved by visibility analysis in combination with photogrammetry geometry. If we know the 3D model of the building, we can do a back-projection of the building's 3D model to the image plane, and then calculate the image coordinates of the building's boundary. In other words, the distance between the building to the perspective centre can be computed by, for example, Point 1,

$$d_{10} = \sqrt{(X_0 - X_1)^2 + (Y_0 - Y_1)^2 + (Z_0 - Z_1)^2} \tag{6}$$

where d_{10} is the distance of Point 1 to the perspective centre 0; X_1, Y_1, Z_1 are the coordinates of Point 1 on the building roof. Similarly, we have the following distances for Points 2 through 4:

$$d_{20} = \sqrt{(X_0 - X_2)^2 + (Y_0 - Y_2)^2 + (Z_0 - Z_2)^2} \tag{7}$$

$$d_{30} = \sqrt{(X_0 - X_3)^2 + (Y_0 - Y_3)^2 + (Z_0 - Z_3)^2} \tag{8}$$

183

$$d_{40} = \sqrt{(X_0 - X_4)^2 + (Y_0 - Y_4)^2 + (Z_0 - Z_4)^2} \qquad (9)$$

where d_{20}, d_{30} and d_{40} represent the distances of Points 2, 3 and 4 to the perspective centre 0.

Theoretically, Points 1 and 2 should have the same position (column and row) in the output image, so do Points 3 and 4. Figure 3 indicates that the d_{20} is greater than the d_{10}, and that the d_{40} is also greater than the d_{30}. Therefore, during the process of ortho-rectification, Point 2 should be rectified to the position of Point 1, and so should Point 4 to the position of Point 3. As a result, a blank area covering Point 4 through Point 3 will be detected as the occlusion, which can be filled by mosaicking the neighbouring slave ortho-images. In order to conduct these operations, we have to generate an ancillary data matrix to hold the distance and the coordinates of the pixels of the source image. When the coordinate of a new pixel n is computed from the above section, we covert the coordinates of the image into the column and row of the pixel in the output image. We then need to check the data matrix to see if there are any pixels sharing the same column and row as this pixel. If we need to compare a pixel with pixel m, the formula below can be used to determine whether or not the two pixels share the same position:

$$d_{nm} = \sqrt{(K_n - K_m)^2 + (L_n - L_m)^2} \qquad (10)$$

where K_n and L_n are the column and row of the pixel n, and d_{nm} is the distance between pixels n and m in the output ortho-image. If $d_{nm} = 0$, the two pixels occupy the same column and row. We then need to further compare the distance of the d_{n0} and the d_{m0}. If $d_{n0} < d_{m0}$, then Point m is occluded by Point n. When the occlusion is detected, we can assign a special value (usually 0) as the background, so that it is distinguished from other pixels.

2.7 Refilling the occlusions

Since occlusions are unavoidable in large-scale aerial imagery, refilling is necessary for the generation of true ortho-images. Usually, an occlusion can only be refilled from neighbouring slave images taken in the same time frame (e.g., the same strip, the identical block of images). In order to refill the occlusion, the occluded area must be visible in one or more slave images.

Sometimes, several slave ortho-images are needed to refill the occlusion. On the other hand, the more slave ortho-images are used, the poorer the true ortho-image quality will appear because of the mosaicking of various images, which may have different radiometric reflectances. Therefore, in order to produce a high-quality ortho-image, the occlusion of buildings in the source images must not be too great. In other words, the proportion between the height of the building and the airbase must be a rational ratio. At the moment, however, there is on consensus on what constitutes

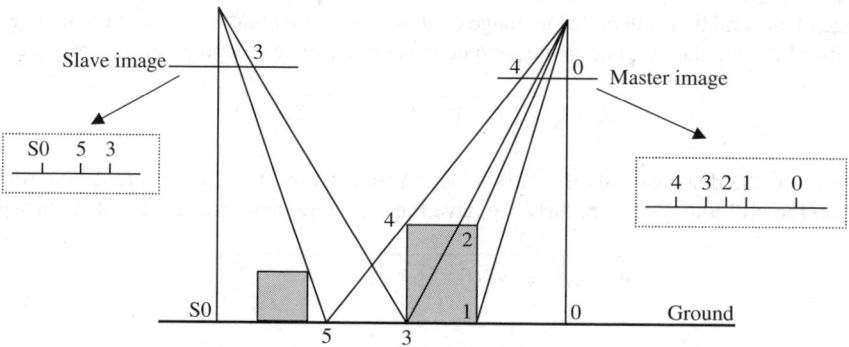

Figure 4.   Visibility of building occlusion in the slave image.

184

a rational ratio. Here, we propose a criterion for practical use: the occlusion must be less than 10% of the height of the building. This means that the distance between Points 5 and 3 must be 10% less than the building's height (the distance between Points 3 and 4). Two factors are critical in determining the occlusion: imaging altitude and the relative location of the building to the nadir point of the image. For the later factor, the imaging scene width is important. Generally speaking, those ground objects closer to the image nadir point have better visibility while those farther from it are more occluded, provided that they are of the same height. This suggests that only the parts of the source image that are close to nadir point of the image are good for the generating of true ortho-images if the imaging altitude is not high and the imaging scene is wide. This is especially important in choosing large-scale aerial images for the generation of true ortho-images.

3 EXPERIMENTS AND ANALYSIS

We have experimented with the proposed method for creating true ortho-images of Denver, Colorado. The source images were acquired on April 17, 2000 at about 11:00am, using an RC30 camera, which has a focal length of 153.022 mm. The flying height was about 1650 m. The aerial photos were originally recorded on film and later scanned into digital format. The DSM, DTM and DBM available to this study only cover a part of the imaging area, the upper quarter of the original image (see Figure 6a and 6b). Due to this fact, we chose a small part of the imaging area for our search. Figure 7a shows the source image, in which there are several high buildings. The heights of the five numbering buildings are respectively 100 m, 103 m, 125 m, 100 m, and 107 m. As seen, the higher buildings have very large occlusions. We have developed the algorithms to generate urban true ortho-images. The flowchart is illustrated in Figure 5. The basic steps and procedures are as follows.

Step 1: Input the source images and corresponding exterior orientation parameters.

Step 2: Input the DTM and DBM, which are available from Wolfgang Schickler of Analytical Survey Inc., Colorado Springs, Colorado. We inputted this data into EARDA/IMAGINE to generate elevation images.

Step 3: Determine the size of the output ortho-image. With the DTM, source image and specified pixel size of the resulting ortho-image, we can determine the size of the resulting image according to the method described in Section 2.2.

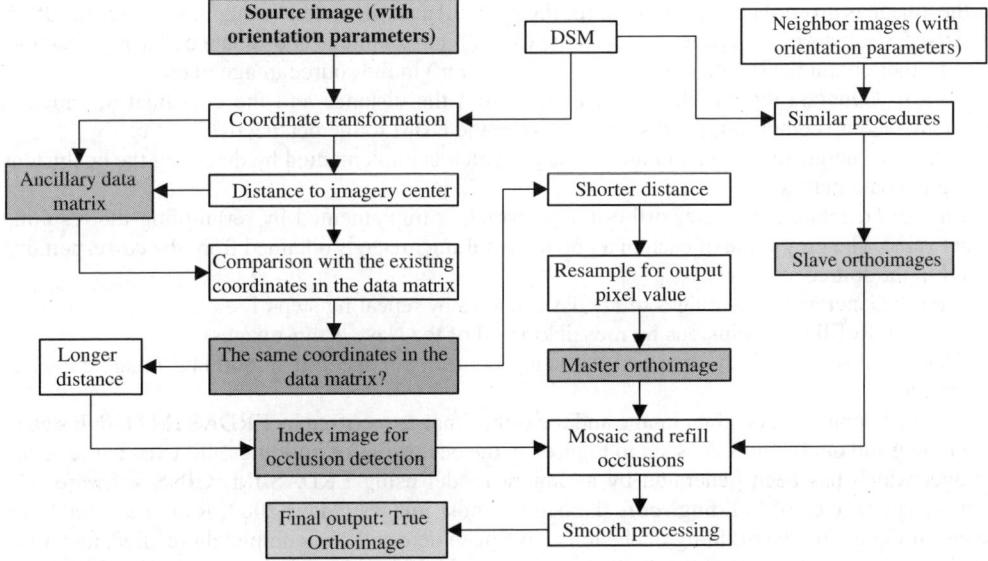

Figure 5.   Procedures for true ortho-image generation.

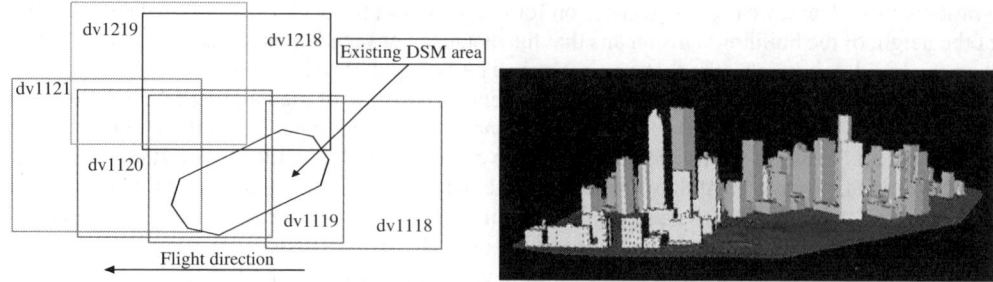

Figure 6. (a) The images acquired and DTM available, (b) DBM in Denver, CO.

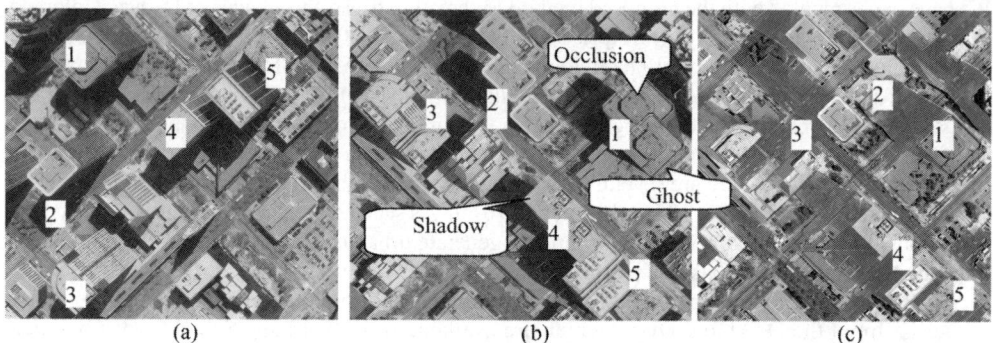

Figure 7. (a) Source image in downtown Denver, CO, (b) ortho-rectified images generated by a camera model using ERDAS/Imagine software, and (c) true ortho-image generated by our method.

Step 4: Compute the geodetic coordinate from the column and row of pixels. The X and Y geodetic coordinates for any pixel can be calculated if the size and resample distance in the resulting ortho-image is given. The Z coordinate for the pixel then is calculated through interpolating DSM.

Step 5: Transform the geodetic coordinate (X,Y,Z) into an image coordinate by back-projection, and further obtain the screen coordinates (row, column) in the source image plane.

Step 6: Generate the ancillary data matrix from the distance and the coordinates, which is implemented by computing the distance between the DBM to the perspective centre.

Step 7: Generate the occlusion index image, which is implemented by detecting the occlusions using the data matrix.

Step 8: Generate the master ortho-image, which is implemented by resampling the resulting pixel value. The grey value of each pixel on the resulting image is obtained from the corresponding pixel in the source images.

Step 9: Generate ortho-images of all slave images by repeating steps 1–8.

Step 10: Refill the occlusions by mosaicking all of the slave ortho-images.

Step 11: Post-process the true ortho-image, by means of spectral smoothing, image enhancement, etc.

Figure 7 depicts the original image and the ortho-images rectified by ERDAS IMAGINE's camera model and our method. As seen in Figure 7b, the occlusions of buildings still exist in the ortho-image, which has been generated by a camera model using ERDAS/IMAGINS software. The double appearance of building roofs dabbed as ghost images (Mayr 2002, Rau *et al.* 2002) are seen. In Figure 7c, the building occlusions have been detected and completely refilled, and all of the buildings are placed in their planimetric position in the resulting ortho-image. The shadows of the buildings are removed and compensated. During the process of generating true ortho-photos,

186

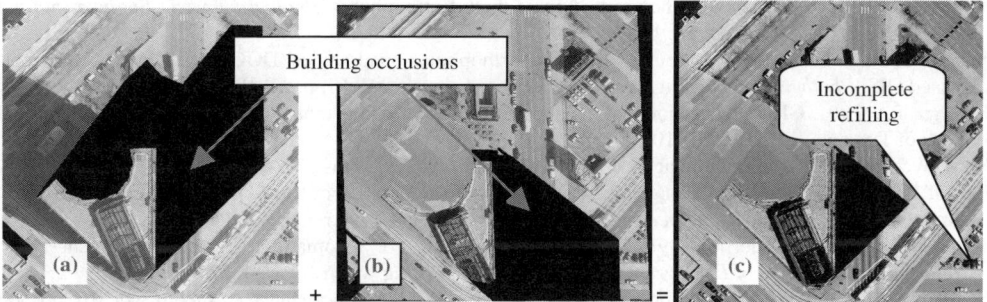

Figure 8.   (a) Building occlusion in the master image, (b) Building occlusion in the slave image, and (c) Result of refilling the occlusion of the master image from the slave image, illustrating the incomplete refilling of the occlusion.

we noticed that sometimes only one slave ortho-image is not enough to refill the occlusion. As illustrated in Figure 8, one master (Figure 8a) and one slave (Figure 8b) ortho-image are still not sufficient to refill the complete occluded area. Therefore, how do we optimize the photogrammetric flying mission for a complete TRUE ortho-image to be investigated in the near future?

4   CONCLUSIONS

Large-scale urban ortho-images taken as urban GIS layer provide decision-makers with the opportunity to optimize the decision-making process. The conventional ortho-photo generation algorithm cannot create satisfactory ortho-images in urban areas because the conventional method, which is based on DTM, is not able to identify building occlusions in urban areas. The method presented in this paper can effectively generate true ortho-images because we used DTM and DBM data, and have developed a method to detect building occlusions for true ortho-rectification. Our experimental results in downtown Denver, CO. demonstrated that our method can effectively and correctly ortho-rectify the distortion of urban buildings. These true ortho-images that have been generated provide accurate information on space and features in their true positions, and can be combined with other urban geospatial information through spatial overlap into the urban decision-making system.

ACKNOWLEDGEMENTS

The project is funded by US National Science Foundation (NSF) under contract number NSF 0131893. The authors sincerely thank Wolfgang Schickler, Analytical Survey Inc., Colorado Springs, Colorado, for his kindness in providing aerial images, building data, and DSMs of Denver, CO.

REFERENCES

Amhar, F., J. Josef, and C. Ries. 1998. The generation of true orthophotos using a 3D building model in conjunction with a conventional DTM, *International Archives of Photogrammetry and Remote Sensing*, vol. 32, Part 4: 16–22.
Baltsavias, E.P. 1996. Digital Ortho-Images – A Powerful Tool for the Extraction of Spatial- and Geo-Information. *ISPRS Journal of Photogrammetry & Remote Sensing*, vol. 51: 63–77.
Baltsavias, E.P. and C. Käser. 2002. DTM and orthoimage generation – a thorough analysis and comparison of four digital photogrammetric systems. Commission IV, Working Group 2.
Chen, L.C. and J.Y. Rau. 1993. A unified solution for digital terrain model and orthoimage generation from SPOT stereopairs, *IEEE Trans. on Geoscience and Remote Sensing*, vol. 31, no. 6: 1243–1252.

Hohle, J. 1996. Experiences with the production of digital orthophotos, *Photogrammetric Engineering and Remote Sensing,* vol. 62, no. 10: 1189–1194.

Joshua, G. 2001. Evaluating the accuracy of digital orthophotos quadrangles (DOQ) in the context of parcel-based GIS, *Photogrammetric Engineering & Remote Sensing,* vol. 67, no. 2: 199–205.

Kyariga, T.A. 2001. GIS as a decision making support tool for urban planning and management: A Practical case of Tanzania, CORP, 2001: 103–106.

Mayr, Werner. 2002. True Orthoimages, *GIM International*, vol. 37, April: 37–39.

Rau, J.Y., N.Y. Chen, and L.C. Chen. 2002. True Orthophoto Generation of Built-Up Areas Using Multi-View Images, *Photogrammetric engineering and Remote Sensing,* vol. 68, no. 6, June: 581–588.

Schickler, W. and A. Thorpe. 1998. Operational procedure for automatic true orthophoto generation, *International Archives of Photogrammetry and Remote Sensing,* vol. 32, Part 4: 527–532.

Sharlatos, D. 1999. Orthophotograph production in urban areas, *Photogrammetric Record,* vol. 16, no. 94: 643–650.

Vassilopoulou, S., L. Hurni, V. Dietrich, E. Baltsavias, M. Pateraki, E. Lagios, and I. Parcharidis. 2002. Orthophoto generation using IKONOS imagery and high-resolution DEM: a case study on volcanic hazard monitoring of Nisyros Island (Greece). *ISPRS Journal of Photogrammetry and Remote Sensing*, 57(1–2): 24–38.

Zhou, G., K. Jezek, W. Wright, J. Rand, and J. Granger. 2002. Orthorectifying 1960's declassified intelligence satellite photography (DISP) of Greenland, *IEEE Geoscience and Remote Sensing,* vol. 40, no. 6: 1247–1259.

Zhou, G., Z. Qin, Susan Benjamin, and Schickler. 2003. Technical Problems of Deploying National Urban Large-scale True Orthoimage Generation, *the 2nd Digital Government Conference*, Boston, May 18–21, 2003: 383–387.

Zhou, G., Z. Qin, Susan Benjamin, Changqing Song, Wolfgang Schickler, Anthony Thorpe, and John Rand. 2003. A comprehensive study on true orthorectification of large-scale urban aerial images, to be submitted to *IEEE Transaction on Geoscience and Remote Sensing*, June 2003.

Advances in Spatial Analysis and Decision Making, Li, Zhou & Kainz (eds)
© 2004 Swets & Zeitlinger, Lisse, ISBN 90 5809 652 1

Studying spatio-temporal patterns of land-use change in arid environment of China

Qiming Zhou
Department of Geography, Hong Kong Baptist University, Hong Kong

Baolin Li[1,2] & Chenghu Zhou[2]
[1] *Department of Geography, Hong Kong Baptist University, Hong Kong.*
[2] *State Key Laboratory of Environment and Resources Information System, Institute of Geographical Sciences and Resources Research, Chinese Academy of Sciences, Beijing, P.R. China*

ABSTRACT: Remotely sensed data have been the most important data source for environment change study in the past 30 years. Large collections of remote sensing imagery have provided a solid foundation for spatio-temporal analyses of the environment and the impact of human activities. This study seeks an efficient and practical methodology for integrating multi-temporal and multi-scale remotely sensed data from various sources with a monitoring time frame of 30 years, including historical and state-of-the-art high-resolution satellite imagery. Based on this, spatio-temporal patterns of environmental change, which is largely represented by changes in land cover (e.g., vegetation and water), were analysed for the given time frame. Multi-scale and multi-temporal remotely sensed data, including Landsat MSS, TM, ETM and SPOT HRV, were used to detect changes in land use in the past 30 years in Tarim River, Xinjiang, China. The study shows that by using the auto-classification approach an overall accuracy of 85%–90% with a Kappa coefficient 0.66–0.78 was achieved for the classification of individual images. The temporal trajectory of land-use change was established and its spatial pattern was analysed to obtain a better understanding of the human impact on the fragile ecosystem of China's arid environment.

1 INTRODUCTION

Land cover change plays a pivotal role in regional social and economic development and global environment changes (Chen 2002). In arid environment, where fragile ecosystems are dominant, the land cover change often reflects the most significant impact on the environment due to excessive human activities.

Remotely sensed data have been the most important data source for environment change study in the past 30 years. Large collections of remote sensing imagery have provided a solid foundation for spatio-temporal analyses of the environment and the impact of human activities. Research has been widely reported on methodologies of detecting and monitoring changes (e.g., Singh 1989, Mas 1999).

For landuse change detection for regional sustainable development, the most commonly used data include Landsat MSS, TM, ETM, SPOT HRV, IRS and AVIRIS. According to most reported research, the change detection was often based on the comparison data acquired on two or three epochs (Michener and Houhoulis 1997, Prakash and Gupta 1998, Sunar 1998, Dai and Khorram 1999, Jensen *et al.* 1999, Houhoulis and Michener 2000, Luque 2000, Ustin and Xiao 2001, Pereira *et al.* 2002, Maldonado *et al.* 2002), and sometimes on four to five epochs (Miller *et al.* 1998, Masek *et al.* 2000). In these studies, it is common to use data acquired from different sensors

(Prakash and Gupta 1998, Luque 2000, Masek *et al.* 2000, Ustin and Xiao 2001). Since the spatial and spectral resolutions of different sensors vary significantly, the ability to distinguish land cover also varies greatly. It is therefore important to address the issue of uncertainty while undertaking a study on detecting changes in land cover using multi-temporal data acquired from different sensors.

When monitoring natural environment and land cover change, three aspects are focused (Singh 1989, Macleod and Congalton 1998):

1. the areal extent of the change, measuring the magnitude of the change;
2. the nature of the change, measuring the temporal trajectory of the change;
3. the spatial pattern of the change, measuring spatial distributions and relationships in the changes.

This study seeks an efficient and practical methodology to integrate multi-temporal and multi-scale remotely sensed data from various sources with a monitoring time frame of 30 years, including historical and state-of-the-art high-resolution satellite imagery. Based on this, the spatio-temporal pattern of environmental change, which is largely represented by the land cover (e.g. vegetation and water) change, has been analysed for the given time frame. Multi-scale and multitemporal remotely sensed data, including Landsat MSS, TM, ETM and SPOT HRV, were used to detect landuse change in the past 30 years in Tarim River, Xinjiang, China.

2 METHODOLOGY

The generic approach of this study is based on post-classification comparison method, which is commonly employed in land cover change detection studies (Miller *et al.* 1998, Larsson 2002, Yang and Lo 2002, Zhang *et al.* 2002). A unified land cover classification scheme was established for classification of images. The classification images were then used for the analysis of temporal trajectory and spatial pattern of landuse change in the past 30 years in the study area.

2.1 *Study area and data*

The study area is centred at about 41°5′N and 85°43′E in Donghetan Township, Yuli County, Xinjiang Uygur Autonomous Region, China. It locates at the middle reach of Tarim River, the longest inland river of China (figure 1). At the fringe of Taklimakan Desert, the 'green corridor' of Tarim Basin is one of the most important habitation areas in aridzone of China. The landscape in Donghetan is typical in Tarim River Valley, with a generally dry and harsh environment, represented by typical desert vegetation and soils. With the increasing land development in recent decades, the fragile environment has experienced quite remarkable change, largely reflecting the general development trend and temporal effects of government policies and administrative measures.

Five multi-temporal remotely sensed images were acquired for change detection of this study (table 1), including Landsat MSS, TM, ETM and SPOT HRV multi-spectral images. In addition, a multi-spectral 4-m resolution IKONOS image was also acquired in September 2000 to assist in field investigations and accuracy assessment of the image classification. The IKONOS image was geo-referenced to a 1:10,000 map using 22 Ground Control Points (GCPs). The other images were then geometrically corrected and registered on the map coordinates using image-to-image registration to the master IKONOS image (table 2). Efforts were made to control registration errors to within half a pixel of the image concerned, so that the errors caused by mis-registration would be less critical.

2.2 *Classification*

To minimize seasonal effects of remotely sensed data, the post-classification comparison method was employed for image processing, since this method is less sensitive to radiometric variations between the scenes (Mas 1999). Supervised classification was employed to classify individual images independently, using a unified land cover classification scheme to ensure that the classifications of the

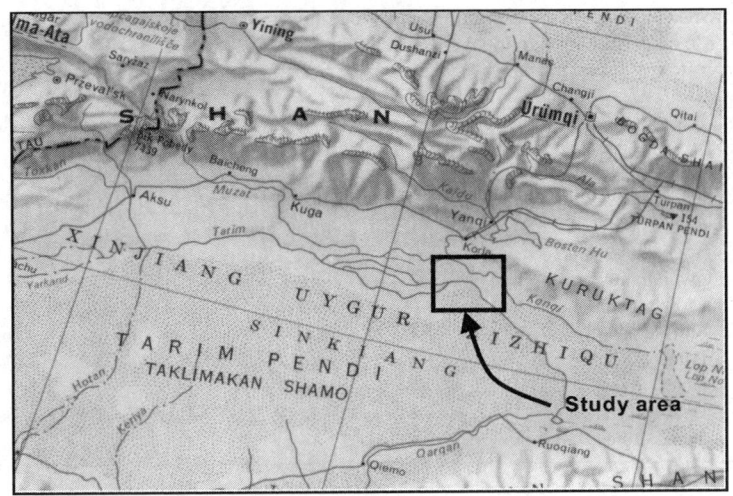

Figure 1. Location map of the study area.

Table 1. Data used in this research.

Satellite	Sensor	Path/Row	Resolution (m)	Acquisition date
Landsat 1	MSS	154/31	57*	3/7/1973
Landsat 2	MSS	154/31	57*	12/10/1976
SPOT 1	HRV	216/266/9	20	20/7/1986
Landsat 5	TM	143/31	30	25/9/1994
Landsat 7	ETM	143/31	30	17/9/2000

* Resampled resolution.

Table 2. RMS errors on the geometric correction and registration of the images.

	RMSE			
	X (pixels)	X (m)	Y (pixels)	Y (m)
MSS (1973)	0.23	13.11	0.35	19.95
MSS (1976)	0.38	21.66	0.49	27.93
SPOT (1986)	0.21	4.20	0.22	4.40
TM (1994)	0.24	7.20	0.20	6.00
ETM (2000)	0.17	4.85	0.16	4.56

multi-scale, multi-temporal images are compatible with each other (table 3). Images were first classified into the Level 2 classes, which were subsequently merged into the five unified classes, as listed in Table 3. Training areas were selected based on image interpretation keys established during the field investigation and from interviews with the local inhabitants. A minimum of five training areas containing at least 100 pixels exhibiting uni-modal distribution for DN were selected for each class. The Jeffries-Matusita (JM) separability test (Richards 1995) was then applied to ensure the correctness of the training area selected. For this study, the JM distances between unified classes were kept greater than 1.8, while the JM distances between Level 2 classes in the same unified class were greater than 1.4. The maximum likelihood classifier (MLC – Lillesand and Kiefer 2000) supported by PCI Geomatica image processing software was used to classify the images.

191

Table 3. Unified land cover classification scheme for multi-scale, multi-temporal images. The numbers denote the land-use class code in individual classifications.

Level 1 classes	Level 2 classes	ETM (2000)	TM (1994)	SPOT (1986)	MSS (1976)	MSS (1973)	Unified classes
Cropland	Cropland	1	1	–	–	–	Cropland (1)
Grass and woodland	Dense grass and woodland	2	2	2	2	2	Grass and woodland (2)
	Sparse grass and woodland	3	3	3	3	3	
	Mowing land	4	–	–	–	–	
	Salty grass	5	5	5	5	5	Salty grass (3)
Water body	Ponds	6	6	6	6	6	Water body (4)
	River	7	7	7			
Unused land	Bare ground and sand dunes	8	8	8	8	8	Bare ground (5)

2.3 Post-classification sorting

Classification results were then merged into compatible groups, as shown in Table 3. This processing was necessary in order to:

1. maintain compatibility and comparability between classes in different multi-scale, multi-temporal images,
2. reduce the number of classes to simplify the analysis of changes in the trajectories of land-use, and
3. minimize potential errors of classification, since it was already known that some of the classes of land use (e.g., croplands) did not exist in early images. A majority filter is also applied on individual classified images to remove isolated pixels.

2.4 Accuracy assessment

In this study we have chosen stratified random sampling scheme for selecting sample points of reference data for classification accuracy assessment. 790 sample points were generated and transferred to GIS. They were then overlaid with the classified images as well as the high-resolution IKONOS multi-spectral image.

Collecting reference data for accuracy assessment on multi-temporal images is always a serious problem for researchers, because simultaneous 'ground truthing' data over a long period of time are very difficult, if not impossible, to find. In this study, we could only acquire a high-resolution IKONOS multi-spectral image that was simultaneous with the 2000 ETM data. Although the IKONOS image has a high enough resolution for 'ground truthing', the 'time gaps' between this 'reference image' and some historical images are large.

In this study, we directly used the IKONOS image as the source of the reference data to assess the results of the 2000 ETM classifications. For the other four historical images, we used the IKONOS image as the basis for comparison for proper interpretation. By this means, obvious land cover changes such as grasslands to water and bare ground to cropland, could be reliably detected by image interpretation. Field visits and interviews were also conducted for sample points where a clear relationship between the present and historical images could not be established.

2.5 Establishing landuse change trajectory

Spatio-temporal landuse change pattern has been an active research field (Roy and Tomar 2001, Weng 2001), and the concept and methodology of change trajectory has been developed (Mertens

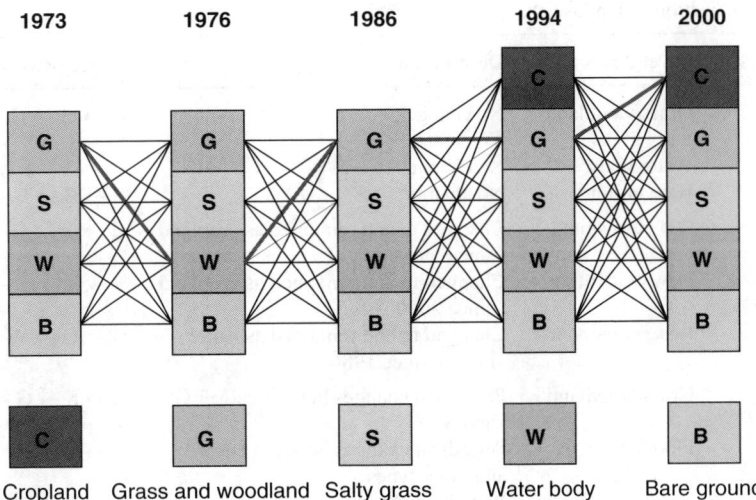

| 1973 | 1976 | 1986 | 1994 | 2000 |

Cropland   Grass and woodland Salty grass   Water body   Bare ground

Figure 2. All possible landuse change trajectory identified for the study area.

and Lambin 2000, Petit *et al.* 2001). In this study the term trajectory of land-use change refers to successions of land cover types for a given sample unit over more than two observations (epochs).

To establish landuse change trajectories, all classified images were integrated in GIS using raster format with ARC/INFO GIS software. Based on the classification scheme as shown in table 3, all possible landuse change trajectories can be shown in Figure 2.

Note that there was no cropland found in this area before the 1990s so that the class 'C' is not included in the classification of the 1973, 1976 and 1986 images. Based on figure 2, a landuse change trajectory can be specified. For example, as highlighted in figure 2, a trajectory can be specified as $G \rightarrow W \rightarrow G \rightarrow G \rightarrow C$, meaning that the land was found to be grass/woodland in 1973, water body (flooded) in 1976, grass/woodland again in 1986 and 1994, and cultivated as cropland in 2000.

For the analysis of temporal human impact on the environment, we have classified all found trajectories into three generic classes, namely, unchanged, human-induced change and natural changes. The *unchanged* class includes trajectories such as $G \rightarrow G \rightarrow G \rightarrow G \rightarrow G$ and $W \rightarrow W \rightarrow W \rightarrow W \rightarrow W$ indicating that the same land cover type was found on the sample point over the past 30 years. Some ambiguous cases (e.g. $S \rightarrow S \rightarrow G \rightarrow S \rightarrow S$ or $G \rightarrow G \rightarrow G \rightarrow C \rightarrow G$) are also included in this type considering that possible classification error may create once-only false class in the trajectory. The *human-induced* change class includes decisive changes due to human activities such as building dam/reservoir and cultivation. These changes are often irreversible so that they represent the major human impact on the environment. The representative trajectories of this class include, e.g., $G \rightarrow G \rightarrow G \rightarrow C \rightarrow C$, $S \rightarrow S \rightarrow G \rightarrow G \rightarrow C$, and $G \rightarrow G \rightarrow W \rightarrow W \rightarrow W$. The *natural* change class includes those indecisive changes due to the natural processes or minor human activities such as light grazing. For example, grassland may be flooded during summer and subsequently dried out as salty grass because of strong evapotranspiration. Examples of trajectories of this class are $G \rightarrow W \rightarrow B \rightarrow G \rightarrow G$ (flooded, eroded and recovered) and $G \rightarrow W \rightarrow G \rightarrow W \rightarrow G$ (repeatedly flooded). The classification of trajectories is summarized in Table 4.

3   RESULTS AND DISCUSSION

3.1   *Image classification*

The five multi-temporal images were independently classified using the unified land cover classification scheme proposed in this study. It is obvious that higher resolution satellite images have

Table 4. Classification of landuse change trajectories.

Level 1 classes	Level 2 classes	Description	Trajectory examples
Unchanged	Grass/woodland	No change*	G → G → G → G → G
	Salty grass	No change*	S → S → S → S → S
	Water body	No change*	W → W → W → W → W
	Bare ground	No change*	B → B → B → B → B
Human-induced	Old cultivation	Changed to and remained as cropland since 1994	G → G → G → C → C
	New cultivation	Changed to and remained as cropland since 2000	S → S → G → G → C
	Reservoirs/ponds	Changed to and remained as water bodies since 1986	G → G → W → W → W
Natural	Grass/woodland	Periodical changes between cover G and S	G → S → G → G → S
	Flooded	Periodical changes between cover W and other types	G → W → G → W → G
	Bare ground	Periodical changes between cover B and other types	G → B → B → G → B

* Some ambiguous cases are also included considering that possible classification error may create once-only false class in the trajectory.

a better ability to separate more classes (i.e. derive more details of land cover). However, these level 2 classes need to be merged in post-classification process so that all multi-platform images can be compared. The classification results are shown in Figure 3.

The overall accuracies for image classifications range from 85.8% to 89.8%, with Kappa coefficients range from 0.66 to 0.78 (table 5). Generally, the classification on images with higher spatial resolution yields better accuracy results. Although the overall accuracies are quite similar for all the classification results, the Kappa coefficients demonstrate much larger range. The Kappa of low-resolution MSS classification were 0.66 and 0.73, in comparison to 0.78–0.83 of those of higher-resolution SPOT, TM and ETM classification.

Kappa coefficients for individual classes over each observation date are listed in Table 6. For croplands, the accuracy of classification of 1994 image is lower than that of 2000. This may reflect the effect of the flood in 1994 when the cropland area was small so that the ratio of misclassified marginal area could be high. For other classes, the accuracy assessment results vary largely, with the lowest Kappa for 1976 salty grass (0.50) and the highest for 2000 bare ground (0.96). In general, the poorer results are shown on early MSS data with lower resolution and poorer reference data. While using better data, the accuracy has raised significantly.

3.2 Area statistics

Table 7 and Figure 4 show the area statistics of land cover types and their change over the 30 year study period. According to the local government record, large-scale cultivation started in this area in 1992 and cropped farmland increased rapidly since then. This change is confirmed by the findings of this study. From the area statistics, it is shown that cropland increased from 4.0% to 12.6% of the total area – three times increase in 6 years. Another significant human impact is the construction of a dam in the north of the study area at the end of 1970s and early 1980s, resulting in flooding vast area of grass/woodlands. The grass/woodland area decreased from 72.2% (1976) to 65.2% (1986) during this period, while water body area increased from 9.6% to 18.0% of the total study area. It is also shown that the area of increasing cropland largely came from salt grass (5.9% (1986) → 2.7% (2000)) and water body (18.0% (1986) → 8.6% (2000)). This is largely because

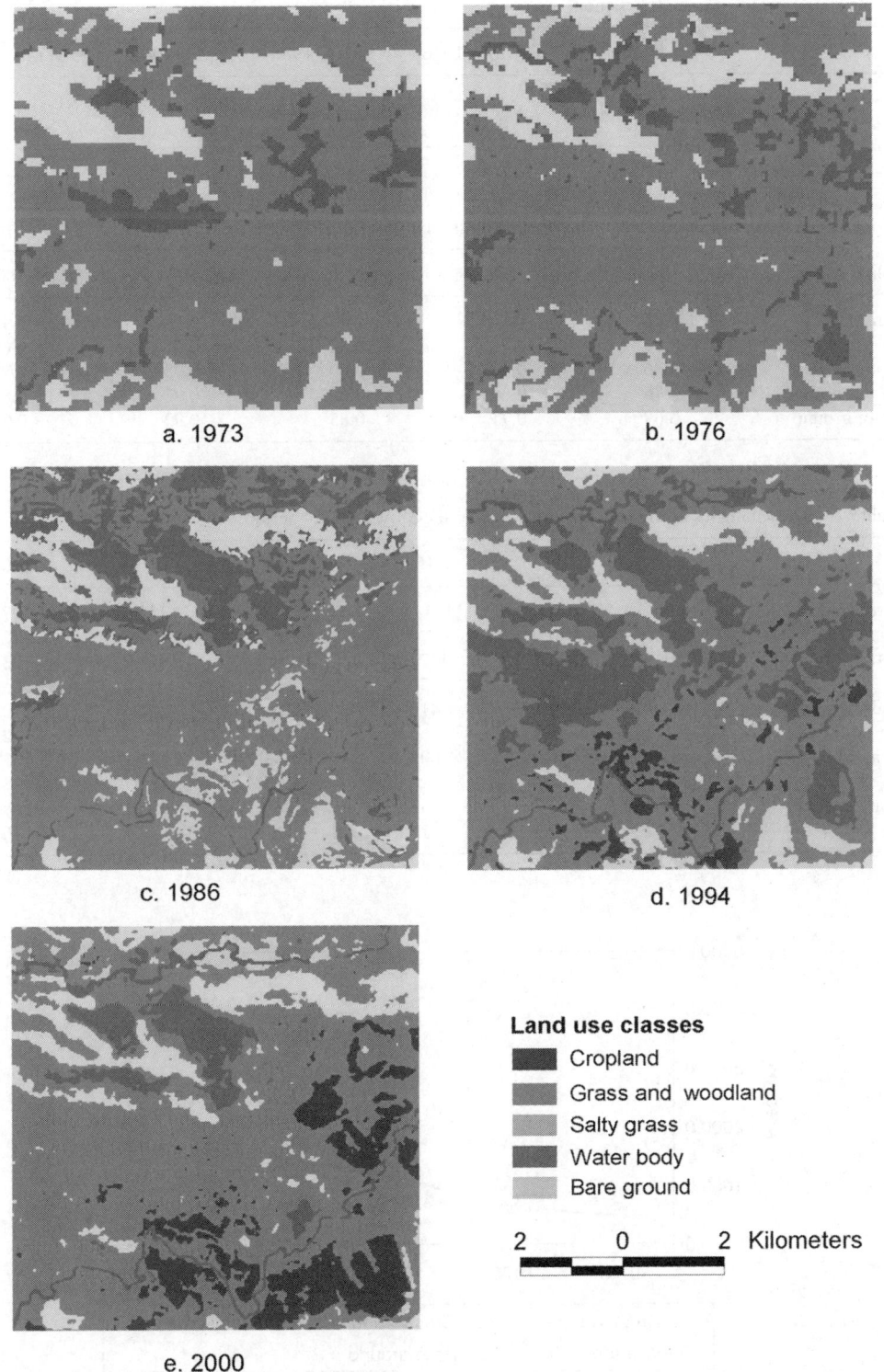

a. 1973

b. 1976

c. 1986

d. 1994

e. 2000

Land use classes
- Cropland
- Grass and woodland
- Salty grass
- Water body
- Bare ground

2   0   2 Kilometers

Figure 3.   Types of land use over the last 30 years.

Table 5. Overall accuracy assessment.

	MSS (1973)	MSS (1976)	SPOT (1986)	TM (1994)	ETM (2000)
Overall accuracy (%)	88.9	85.8	88.9	89.8	87.1
Kappa	0.73	0.66	0.79	0.83	0.78

Table 6. Kappa for individual classes for each classification result.

Class name	1973	1976	1986	1994	2000
Cropland	–	–	–	0.70	0.91
Grass/woodland	0.70	0.73	0.86	0.75	0.66
Salty grass	0.86	0.50	0.89	0.93	0.70
Water body	0.76	0.60	0.60	0.91	0.80
Bare ground	0.69	0.71	0.85	0.92	0.96

Table 7. Area statistics of the land cover types over the 30-year study period.

		1973	1976	1986	1994	2000
Cropland	(ha)	–	–	–	254.3	797.0
	(%)	0	0	0	4.0	12.6
Grass/woodland	(ha)	4746.8	4577.2	4129.9	3811.9	4153.2
	(%)	74.9	72.2	65.2	60.2	65.6
Salty grass	(ha)	331.1	416.4	376.1	202.6	170.7
	(%)	5.2	6.6	5.9	3.2	2.7
Water body	(ha)	476.9	608.8	1143.3	1441.2	547.8
	(%)	7.5	9.6	18.0	22.7	8.6
Bare ground	(ha)	781.3	733.7	686.7	626.1	667.1
	(%)	12.3	11.6	10.8	9.9	10.5

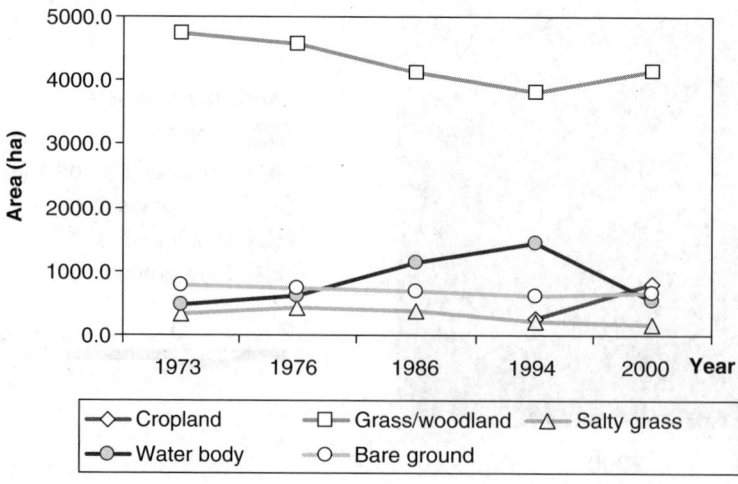

Figure 4. Change of land cover types in 30 years.

that the most cultivated lands are located along the river bands of Tarim River in the south and east part of the study area, with the construction of dykes along the river bed to prevent flood. Cover types with less favourite cultivation conditions (particularly lack of water supply), such as grass/woodland and bare ground , were hardly changed during the 30 year period. The flood effect showed by significant temporary increasing water body and decreasing grass/woodland on 1994 image should also be noted.

3.3   Land cover change trajectories

Table 8 shows the area statistics of land cover change trajectories. During the 30-year study period, the unchanged area occupies 42.8% of the total area, human-induced changes occupied 17.5%, and natural change area occupies 39.7%. For unchanged area, grass/woodland constitutes 80.3%. For human-induced change, new cultivation occupies 60.8%. For natural change area, flooded area obviously dominates, constituting 64.3%. Given the fact that the study area is quite remote and human activities appear to be quite limited, it is understandable that most of environmental change is due to natural force (e.g. flood) rather than human impact. However, it should also be noted that since early 1990s, human activities start to play an important role on the environmental change, by altering the natural courses of water and surface materials. Although the total area of human-induced change is still relatively small (<20%), its pace of growth is alarming.

3.4   Spatial distribution of landuse change trajectories

Figure 5 shows the spatial distribution of landuse change trajectories in the past 30 years. In the study area, the human-induced changes mainly appeared in two areas, namely, cultivation in south and southeast along Tarim River, and dam/reservoir construction in the north. In between the two areas, human impact was limited and natural change was predominated by flood. It should also be noted that areas of natural change with bare ground most likely appeared at the fringe of unchanged bare ground areas, indicating the temporary cover changes in those bare ground areas. Since the total bare ground area had very little change (12.3 (1973) → 10.5 (2000), table 7), there is no evidence that shows expansion or shrinking of the bare ground area (mostly sand dunes), nor desertification processes in the study area in the past 30 years.

Table 8.   Classification of land-use change trajectories.

Level 1 classes	Level 2 classes	Area (ha)	%
Unchanged	Grass/woodland	2177.5	34.5
	Salty grass	7.7	0.1
	Water body	47.0	0.7
	Bare ground	478.0	7.5
	Sub-total	2710.2	42.8
Human-induced	Old cultivation	124.0	2.0
	New cultivation	675.4	10.7
	Reservoirs/ponds	311.8	4.9
	Sub-total	1111.2	17.5
Natural	Grass/woodland	318.4	5.0
	Flooded	1618.0	25.5
	Bare ground	578.2	9.1
	Sub-total	2514.6	39.7
Total		6336.1	100.0

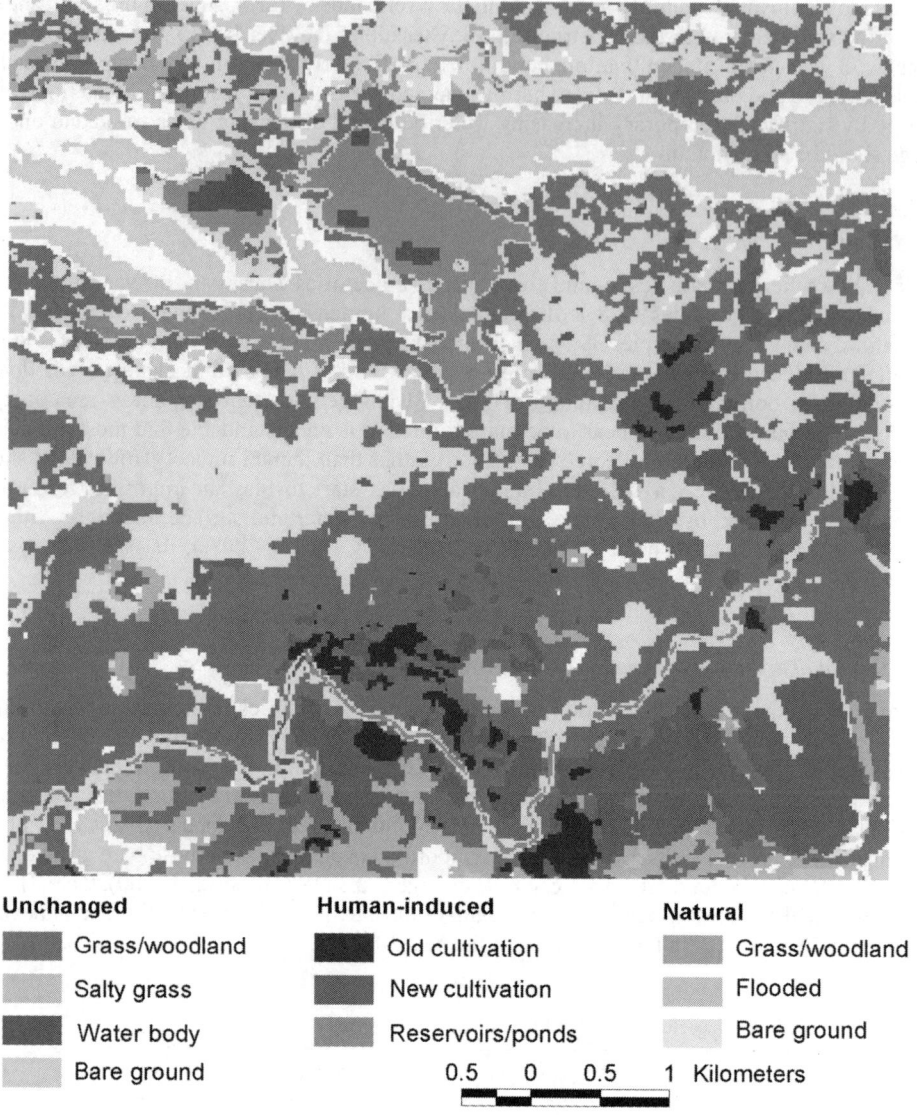

Figure 5.   The trajectories of land-use change in the past 30 years.

4   CONCLUSION

In this paper, we have demonstrated a project that studies the spatio-temporal pattern of land-use change in the fragile ecosystem of arid environment of China. Multi-scale and multi-temporal remotely sensed images, including Landsat MSS, TM, ETM and SPOT HRV, were used to analyse the land-use change trajectories and to identify the human-induced environmental change, such as cultivation and dam/reservoir construction. The study shows that using the auto-classification approach an overall accuracy of 85%–90% with a Kappa coefficient 0.66–0.78 was achieved for individual image classification. The temporal trajectories of land-use change were established and their spatial distribution was analysed. It is shown that the natural forces such as flood were still dominating the environmental processes of the study area, with unchanged (42.8% of the total area) and indecisive changes between land cover types (39.7%). However, human-induced changes, constituting 17.5% of the total area, started to play more important role in the environmental change. This was demonstrated by the triple increase of croplands in the last 6 years and overall 5% of the area was permanently submerged due to the construction of dam and reservoirs since the 1980s.

The study has demonstrated an efficient and practical approach for analysing and modelling environmental changes in an arid environment by integrating multi-temporal and multi-scale remotely sensed data from various sources. In this study, we extended the concepts and methodology of change trajectory analysis to the study of an arid environment. With this approach, human-induced changes, which represent an irreversible impact on the environment, can be extracted from changes caused by natural forces. Regional management can therefore focus on the human-induced changes that have been identified, and on their impact, to achieve the goals of environmentally sound land management practices and sustainable economic development.

Further studies will be focused on the methodology for assessing and modelling the uncertainty of the change trajectory analysis and its impact on environmental change applications. Through further studies on the spatio-temporal patterns and relationships between cover types and local land-use practices, the ultimate goal of this study is to assess human activities and predict environmental responses, particularly the long-term irreversible consequences, so that better, science-based management decisions on land management and development can be made.

ACKNOWLEDGEMENT

This study is supported by the Earmarked Research Grant (Project No. HKBU 2086/01P), Research Grants Council, Hong Kong. The authors would like to thank the staff of the Institute of Geography, Chinese Academy of Sciences, particularly Mr Alishir Kurban for their support during the fieldwork. The constructive criticism and comments of anonymous referees are also acknowledged.

REFERENCES

Chen, X., 2002, Using remote sensing and GIS to analyse land cover change and its impacts on regional sustainable development. *International Journal of Remote Sensing*, 23(1), 107–124.

Dai, X.L. and Khorram, S., 1999, Remotely sensed change detection based on artificial neural networks. *Photogrammetric Engineering & Remote Sensing*, 65(10), 1187–1194.

Houhoulis, P.F. and Michener, W.K., 2000, Detecting wetland change: a rule-based approach using NWI and SPOT-XS data. *Photogrammetric Engineering & Remote Sensing*, 66(2): 205–211.

Jensen, J.R., Cowen, D.J., Althausen, J.D., Narumalani, S. and Weatherbee, O., 1999, An evaluation of the coastwatch change detection protocol in South Carolina, in Lunetta, R.S. and Elvidge, C.D. (eds.), *Remote Sensing Change Detection: Environmental Monitoring Methods and Applications*, London: Taylor & Francis, pp75–87.

Larsson, H., 2002, Analysis of variations in land cover between 1972 and 1990, Kassala Province, Eastern Sudan, using Landsat MSS data. *International Journal of Remote Sensing*, 23(2): 325–333.

Lillesand and Kiefer, 2000, *Remote Sensing and Image Interpretation*, 4th Edition, New York: John Wiley & Sons.

Luque, S.S., 2000, Evaluating temporal changes using Multi-Spectral Scanner and Thematic Mapper data on the landscape of a natural reserve: the New Jersey Pine Barrens, a case study. *International Journal of Remote Sensing*, 21(13&14): 2589–2611.

Macleod, R.D. and Congalton, R.G., 1998, A quantitative comparison of change-detection algorithms for monitoring eelgrass from remotely sensed data. *Photogrammetric Engineering & Remote Sensing*, 64(3): 207–216.

Maldonado, F.D., dos Santos, J.R. and de Carvalho, V.C., 2002, Land use dynamics in the semi-arid region of Brazil (Quixaba, PE): characterization by principal component analysis (PCA). *International Journal of Remote Sensing*, 23(23): 5005–5013.

Mas, J.F., 1999, Monitoring land-cover changes: a comparison of change detection techniques. *International Journal of Remote Sensing*, 20(1): 139–152.

Masek, J.G., Lindsay, F.E. and Goward, S.N., 2000, Dynamics of urban growth in the Washington DC metropolitan area, 1973–1996, from Landsat observations. *International Journal of Remote Sensing*, 21(18), 3473–3486.

Mertens, B. and Lambin, E.F., 2000, Land-cover-change trajectories in southern Cameroon. *Annals of the Association of American Geographers*, 90(3): 467–494.

Michener, W.K. and Houhoulis, P.F., 1997, Detection of vegetation changes associated with extensive flooding in a forested ecosystem. *Photogrammetric Engineering & Remote Sensing*, 63(12): 1363–1374.

Miller, A.B., Bryant, E.S. and Birnie, R.W., 1998, An analysis of land cover changes in the Northern Forest of New England using multitemporal Landsat MSS data. *International Journal of Remote Sensing*, 19(19): 245–265.

Pereira, V.F.G., Congalton, R.G. and Zarin, D.J., 2002, Spatial and temporal analysis of a tidal floodplain landscape-Amapá, Brazil-using geographic information systems and remote sensing. *Photogrammetric Engineering & Remote Sensing*, 68(5): 463–472.

Petit, C., Scudder, T. and Lambin, E., 2001, Quantifying processes of land-cover change by remote sensing: resettlement and rapid land-cover changes in south-eastern Zambia. *International Journal of Remote Sensing*, 22(17): 3435–3456.

Prakash, A. and Gupta, R.P., 1998, Land-use mapping and change detection in a coal mining area–a case study in the Jharia coalfield, India. *International Journal of Remote Sensing*, 19(3): 391–410.

Richards, J.A., 1995, *Remote Sensing Digital Image Analysis*, 2nd Edition, Berlin: Springer-Verlag.

Roy, P.S. and Tomar, S., 2001, Landscape cover dynamics pattern in Meghalaya. *International Journal of Remote Sensing*, 22(18): 3813–3825.

Singh, A., 1989, Digital change detection techniques using remotely-sensed data. *International Journal of Remote Sensing*, 10(6): 989–1003.

Sunar, F., 1998, An analysis of changes in a multi-date data set: a case study in the Ikitelli area, Istanbul, Turkey. *International Journal of Remote Sensing*, 19(2): 225–235.

Ustin, S.L. and Xiao, Q.F., 2001, Mapping successional boreal forests in interior central Alaska. *International Journal of Remote Sensing*, 22(6): 1779–1797.

Weng, Q., 2001, A remote sensing–GIS evaluation of urban expansion and its impact on surface temperature in the Zhujiang Delta, China. *International Journal of Remote Sensing*, 22(10), 1999–2014.

Yang, X. and Lo, C.P., 2002, Using a time series of satellite imagery to detect land use and land cover changes in the Atlanta, Georgia metropolitan area. *International Journal of Remote Sensing*, 23(9): 1775–1798.

Zhang Q., Wang, J., Peng, X., Gong P. and Shi, P., 2002, Urban built-up land change detection with road density and spectral information from multi-temporal Landsat TM data, *International Journal of Remote Sensing*, 23(15): 3057–3078.

Spatial analysis and decision support systems

Advances in Spatial Analysis and Decision Making, Li, Zhou & Kainz (eds)
© 2004 Swets & Zeitlinger, Lisse, ISBN 90 5809 652 1

Development of a spatial decision support system to support road network traffic transport analysis and management at three gorges hydroelectric project

Quan Bao & Vincent C. Tao
GeoICT Lab, York University

Fulin Bian
Wuhan University, China

ABSTRACT: To improve the usefulness of GIS as a decision support tool, two needs are apparent, i.e. (a) methods allows decision-makers to easily select alternatives most closely aligned with their priorities across a number of relevant criteria and (b) recognition of multiple participants in most decision-making processes. Based on these requirements, a framework of spatial decision-support system is designed for traffic analysis and management. Three kinds of decision functions are provided, i.e. best path analysis, least transport expenditure analysis and break-down alarm analysis.

1 INTRODUCTION

Nowadays, Geographical Information Systems (GIS) are powerful and useful as means of information, visualization and research tools. However, beyond basic levels of decision support, GIS remains largely external to the decision-making process. This suggests that despite increased analytical sophistication, most GIS software is more suited to providing limited types of output (maps, tables, etc.) than as a tool to support, at anything other than a superficial level, tactical or strategic decision-making processes. To improve the usefulness of GIS as a decision support tool, two needs are apparent. First, decision-makers require methods that allow them to easily select alternatives most closely aligned with their priorities across a number of relevant criteria. Second, it is necessary to explicitly recognize that most decision-making processes involve multiple participants. Since problem solving is often characterized by multiple and conflicting objectives, methods that contribute toward more a completely systematical SDSS system are required.

The Road Network Traffic Transport Analysis and Management System (RNTTAMS) of Three Gorges Hydroelectric Project (TGHP) processes an enormous amount of data (both historic and current) related to all aspects of road traffic. The data are generated by many different agencies such as the planner and designer, and stored in just as many different formats. The road management department is beginning to explore the use of the GIS in their decision-making processes by generating maps that convey information gleaned from their respective databases. As they try to use data from other departments, they increasingly encounter problems in accessing and interpreting the data. For example, when huge and heavy hydroelectric equipment or materials for TGHP need to be carried from outside the construction sites, the Road Traffic Transport department has to make a decision about how to transport the equipment. That means that the delivery route selected must be the best at the least cost, and the procedure must be safe, and so forth. All of these decisions are based on the information on the load of the road surface, complicated topographical information, road planning and design parameters, road accessories, etc. The Road Traffic Transport department has tremendous difficulty in accessing and integrating the data in a manner consistent with their existing tools and processes. They would like to integrate data on climate, topography, geology and road project planning and design parameters in their spatial analysis of the Road Traffic Transport Management and Maintenance System, as well as decision-making. Often, the applications of this multidisciplinary database require a whole corps of domain specialists to extract meaningful interpretations.

In this paper, we present an extensible SDSS software architecture and demonstrate the architecture used by researchers, analysts and public officials working on traffic transport analysis, traffic transport management, traffic system maintenance, planning and design.

2   FRAMEWORK FOR SPATIAL DECISION SUPPORT SYSTEM (SDSS)

A Spatial Decision Support System (SDSS) could be viewed as an integrated computer software and hardware package realized, in spatial context, using GIS functionality on database management, knowledge and criteria management, analytical modelling and spatial-tabular display, together with a framework for adopting the expert knowledge of decision-makers.

There are five key components in the SDSS, which differ quite considerably from a GIS. There are:

- Data base management
- Analysis logic (objectives, constraints, decision rules)
- Spatial display
- Report generation
- User's interface

Whereas a GIS offers the above functionality as loosely coupled set of tools or primitives, in a SDSS all of these functionalities appear to be a seamless entity.

We have developed a logical architecture for spatial decision support systems. The decision-making process begins by combining and organizing data into pieces of information. Multiple pieces of information are then examined and combined to discover or create knowledge, which is the basis upon which a decision should be made.

The abstract, five-system architecture of Figure 2 shows the logic architecture we have created to support this view of the decision-making process.

Data management system   The DBMS used in general management is not adequate for managing spatial data directly. The reasons are: (1) it cannot provide the function for querying spatial location; (2) it does not have a complicated graphic display function. The results of a spatial query

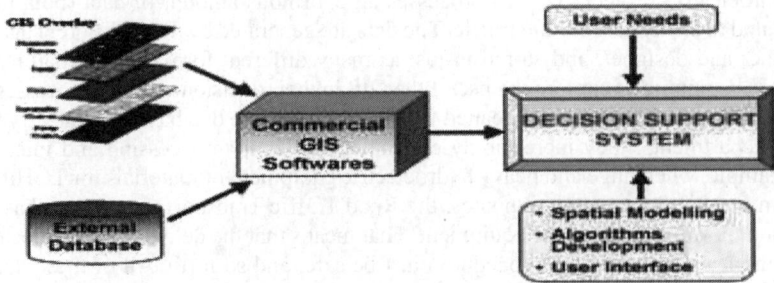

Figure 1.   Components of SDSS for RNTTAMS.

and analysis are not only a text report, but also a display of graphics. General DBMS do not have such capabilities.

In this data management system, we used general DBMS (Foxpro) to manage attribute data and used special commercial software to manage spatial data (ArcView). Users can both access these two kinds of database and access the database separately by different uses. Figure 3 shows the structure for the database management model.

The data management system contains distributed spatial, constraint and relational databases. The purpose of this system is to provide transparent access to either local or remote data without concern for data formats. Figure 4 shows the structure of the system's database.

Knowledge management system   Knowledge is created or discovered by combining information in new ways. Systems that provide or discover domain-specific knowledge are implemented in the knowledge management sub-system. Examples might include data mining and knowledge discovery algorithms as well as simulation models. Systems at this level might also provide a more traditional domain-specific regression analysis of information (or data) generated (stored) at the data management system. The intention is that decision-makers will interact with this system, via the User Presentation interface, to build and gather domain-specific knowledge. The tools in

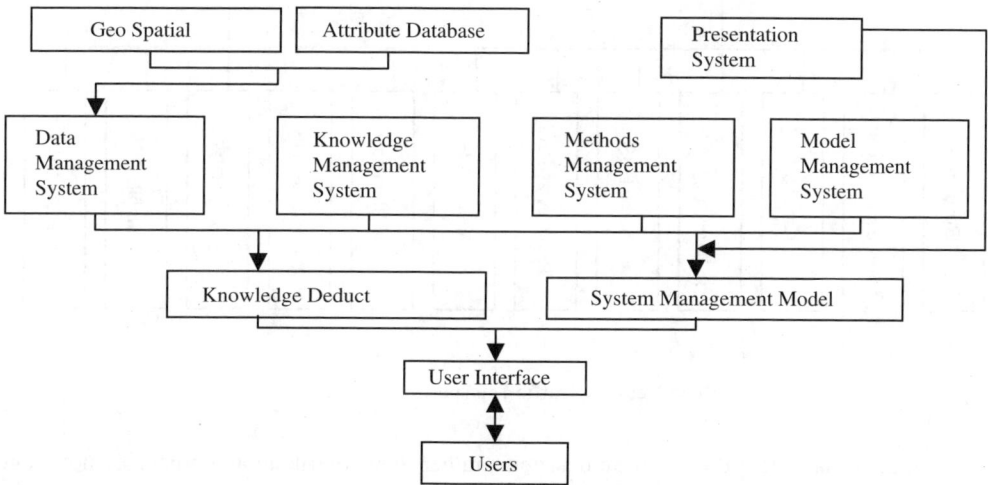

Figure 2.   Five-system software architecture for a Geospatial decision support system.

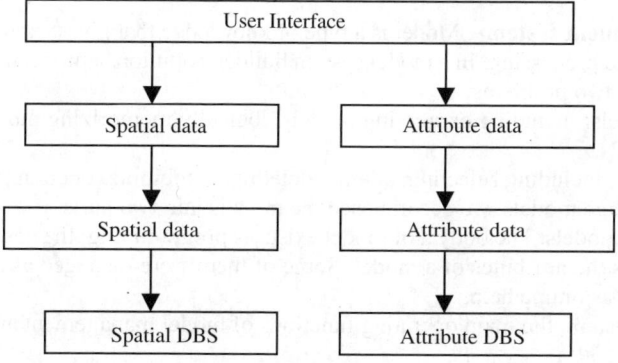

Figure 3.   Database management.

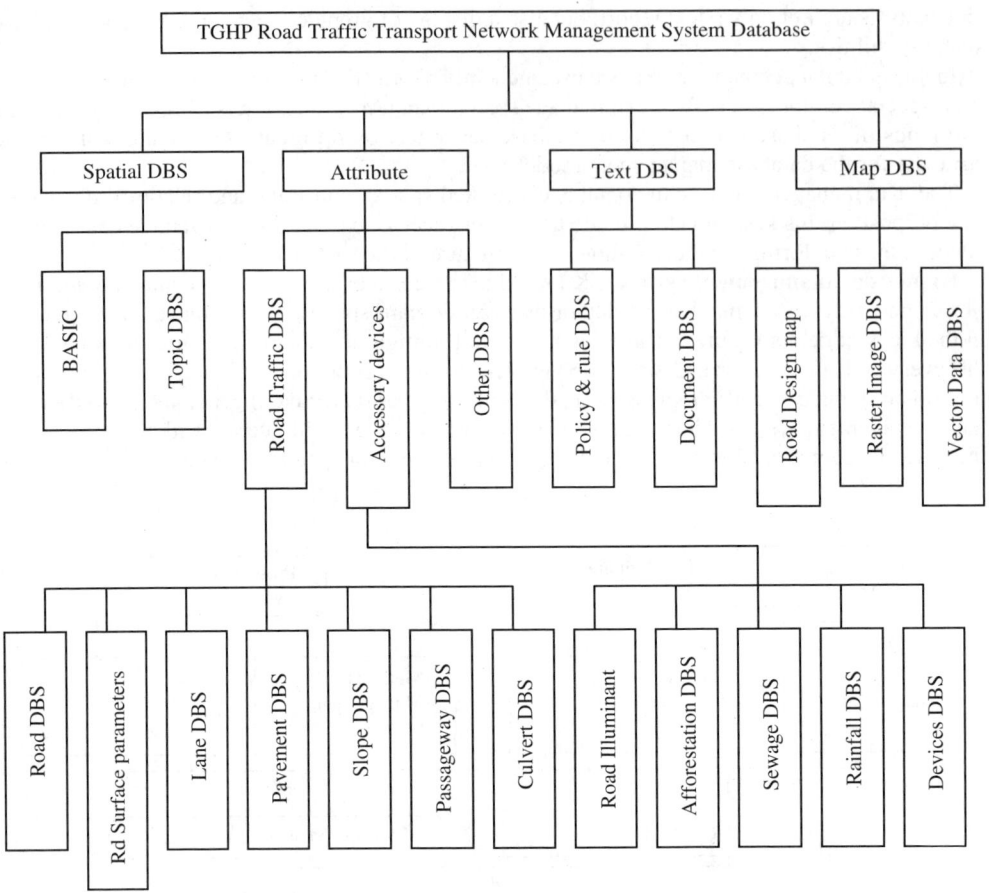

Figure 4.   The database of a Road Network Traffic Transport System.

knowledge management do not make decisions; rather they contribute and organize knowledge that is used in the decision-making process. (An additional system could be placed on top of the knowledge management system to create an expert system, but that is not the goal of this paper).

Methods management system   Method means the criteria or criterion for processing data and models. We realize the system with an interacting operation to dynamically restructure and manage models.

Model management system   Model is a type of knowledge that gives expression to the procedure of analysis and processing. In a model system builder's opinion, a model management system must mainly solve two problems:

Memorize models: including expressing models, logically memorizing models and physically memorizing models.

Operate models: including selecting, adding, deleting, combining, operating, and so forth.

In order to manage models, we decomposed the models into two parts: the body of models and the description of models. The body a of model exists as programming; the description of a model can be regarded as the attributes of a model. Some of them were managed as a cell in a database and others existed as online help.

In this SDSS system, the main operating functions of model management are:

- Create and construct new models
- Modify models in order to suit the change of data range

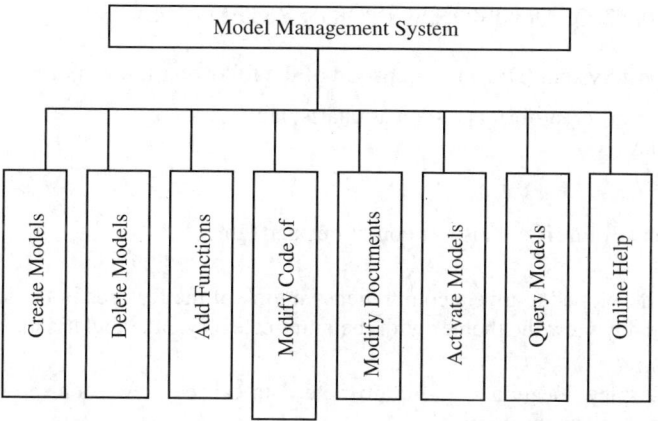

Figure 5. Model management structure.

- Redefine or recreate models
- Query models in order to get some relative information
- Make fuzzy queries in order to get model assemble of which support some functions
- Model delete
- Model active
- Model online help.

A model management system helps in building operational models that satisfy a range of criteria in multiple stages until the desired outcome or a satisfactory result is achieved. At any stage, if the user is not satisfied with a certain result, he can simply go one step behind and modify the earlier criteria and then can proceed to next step. At each step, the results can be viewed on the screen, allowing for an interactive approach. Figure 5 shows the model management structure for the system.

Presentation system   A Presentation System has to comply with the requirements of the interactive, iterative and participative involvement of a decision-maker. A decision-maker, who could be an expert in his sphere of activity, need not always be a GIS expert. The system should provide an interface which

- Is easy to use in order to be effective;
- Has communication systems capability;
- Facilitates the easy selection of parameters, data output, and so forth, without forcing a user to refer documents;
- Is transparent, to facilitate visualization of the process represented within the analysis model.

Aim   The aim of the research is to provide a Spatial Decision Support System in a GIS environment to make the decision-making process qualitatively effective.

Objectives
- To provide the users with an accurate spatial view of the road and its related information, such as the design blueprint, parameters, accessories, etc.
- To provide the users with detailed road data and road-related data on a desktop in a GIS environment.
- To assist the users in finding the possible path and methods depending on several parameters, such as the road's load, road surface condition, slop, culvert, road illuminant, road maintain, etc.

3   DEVELOPMENT OF DECISION SUPPORT SYSTEM (DSS)

A Decision Support System (DSS) is comprised of the following five major modules:

- Data displays in different formats, such as charts, tables, etc.
- Attribute displays
- Query builder
- Model builder
- Dynamic displays of traffic transport conditions and jams.

Data display
This module, developed to give a comprehensive look of the features in the study area, allows the user to selectively view the themes with their sub-category and tabular data simultaneously.

Attribute display
In Attribute Display, location-specific attribute data can be viewed on the screen by simply placing the cursor over the position.

Query builder
Three kinds of query shells were developed; namely, Simple, Complex and Compound query shells. Simple query allows building a condition on a particular theme alone. Complex query allows further conditions with connectors like 'AND', 'OR' on the same theme. Apart from all of the capabilities of a complex query, compound query allows a condition to be built on more than two different themes with connectors.

Analysis and decision-making functions
With the help of analysis software, we combined an attribute DBS (Database) with a spatial DBS, and generated a comprehensive analysis and decision support sub-system for the Three Gorges Road Network Traffic Transport.

3.1   Best path analysis

The Three Gorges Hydroelectric Project is a huge hydroelectric project. Limited by characteristics such as provisionality, seasonality, the need to deliver hugely heavy equipment, the use of one-way heavy container trucks, a great number of deliveries of concrete, and so forth, it is difficult to select a path to delivery and road management. In the Three Gorges construction site, there are many bridges and tunnels instead of roads. According to the different kinds of equipment, users have to take into consideration both the minimum amount of expenditure and most executable path. Meanwhile, a decision-maker has to consider the load and width of the roads, the load and width of the bridges; the width and height of the tunnels, etc. to ensure that the equipment will be delivered safety.

3.2   Least transport expenditure analysis

During the constructing process of Three Gorges Project, there were some problems concerned sending a great deal of basic materials to several dam constructing sites. The main function of this model is to figure out the least transport expenditure path for user by spatial analyst, query, and calculation. The result was shown at screen or printed out for user as reference.

3.3   Break-down alarm analysis

When there is a traffic jam, this model is used to determine the location. It can rapidly find a dispersed transport line and assist the control centre to control the transport.

4   CONCLUSION

The DSS is effective and satisfies the stated objectives. However, it can certainly be further refined. The presented framework for Spatial Decision Support Systems has been adopted to facilitate

Spatial Decision Support for Road Traffic Transport Management. The most important aspect was the knowledge and methods, or the modelling of the impact, and the customization of the GIS tools to realize such a system. The approach thus presented can be very successfully adopted for a wide spectrum of planning and decision support problems in a spatial context.

REFERENCES

Chomicki, J. and Revesz, P. Z. 1999. Constraint-based Interoperability of Spatiotemporal Databases, *Geoinformatica*, 3(3): 211–243.

Densham, P. J. 1991. Spatial Decision Support Systems, In D. J. Maguire, Goodchild, M. F. and Rhind, D. W. ed., Geographical Information Systems: Principles and Applications, 403~412, New York: John Wiley.

ESRI, http://www.esri.com/software/

Geographic Resources Analysis Support System (GRASS), http://www.geog.uni-hannover.de/grass, accessed March, 2002.

Goodchild, M. F., Haining, R. P. and Wise, S. M. 1992. Intergrating GIS and Spatial data analysis: problems and possibilities. *Int. J Geographical Information Systems*, 6(5).

Goodchild, M. F. 2000. Perspective: Browsing metadata, where do we go from here? *GeoInfo Systems* 10(5): 30–31.

Jankowski, Piotr. 1993. Integrating GIS and multi-criteria decision-making methods. *Journal of International Geographic Information System* 7(5).

Turban, E. 1990. Decision Support and Expert Systems: Management Support System. 2nd ed. New York: MacMillan Publishing Company.

Oil and gas pipeline route planning using geo-spatial information system analysis

Mahmoud Reza Delavar & Fereydoon Naghibi

Department of Surveying and Geomatic Engineering, Engineering Faculty, University of Tehran, Tehran, Iran

ABSTRACT: A prototype of least-cost pipeline routing was performed using various data and GIS analysis. Ahvaz-Marun oil pipeline in the southwest of Iran was chosen for the development of the prototype. The pipeline carries Ethan gas from Marun Petrochemical Company (MPC) to Ahvaz. A model was developed incorporating the length of the pipeline, topography, geology, land use, streams, wetlands, roads and railroad crossings to identify a least-cost pathway. Geo-spatial Information System (GIS) analysis based on a cartographic modeling approach was used for spatial modeling and overlay. Costs associated with terrain conditions, geology and land use were given from actual costs in the section of design of pipeline Iranian National Oil Company (NIOC). The length and cost associated with existing pipelines made using traditional approaches were compared with those of the least-cost pathway obtained through GIS best path analysis using fuzzy logic, index overlay and Boolean overlay techniques. The best results were those obtained using the fuzzy logic approach. The existing path of the pipeline is 34.5 km long, and the least-cost pathway is 35.2 km long. Although longer in length, the least-cost pathway is 29% cheaper than the existing pipeline path. The results of this analysis demonstrate the benefits of integrating geo-spatial data within a GIS environment that has been used as a spatial decision support system for pipeline routing.

1 INTRODUCTION

GIS is a science and technology which can integrate different geo-spatial data from various sources for route design processes through spatial analysis. It is also used to unify project processes, including environmental parameters and project team decision-making (Naghibi, 2003). A number of studies have already been performed on designing pipeline routes using GIS. One of the optimal routing methods based on the discrete grid-net and Dijkstra's algorithm was developed at the University of Helsinki. This method had been applied to the routing of the gas pipeline to western Finland (Sarkka and Esko, 1999).

Another study was conducted by the TransCanada Pipeline Ltd. TransCanada selected and used a GIS approach to automate the route selection for an expansion line. The process had been benchmarked against TransCanada's standard route planning approach for the selection of alternative routes, and was chosen to improve the speed and consistency of route selection (Montemurro and Gale, 1996). Also, a prototype for least-cost analysis was performed for pipeline routing using remotely sensed data and GIS analysis. A small section of Caspian oil pipeline was chosen for development of the prototype (Hicken and Krumbach, 1998).

This paper describes one of the efforts toward the investigating innovative approaches to pipeline routing. The present study was initiated to demonstrate the use of various data from different sources and GIS analysis for developing a least-cost design for pipeline placement using Ahvaz-Marun pipeline as an example (Delavar and Nagibi, 2003 and Naghibi, 2003). The study area selected is in

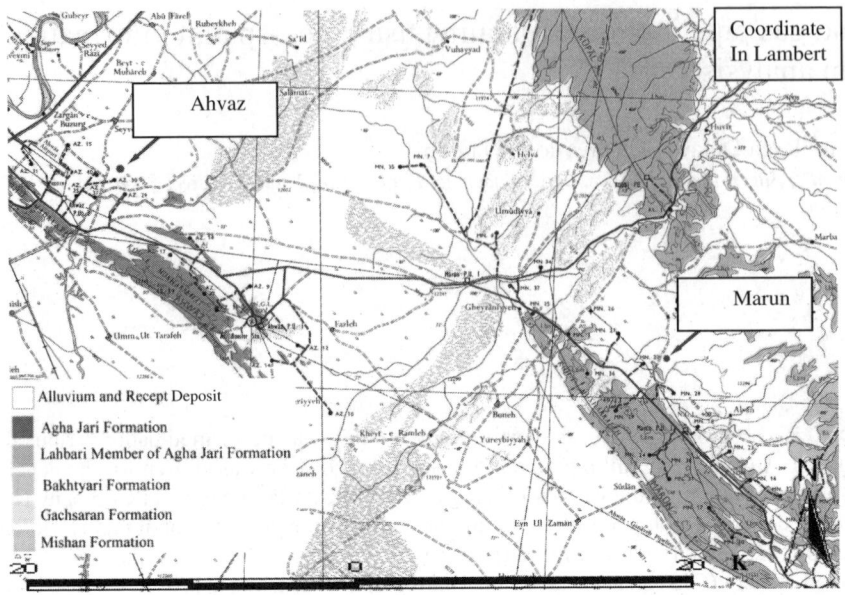

Figure 1.   Study area on a geologic map at 1:250,000 scale, Khuzestan Province in southwest Iran.

Khuzestan Province in southwest Iran. The site lies in the southern Zagros Mountains near the Persian Gulf and is an area of low relief along the proposed pipeline route (Figure 1).

2   LEAST-COST PATHWAY ANALYSIS

The computation of least-cost paths is perhaps one of the most potentially useful applications of GIS. Finding a minimum path over a surface partitioned into regions of different frictions to movement has two aspects (DeMers, 2002 and Douglas, 1994):

– Creation of an accumulated cost surface; and
– Tracing a slope line down the accumulated cost.

The cost surface in a GIS environment is a grid where the values associated with the cells are used as weights for calculating the least-cost paths. These weights represent the friction or difficulty involved in crossing the cell and may be cost, time, distance or risk (Stefanakis and Kavouras, 1995).

The integration of an accumulated cost surface from a cost of passage surface requires a spreading function that begins at the previously defined destination of the path. At the start of the procedure, only the destination point cell has a defined value of accumulated cost and the spreading algorithm searches its eight neighbor cells, stopping on the first one not previously assigned a value. From that first cell, the algorithm begins another search of its eight neighbors to find those with defined values of the accumulated cost. For each of them the cost of the cell is added and the smallest cost is recorded as the accumulated cost of the cell. This procedure is repeated until every cell has been assigned a cumulative cost value.

Once a cumulative cost surface is created, a new surface can be created, assigning each cell a number indicating the least-cost direction to the ending point, following slope lines over the cumulative cost surface. This best-direction surface is used to draw the least-cost path. This procedure is an adaptation of algorithms, such as Dijkstra's, which are traditionally applied to solve network problems (Berry, 1993; DeMers, 2002; Dijkstra, 1959 and Tomlin, 1990; Naghibi et al., 2003).

Figure 2 shows the optimum path determination process in this research. These steps are elaborated in the next sections.

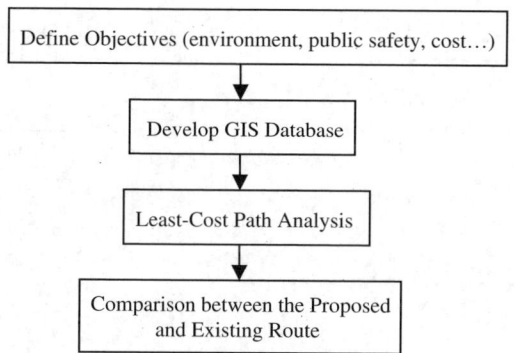

Figure 2.   Route selection process implemented.

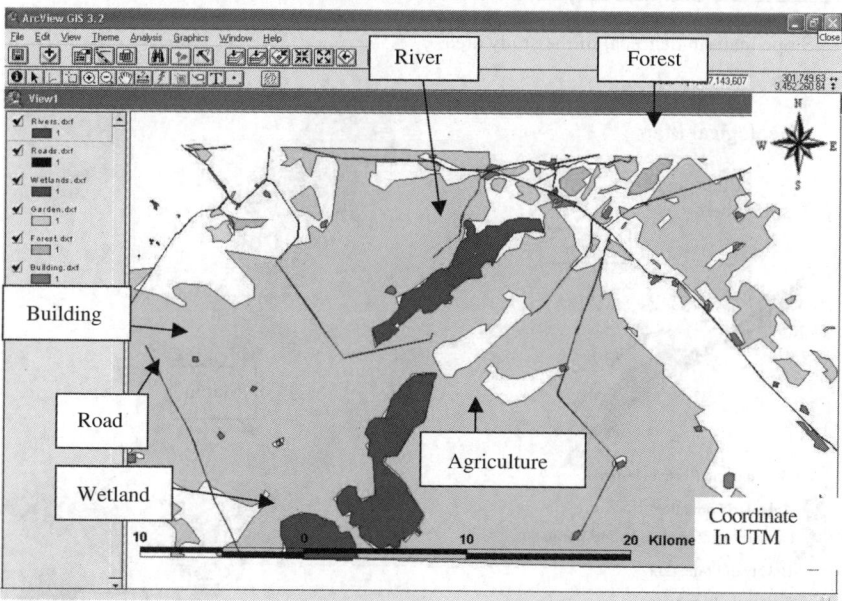

Figure 3.   Land use/Land cover map of the study area.

3   DATA INPUT

Maps and field work are required for pipeline routing, design and construction. For this route, topographic maps at a scale of 1:25,000 produced by Iranian National Cartographic Center (NCC) were used. The available geologic maps of the area are at a scale of 1:100,000 and 1:250,000 produced by NIOC. The area of interest for this analysis is shown in Figure 3. The least-cost pathway analysis, using various data and GIS analysis, was intended to confirm the best pipeline route within this site.

3.1   *Input of topographic and geologic data*

Topographic and geologic data of the Ahvaz-Marun pipeline area were prepared in a GIS-ready format and used as input to the GIS database. The locations of roads, railways, wetlands, forests and drainage features were derived from the topographic map layer. This layer that was produced by NCC is the base for a National Topographic Database (NTDB) in the development of a national GIS for Iran (Figure 3).

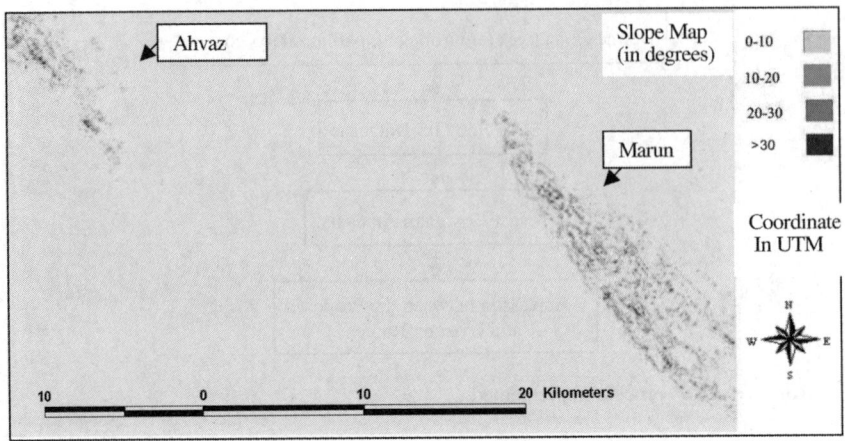

Figure 4.   Slope map (in degrees) of the study area.

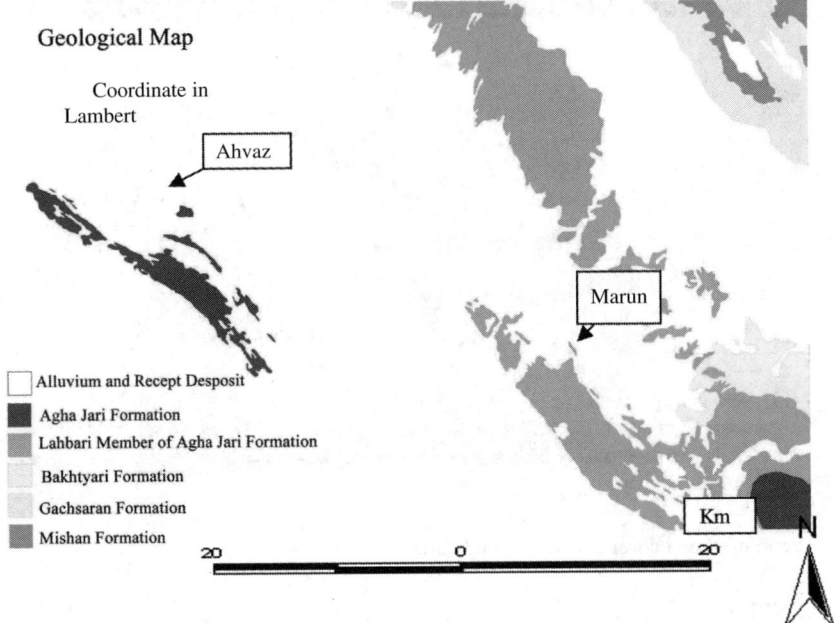

Figure 5.   Formations derived from the geological map.

In this project, a Digital Elevation Model (DEM) was produced from the elevation data. The slope map, derived from the DEM, is shown in Figure 4. It was used as input to the least-cost pathway analysis. Lighter tones indicate steeper slopes. Most of the steeper slopes found near Marun and Ahvaz and other areas have low slopes of between zero and five degrees.

Geological formations in the area under study are shown in Figure 5. These boundaries between geologic units were extracted in digital form from the geological map and incorporated into the GIS database. These geologic units were divided as very hard rock, hard rock, soft rock and very soft rock based on the descriptions in the geological map legend.

Faults were also considered in the analysis (least-cost pathway) because the crossing of pipelines over them must be avoided (Figure 6).

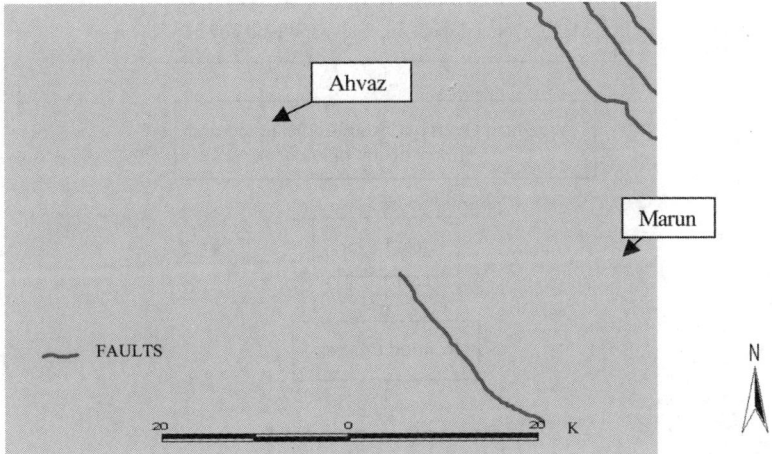

Figure 6.   Location of faults derived from the geological map.

4   PIPELINE ROUTING CRITERIA

Traditional route selection begins with the definition of a straight line connecting a start point to an end point. This line is then modified to accommodate physical, environmental and economic features that meet the objectives of the project.

Therefore, the factors influencing the selection of pipeline routes are technical and engineering requirements, environmental considerations and population density. However, these factors are chosen to balance engineering and construction costs against environmental costs and future liability (Montemurro and Gale, 1996).

The engineering and technical considerations used in this research include pipeline length, topography, surface geology, river and wetland crossings, road and railroad crossings and the proximity to large population centers. High-relief terrain would result in higher construction costs and increase the need for pump stations (Hicken and Krumbach, 1998).

Cost factors used in the least-cost path analysis were calculated from a pipeline project of NIOC and its normalized baseline cost. Using the costs associated with the pipeline project, percentages over the baseline cost were calculated for construction in rock, the clearing of brushes and trees, the crossing of rivers and railroads and passing through agricultural land and wetlands. Estimates were made of the slope ranges that are associated with four categories of terrain including flat, rolling, sharp, choppy and rough that are commonly used by pipeline estimators. The cost of a pump station, however, has not been considered in this analysis. As it is not generally considered desirable to route pipelines through urban and industrial areas, these areas were assigned high costs above the baseline value.

5   IMPLEMENTATION

The pipeline project had no constrained point for passing except in Ahvaz (destination) and Marun (source). The objective of the cost pathway analysis was to compare the cost of the existing pipeline route to a least-cost pathway between the two points. The analysis was performed using 50 m resolution cells (for raster layers). In other words, in each GIS layer, a value was calculated and assigned for each 50 m × 50 m cell in the study area (Naghibi, 2003). Figure 7 shows the implementation phase, which is discussed below.

The analysis was accomplished by inputting map data into a GIS. The GIS analysis is used for spatial modeling and data overlay. The GIS provided a framework for developing and overlaying

215

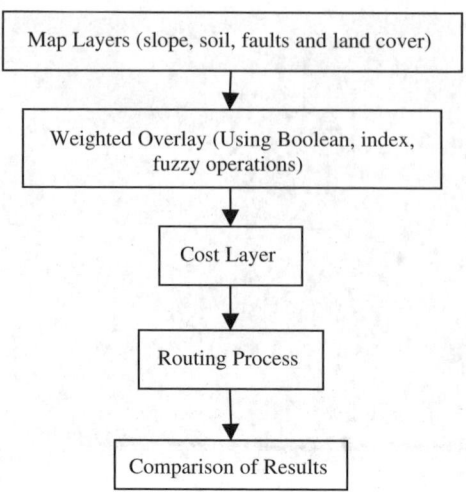

Figure 7.   The implementation phase.

all of the input layers and for carrying out the spatial analysis. ArcView and Arc/Info software were used for the display and the spatial analysis, respectively.

The topographic, geologic and land use data were used to develop a least-cost pathway for the design of the pipeline. The least-cost analysis was performed by assigning cost factors associated with the crossing of slopes, streams, wetlands, roads, railroads, rock, agricultural land, urban and industrial areas; developing a cumulative cost surface; and then calculating a path of least resistance across the surface. The location of the stream, road, and railroad crossings were digitized from the topographic map. The areas where rock was likely to be encountered were defined from the geologic map. A land use map, produced from NTDB, was used to identify agricultural land and urban areas. Pipeline construction costs associated with terrain conditions, geology and land use were calculated from actual pipeline construction projects. Very high values were assigned to urban and industrial areas; to the crossing of road, railway and river features; and to areas outside the defined boundary of the layers.

Least-cost pathway analysis is a modification of algorithms that are traditionally used in GIS for drainage basin analysis, with the path constrained to flow through specific nodes to specified endpoints. The first step in the least-cost pathway analysis is the generation of a single weighted surface layer, based on input layers generated from a table of weights that contribute to the cost of traversing between two points (cells) (Berry, 1993; DeMers, 2002; Dijkstra, 1959; Tomlin, 1990; Stefanakis and Kavouras, 1995).

The combined weighted surface is analogous to the topographic surface, in that it has peaks (areas of relatively high cost) and valleys (areas of relatively low cost). The cost surface is shown in Figure 8. The cost layer has been created using different weighted overlay approaches including Boolean, fuzzy and index overlay (Tables 1 to 4). The cost surface created from the fuzzy logic approach presented an optimum area for a pipeline route compared to the other approaches.

The darkest tones show the areas with the highest costs. The highest costs are in urban areas and in large bodies of water and roads. Moderate costs are in forests and wetlands with steep slopes. The lowest costs are in areas with bare ground, dry grass, less dense native vegetation and agriculture. From the weighted surface, an accumulative cost surface was generated. Least-cost pathways could follow the resulting surface. Accumulative cost surface utilizes a single point of origin and accumulates the sum of cells as one moves from the origin. The accumulative cost surface consists of the sum of cells between any location on the surface and the point of origin.

Sums accumulate as they radiate out from the point of origin. The accumulative cost surface does not show which cell to return to or how to get there. A separate surface has to be generated

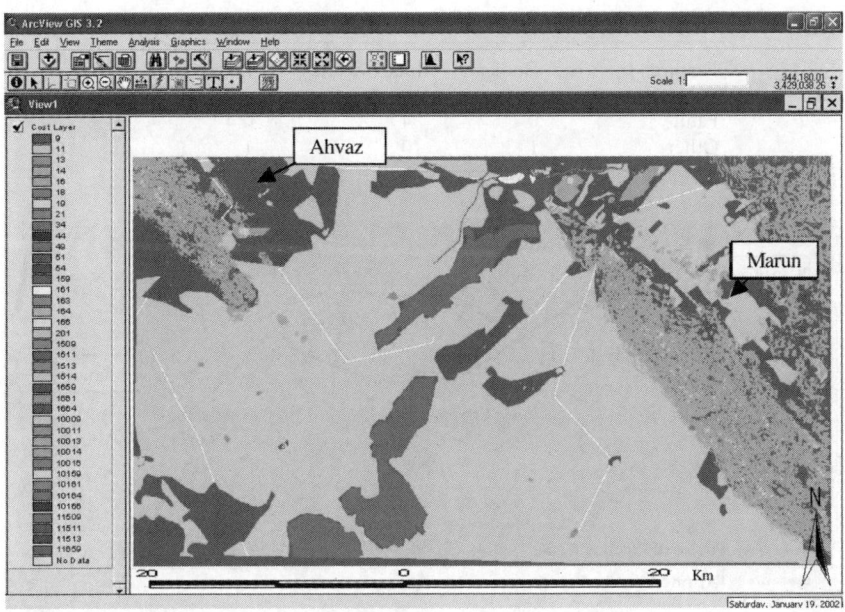

Figure 8.   The cost surface.

Table 1.   Weight of the geologic units.

Feature	Boolean	Index overlay	Fuzzy logic
Very soft rock	1	10	1
Soft rock	1	8	0.8
Hard rock	1	5	0.5
Very hard rock	0	4	0.4

Table 2.   Weight of the land cover features.

Feature	Boolean	Index overlay	Fuzzy logic
Wasteland	1	10	1
Agriculture	1	9	0.9
Forests	1	7	0.7
Wetlands	0	5	0.5
Roads & railways	0	3	0.3
Rivers	0	2	0.2
Population centers	0	1	0.1

Table 3.   Weight of the slope map.

Feature	Boolean	Index overlay	Fuzzy logic
0–10	1	10	1
10–20	1	8	0.8
20–30	1	6	0.6
>30	0	0	0

Table 4. Weight of areas with faults.

Feature	Boolean	Index overlay	Fuzzy logic
Fault	0	4	0.4
Other	1	10	1

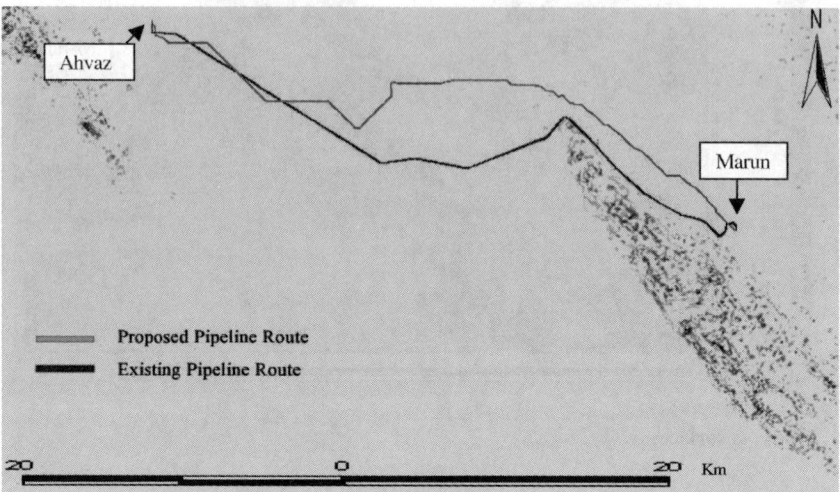

Figure 9. Comparison between proposed and existing route.

Table 5. Number of cells of the existing and the least-cost pathways that are crossing various features.

Feature	Existing route	Least–cost pathway
Urban and industrial	8	5
Roads	23	9
Rivers	0	0
Wetlands	46	12
Forest	0	0
Very hard rock and hard rock	203	175
Slope < 20°	823	952

to return a surface with a value ranging from 0 to 8 that can be used to reconstruct the route to the origin. Each value (0 to 8) identifies which neighboring cell to move into to get back to the origin. Once the accumulative cost and direction surfaces are created, a least-cost path route can be derived from any designated destination cell or from the endpoint (Berry, 1993; DeMers, 2002; Dijkstra, 1959; Sarkka and Esko, 1999).

There are a total of 1034 cells along the existing pipeline route and 1084 cells along the least-cost pathway (Figure 9). Table 5 shows that the existing pipeline route traversed a larger number of urban area, roads and wetlands cells than did the least-cost pathway.

6 CONCLUSION

Incremental costs resulting from terrain, geology and land use were accumulated for these routes (the existing and proposed pipelines) along the cost surface. The existing pipeline path was 34.5 km

218

long and the least-cost pathway was 35.2 km long. Although the least-cost pathway using fuzzy logic overlay was longer, the analysis indicated that it would be 29% cheaper to construct than the existing pipeline path. The results indicated that the proposed pipeline route is the most cost-effective.

Most of the cost difference between the existing pipeline route and that obtained through the least-cost analysis can be attributed to greater costs associated with the larger number of urban, road and river crossing cells.

Methods also need to be developed to eliminate sharp angles from the least-cost pathway. Actual costs should include costs incurred by construction and environmental considerations. When developing a database including topography, geology and land use from satellite imagery and available maps for an area of interest, additional data can be incorporated to refine the model.

REFERENCES

Anonymous 1993. GIS lead to more efficient route planning, Oil Gas Journal. Apr. 26, 1993, p.81.

Berry, J. K. 1993. Cartographic Modeling: The Analytical Capabilities of GIS, In Environmental Modeling with GIS, Oxford University.

DeMers, M. 2002. GIS Modeling in Raster, New Mexico University, John Wiley & Sons Inc.

Dijkstra, E. W. 1959. A note on two problems in connection with graphs: 269–271, Numerische Mathematik.

Douglas, D. H. 1994. The parsimonious path based on the implicit geometry in grid data and on a proper slope line generated from it , Proceedings of the International Symposium on Spatial Data Handling: 1133–1140. Edinburgh, Scotland.

Douglas, D. H. 1994. Least cost path in GIS using accumulated cost surface and slope lines, Cartographica 31(3): 37–51.

Hicken, J. and Krumbach, Y. 1998. Use of high resolution remote sensing for route selection, Environmental Remote Sensing Centre, University of Wisconsin-Madison. Series ARC-UWM-004-97.

Montemurro, D. and Gale, T. 1996. GIS-based process helps TransCanada selection route for expansion line, June 22, Oil and Gas Journal: 63–71.

Naghibi, F. 2003. Optimum Oil and Gas Pipeline Routing Using GIS, MSc. Thesis, Engineering Faculty, University of Tehran, p105.

Naghibi, F. Delavar, M. R. and Rahmanizadeh, A. 2003. Extension of the Dijkstra algorithm to determine the shortest path in GIS emphasizing a large volume of data, (in Persian), Proc. Conf. Geomatics 82 (2003), May 11–12, 2003, NCC, Tehran, Iran, CD-ROM, 80.

Sarkka, P. and Esko, L. 1999. Optimal routing of pipeline, Helsinki, University of Technology, GIM, Feb. 6, p.6–9.

Shafiee, S. 2000. Evaluation of Usabilities of Raster-Based GIS for optimum path determination , MS Thesis, Dept. of Surveying and Geomatic Engineering, Engineering Faculty, University of Tehran.

Stefanakis, E. and Kavouras, M. 1995. Determination of the optimum path on the earth surface, Proceedings of the 17th International Cartographic Association Conference, September 1995. Barcelona, Spain.

Tomlin, C. D. 1990. Geographic Information System and Cartographic Modeling, Englewood Cliffs, NJ: Prentice Hall.

Advances in Spatial Analysis and Decision Making, Li, Zhou & Kainz (eds)
© 2004 Swets & Zeitlinger, Lisse, ISBN 90 5809 652 1

A real-time geo-spatial information system: an integration of GIS technologies for ice forecasting

Awtar Koonar
Services Clients and Partners Directorate, Meteorological Service of Canada

Ziqiang Ou
Canadian Ice Service, Services Clients and Partners Directorate, Meteorological Service of Canada

Brian Scarlett
GIS Consultant

ABSTRACT: Each year, ships face many challenges travelling through the ice- and iceberg-covered waters of North America. The formulation and delivery of timely ice hazard warnings have become very important for the safety and security of ships, marine and Northern Canadian communities. The ice forecast information is used by ships and the marine community in decision-making and route-planning. Transportation, oil exploration, environmental protection and research institutions rely on ice information for purposes of planning, decision support and research. The Canadian Ice Service (CIS) works closely with the Canadian Coast Guard (CCG), the US Coast Guard (USCG) and the US National Ice Centre (NIC) to provide information on ice and icebergs to national and international clients. CIS relies on geo-spatial information technology and information management technologies to produce information about ice conditions and disseminate it to clients. The production, mapping and dissemination systems, based on client/server and parallel computing architecture, use the latest remote-sensing, GIS, web-mapping and data-warehouse technologies. The servers acquire over 6 GB/day of data in real-time from radar and camera systems on board aircraft, polar orbiting and geo-stationary satellites. These servers also receive ice observation charts from CCG icebreakers and ice reconnaissance aircraft, environmental and weather information from the Canadian Meteorological Centre, ice and GIS information from the US National Ice Centre, and spatial data from other national and international partners. The data received is automatically processed and stored in a central geo-spatial data repository. The ice forecasters and analysts use highly sophisticated graphic workstations to analyse and integrate information to generate products and digital maps. These products and maps are disseminated to partners and clients using satellite and terrestrial telecommunications systems. The clients and partners can also access this geo-spatial data repository using the Internet and geo-spatial data networks. In the near future, a user will be able to generate maps interactively and on demand. The CIS clients and ice service specialists on board the CCG ships will be able to browse the data repository using standard browsers or Java-based spatial browsers, select and download the required information or retrieve digital maps that have been ordered. The Integrated Spatial Information System (ISIS) is a decision-support system built upon technologies such as ESRI ArcGIS and ArcSDE, ERDAS Imagine, Oracle Relational Database and Blue Angel – Metastar. Microsoft Windows, Linux and HP-UX platforms are supported.

1   INTRODUCTION

The mandate of the Canadian Ice Service is to generate and disseminate daily products to aid navigation in Canadian waters. To fulfil this obligation, the CIS is dependent upon highly integrated geo-spatial and supporting information technologies for the acquisition and processing of data from satellites, airborne sensors, ice and weather models and many other sources of data; and for the dissemination of products.

A typical day at CIS involves the near real-time processing and integration of observed reports, remotely sensed imagery and forecasts from model simulations. Charts and bulletins based upon these analyses are then automatically disseminated in a variety of formats to clients and partners. All analyses and products are archived in a spatially enabled repository and are made available to the public via the Web.

The products routinely provided by CIS are diverse. They include:

– Daily ice hazard warnings and charts describing ice conditions in navigable waters
– A warning service for extreme ice events
– Planning support for transportation in Canadian waters beset with sea ice and icebergs
– Regular analyses for climate monitoring and sustainable development
– Archiving of ice and iceberg data for climatological purposes
– Contribution of Canadian ice data to the World Data Centre for Glaciology
– Publication of cartographic-quality seasonal atlases for the Canadian Arctic.

The Integrated Spatial Information System (ISIS) satisfies the requirements for a system that is reliable and efficient in a production environment, yet flexible and scalable enough to adapt to new data sources, products, platforms and evolving business requirements. ISIS is built around commercially available software packages that have made it viable to create, integrate and distribute geo-referenced imagery and other GIS based products within the constraints of the daily CIS production schedule.

Interactive and on-demand web mapping are also being introduced to allow clients and partners the ability to generate and deliver custom products in the most rapid fashion possible.

2   ISIS ARCHITECTURE

ISIS can be viewed as being comprised of six sub-systems, each providing the following functions:

1. Data Acquisition and Processing, in which data is automatically received from different sources and platforms, raw imagery is imported using various de-compression, geo-referencing and enhancement modules, vector model outputs are converted and contoured, and all results are inserted into a central data repository. Load balancing, distributed databases and parallel processing are necessary for this sub-system to meet the requirements dictated by strict production and delivery deadlines.
2. Central Data Repository is composed of a relational database management system (RDBMS) containing data and metadata. The repository is streamlined for the demands of a production environment so that expired data and products are automatically cycled to a long-term archive that supports the analyses of trends.
3. Production Sub-system is implemented on workstations that provide access to all data provided by the Central Data Repository. Custom tools are available for sorting, displaying, analysing, and integrating data in order to reveal the characteristics of ice and iceberg targets. Maps, bulletins and other spatial products are generated with a minimum of effort to enable analysts to extract the maximum amount of information from the available data. Components of this sub-system make extensive use of ERDAS Imagine and ESRI ArcGIS software and include the following modules:
   – Geo-Spatial Metadata Browser is a lightweight visualization utility that is the primary user-interface to the Central Data Repository. An analyst can preview in tabular format and can also superimpose the 'footprint' of the data on reference layers. Temporal and spatial filters assist in the rapid discovery and extraction of data.

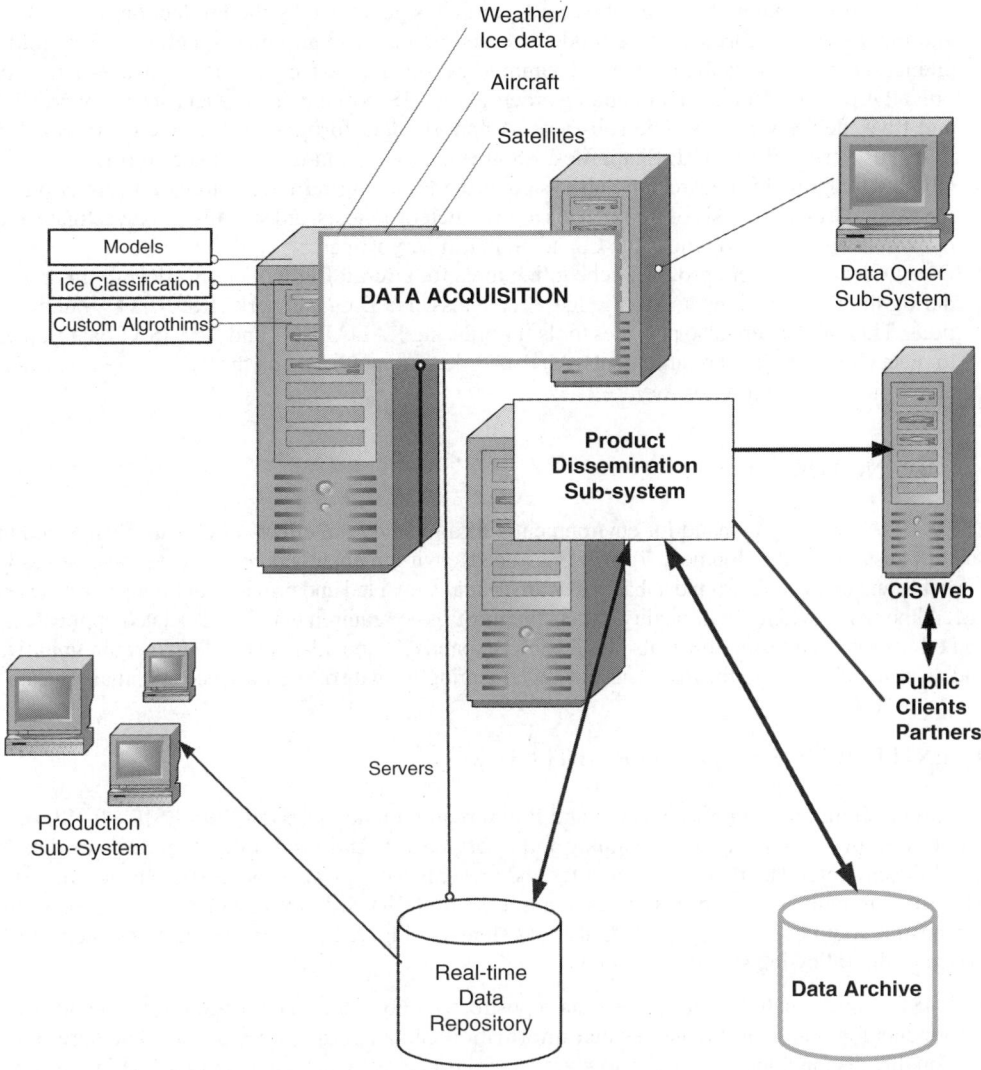

Figure 1.   ISIS architecture – today.

- Image Analysis Module allows the analyst to manage and manipulate numerous layers of raster and vector data. Tools for geographic adjustment, histogram enhancement and mosaicking assist in the annotation of imagery and provide the basis for final analyses. ERDAS Imagine is the engine for this module.
- Data Integration and Product Generation Module unites spatial data from various sources to generate final analyses and products. Data sources include the Image Analysis Module, model simulation results, output from partner organizations and archived analyses. Integrated data entry and relationship rules based on World Meteorological Organization standards ensure the correctness of the topology and integrity of the data.
- Berg Analysis and Prediction Module combines model simulation results and observation reports from ships, aircraft and remote offices to chart the annual drift and decay of thousands of icebergs. Automatic iceberg detection from high-resolution imagery is also used to validate model predictions on the location and size of icebergs near busy shipping channels.

223

4. Product Dissemination Sub-system receives products generated by the Production Sub-system and other systems, formats these products based on client requirements and delivers them to clients and partners via their preferred communication methods; e.g., FTP mailbox, e-mail, Fax or FTP. A product database and standing orders allow CIS to maintain product lists, client profiles and track delivery results. The sub-system supports data formats such as Geo-Tiff, MrSID, PDF, PostScript, JPEG, GIF, Shapefiles, ESRI e00 exports and ERDAS raster formats.

5. Climate Integrated Data Archive Sub-system provides a long-term data store of information on ice to support the analysis of trends. This allows meteorologists and scientists to investigate and document changes throughout the Canadian marine cryosphere.

6. Data Order Sub-system provides scheduling tools for aircraft flight plans and downlinking data during the flight to ships and to the Ice Reconnaissance Data Network (IRDNET) ground stations. This sub-system also provides tools for ordering RADARSAT and ENVISAT data frames to meet CIS operational requirements. All data orders are published on the web site and are available from the central repository.

3 MOVING FORWARD

There is an increasing demand for environmental data to study the effects of climate change and to support sustainable development. To meet this need, environmental and coastal data must be easily locatable and extractable from databases that are spatially enabled and provide continuous data coverage. Although providing high-quality data is the most basic requirement, the associated applications and e-business architecture must also be flexible, expandable and adaptable. Clients require intuitive and on-demand access to the many current and emerging formats of vector, video and imagery data.

4 ENTERPRISE ISIS (E-ISIS) ARCHITECTURE

The architecture must support existing and future requirements. Derived from ESIF[1], E-ISIS is a framework for decision-support, mapping and e-service applications. It integrates complex details about the environment and human activities and presents them in ways that assist decision-makers. The efficient management of resources and mitigation of hazards are key elements in promoting responsible socio-economic growth. E-ISIS, as shown in Figure 2, can be viewed as a system comprised of the following six environments:

1. E-ISIS Core – includes the spatial data repository and provides a collection of tools and components for accessing the geo-spatial information contained in the repository. The core functionality also includes the ability to search and retrieve metadata using a visual spatial browser. The data acquisition and processing component acquires and processes both real-time and non-real time data, and catalogues the derived metadata and data.

   Through the E-ISIS Core, spatial data can be exchanged with other spatial data providers using metadata standards and exchange protocols. The metadata catalogue, a component of the repository, can also provide access to data stored in local or departmental data warehouses or international data warehouses.

   – Data environment – comprised of data acquisition, processing, managing and archiving functionalities. Data is stored in a central repository and is accessed via a metadata catalogue by other environments that include e-mapping, dissemination and climate analysis sub-systems.

   – Data processing – performs automatic geo-location, image compression, enhancements and the contouring of satellite, imagery and model predictions; and stores the output in a central repository.

   – Data management – using common interface components, stores spatial data records in a central spatial data repository. It also invokes processes to merge data sets to support databases with continuous spatial coverage. Also supports the archiving of data and e-service delivery.

   – Spatial data repository – composed of a relational database management system implemented with a schema based on the Content Standard for Digital Geo-spatial Metadata (CSDGM).

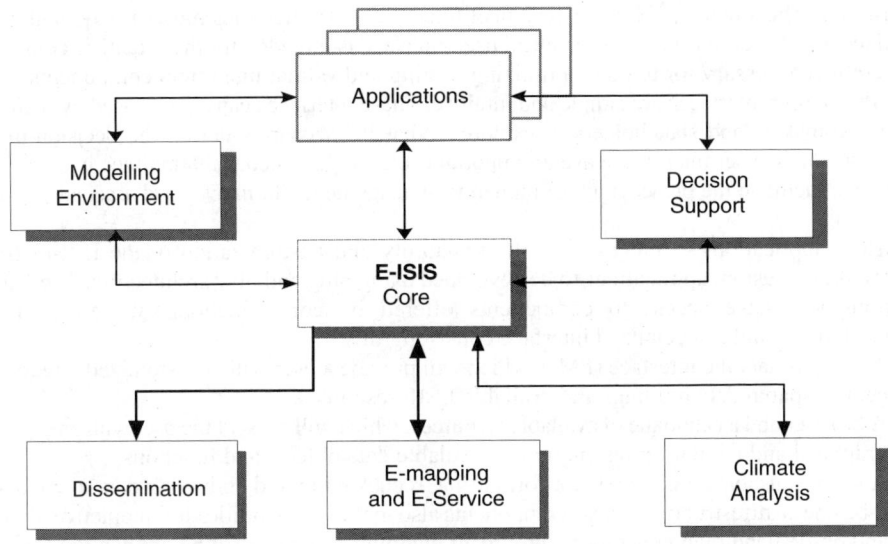

Figure 2.   E-ISIS architecture.

Configurations support either production or long-range workflows. Based on ESRI ArcSDE and Oracle Relational Database technologies.

2. Modelling environment – provides a common interface for running different simulations such as hydrographical, hydrological, environmental or ice models. The core engine provides the necessary input and manages outputs through the data acquisition component. The modelling environment provides its own visualization capabilities due to the sophisticated requirements of 3-D display and time series animation.

   The modelling environment simplifies the task of preparing, managing and analysing numerical modelling data through a consistent user interface. Data can be visualized in either 2-D or 3-D and in either a static or animated mode. More significantly, the design is based on encapsulation and event-driven triggers, which permit the concurrent simulation of environmental, atmospheric, hydrologic, and other ecosystem processes.

3. Decision-support environment – provides tools, visualization, analysis, automation and product generation to assist in tactical and strategic decision support at the enterprise or business level. Ice forecast, flood forecast and management systems are a few examples of applications for this environment. Spatial support is comprised of GIS, remote sensing, data integration and visualization functions.

4. Climate support and analysis – provides access to various analytical tools for analysing meteorological and ice data trends and tools for supporting forecasters who produce long-range forecasts. This environment has direct access to archive repositories based on FGDC Metadata standards. It supports other tools for producing climate reports and atlas products.

5. Dissemination and e-mapping – provides common tools, components and integration with off-the-shelf software to provide on-demand mapping, interactive functionality for custom maps, data format conversion and delivery mechanisms. The gathering of statistics, analysis of performance, and management and invoicing of orders are also supported.

   Using e-mapping functionality, a client will be able to browse spatial data from the central repository, request a default template, choose data sets based on geographical and temporal needs and then generate a cartographic product. This environment is based upon OGC and XML standards, with Scalable Vector Graphic (SVG) used for interactive displays and Geographic Mark-up Language (GML) used for simplified data exchanges.

225

6. Custom applications – public and Technical Interfaces –Applications must provide both public and technical user interfaces. A technical user interface is intended for the scientific community where it is necessary for tuning, optimizing, testing and validating system components.

   Public user interfaces are simple and intuitive. These interfaces can also be used by managers, decision-makers and stakeholders to explore "What if" scenarios during the decision-making process. Public user interfaces have an important role to play in consultation meetings, in which they can facilitate the presentation of numerous management alternatives and associated courses of action.

7. Custom applications – Integration – the scalability and customization of the E-ISIS framework allows custom applications to be developed using plug and play architecture. These applications will have access to components offered by core functionality, various E-ISIS environments and a specialized interface consisting of:
   - A person-machine interface (PMI), which will provide a user with a customized window into central spatial data holdings and available ESIF resources.
   - A browser and a catalogue of available resources, which will present the user with an organized graphical and tabular representation of available data objects and functions.
   - A set of reusable components for formatting, repackaging and delivering data to clients and e-business infrastructure. These components also include capabilities for interactive, dynamic and on-demand map generation, intelligent mapping and data drilling applications.
   - A spatial data engine and a central metadata and data repositories, both local and remote and capabilities to enable batch processing, e-service delivery and e-mapping.
   - The E-ISIS engine will manage all of the above functions and processes including data acquisition, data management, data cataloguing, security, service delivery tracking, billing and order management.

5 INFORMATION AND SERVICE DELIVERY IMPROVEMENTS

CIS has installed or is in the process of installing high-speed Express-Vu based satellite direct links on major Canadian Coast Guard ships. This will provide cost-effective transfers of imagery, GIS and other datasets with significantly improved performance. The satellite links will increase the amount and quality of information provided to navigators who are often restricted by unreliable and slow analogue communications.

As well, a new On-Board Analysis Sub-system is in the conceptual stage. Data entry will be based on Pen computers that will use GIS technology for attribute and feature inputs. The Canadian Coast Guard will be installing Pen computers on board ships and helicopters to generate ice observation charts.

To assist the CIS personnel on the aircraft, the On-Board Analysis Sub-system will be configured with a small network linking server, workstation(s) and Pen computers. This will allow the analysts to analyse newly acquired SAR/SLAR imagery in real-time, generate an analysis enhanced by visual and other sensor data and then transmit the product directly to coast guard ships, commercial ships or satellite broadcast stations. This sub-system will be based on Windows 2000 and ArcGIS.

CIS production is currently based on an Area of Interest (AOI) paradigm, in which products are generated for specific and discrete geographical areas. With the geo-spatial data repository, CIS will also have the ability to generate products based on product and regional templates. This in turn will provide clients with the ability to generate custom maps on a demand basis.

In conclusion, the open and scalable architecture of E-ISIS will enable CIS and our partners to offer new and cost-effective products to meet the demands of clients, partners and the public. As part of e-government initiatives, the delivery of these products will be via the Web. The E-ISIS framework can be easily adapted to implement decision-support systems in such areas as pollution control, agriculture administration, coastal-resource management, flood forecasting, and many others.

REFERENCES

Awtar Koonar and Brian Scarlet, Integrated Spatial Information System, An integration of GIS technologies for ice hazard warnings, ESRI Conference, 2002.

Koonar, Awtar, Navigating Icy Waterways with GIS, ArcNorthNews, 2002.

Koonar, Awtar and B. Scarlett, GIS Technology: Canadian Ice and Marine Service, 2001.

Koonar, Awtar, and ISIS Development Team, ISIS Conceptual Overview, 1997.

Awtar Koonar and Ziqiang Ou – Real Time and Near Real Time Tactical Support Using Ice Reconnaissance Aircraft Systems, Fourth International Airborne Remote Sensing Conference and Exhibition, June 1999, Ottawa, Canada.

Ziqiang Ou, Mac McGregor, Awtar Koonar and Bob Zacharuk – Automated Image Georeference and Tactical Support to Canadian Coast Guard Using Downlink IceVu Systems, Proceedings of ISPRS Symposium Integrated Systems for Spatial Data Production, Custodian and Decision Support, Xi'an China, August 2002, pp. 365–368.

Prins, Mark, Is GML only for Internet GIS? Directions Magazine, 2003.

Seff, George, SVG and GIS, Directions Magazine, 2003.

Advances in Spatial Analysis and Decision Making, Li, Zhou & Kainz (eds)
© 2004 Swets & Zeitlinger, Lisse, ISBN 90 5809 652 1

Spatial modelling in a GIS for an environmental decision-support system

E.S. Malinverni & G. Fangi
DARDUS – Università Politecnica delle Marche, Ancona, Italy

P. Salandin
Istituto di Idraulica – Università Politecnica delle Marche, Ancona, Italy

ABSTRACT: Information System (IS) and Decision-Support System (DSS) are important tools in environmental management. They are expected to play an even more important role in the future. In particular, the introduction of Geographic Information System (GIS) as a desktop tool provides an easily understandable way of presenting data, making these tools more applicable. The use of GIS, as the platform for the discussions, pushes the process of bridging the gap between the scientists and environmental managers, and hence improves the basis of making better decisions for the use, protection and planning of the environment. Environmental management and environmental planning are improving daily, thanks to better techniques and approaches, larger processing power, cheaper computer memory, more accurate and available data, and so forth. This trend, together with the rapid progress in research and technology, is facilitating all kinds of decision-making by using a decision-support system. In this research we applied innovative mapping techniques using HEC-GeoRAS with *ArcView* to efficiently develop a spatially referenced floodplain analysis referring to a small area near Ancona (Italy). The project would design an approach, a methodology and an integration process among software to deal with the complexities of such a situation; that is, it will develop the first step in an interactive, iterative and participatory spatial decision support system (SDSS).

1   INTRODUCTION TO ENVIRONMENTAL MANAGEMENT BY GIS

Problems of environmental management are increasing in complexity. When we refer to environmental management, the word *environment* is used to describe the interrelationship between people and nature. In environmental management and environmental planning, there has been rapid progress in research and technology and in facilities of all kinds for decision-making using a decision-support system (Batty and Densham, 1996). Furthermore, new models and modelling techniques continue to influence environmental planning. Environmental management is facilitated by these techniques and will continue to rely on decision-support systems to obtain scientifically verifiable decisions.

To illustrate a typical situation where a spatial decision support system might be needed, we briefly present the example of a project regarding an analysis of a floodplain. For millennia, river systems have been a focus for travel, industry and human culture. Because of their importance for human activity and quality of life, engineers and scientists have tried to understand their physical properties and behaviour. Watershed studies have used various sources of data such as paper maps, reports on existing and future land use, and hydrological database tables to model and delineate floodplains. These data sources were often difficult to use because they were produced at different scales and accuracies. Furthermore, the data were difficult to track with model input, making it almost impossible to completely reproduce the study. For these reasons, watershed studies are rarely used as a long-term planning tool that can be easily updated as development occurs in the watershed. To improve the efficiency, consistency and usability of watershed studies, innovative methods were developed including the use of continually expanding geographical information system (GIS) data sources. It is possible to take full advantage of GIS tools, which allow model input and output data to be stored and reproduced in a GIS format and graphical user interfaces to be effectively used.

The use of GIS technology in water management is not a new concept as HEC-RAS, probably the most popular river analysis model, demonstrates. The novelties are the interactive application of this software in a GIS and the possibility of exchanging the data. Methodologies and procedures, developed for spatially referenced floodplain studies, incorporate innovative GIS techniques to integrate and visualize detailed hydraulic simulations performed by HEC-RAS. Model input and output data can be generated in a spatial format by the GIS database, which provides uniformity and consistency of model parameters for HEC-RAS models. The final floodplain boundaries can be produced in a geo-referenced format that can then be added to the GIS database. *ArcView*'s coverage of land use, urban areas, slopes and elevations are suitable for determining basin flow paths, slopes and drainage areas. Therefore, in this study innovative mapping techniques using HEC-GeoRAS with *ArcView* have been applied for the efficient development of a spatially referenced floodplain analysis referring to a small area near Ancona (Italy).

2   THE USE OF GIS IN DECISION-SUPPORT SYSTEMS

Information System (IS) and Decision-Support System (DSS) are important tools in environmental management. They are expected to play an even more important role in the future. In particular, the introduction of GIS as a desktop tool provides an easily understandable way of presenting data, making these tools more applicable. The use of GIS as the platform for discussions helps to bridge the gap between scientists and environmental managers, and hence improves the basis for making better decisions on the use, protection and planning of the environment (Armstrong, 1992).The two terms Information System and Decision Support System are frequently used, but not very clearly defined.

The main differences between IS and DSS are, according to the author's knowledge, that DSS seems to be more focused on handling different scenarios/alternatives than IS. In other words, some types of modelling tools are operating and hence producing the consequences/effects of different measures/actions. IS seems to be more focused on the need to provide (aggregated) information to the general public. An IS becomes GIS when it uses computers and geography to help people to better understand the world in which they live and to solve problems (ESRI, 1996). A GIS only provides rudimentary support for decision-making, but when it is combined with the relevant data, spatial modelling tools and appropriate interfaces for decision-makers, more sophisticated decision support can be achieved. The ability to integrate information and support decision-making is the true power of a GIS.

Decisions are based on people's understanding of a certain situation. Computer systems are famed for their ability to take the data and process it into information. If a decision is required, one needs to take the process one step further, so this information can be turned into a process of understanding problems.

Hargrove (1995) notes that '*Decisions, by definition, are based on sufficient data; if enough information was available, the correct answer would become obvious and no decision would be required.*' Conger (1994) describes a system that supports decision-making, such as a system that seeks to identify and solve problems. On the issue of geography in decision-making, Petch (1993) explained that there are certain management decisions for which spatial information, analysis and cognition are essential, and the use of maps influences how we look at or conceive of things and, therefore, what we decide. Thus, the GIS can be used as a means to respond to a series of 'what if' scenarios, in which any single map solution is not important. Every GIS helps the user to make more informed decisions. The final step is where GIS is used as a Decision Support System.

Densham (1991), giving a definition of Decision Support System, suggested six characteristics: 1) explicit design to solve ill-structured problems; 2) powerful and easy-to-use user interface; 3) ability to combine flexibly analytical models with data; 4) ability to explore the solution space by building alternatives; 5) capability of supporting a variety of decision-making styles; and 6) allowing interactive and recursive problem-solving.

As an extension of DSS, the Spatial Decision-Support System is a computer-based information system used to support decision-making where complex spatial problems are ill or semi-structured and decision makers cannot define their problems or fully articulate their objectives. The Spatial Decision-Support System (SDSS) has been extensively and adequately covered in the literature (Craig and Moyer, 1991; Densham, 1991; Goodchild et al., 1992; Moon, 1992; NCGIA, 1996).

In general, the technology at the basis of DSS is summarized in the acronym DDM: Dialogue, Data and Modelling (Sprague, Watson, 1996). In the SDSS, these three aspects have to be well organized in three fundamental components:

– Data Base Management System (DBMS): substantially, the software managing the geographical data, related to the types of data (position, topology, attributes) or to their logic relationships and hierarchies.
– Model Base Management System (MBMS): the motor producing the decision, has some functionality of spatial analysis, of modelling and evaluation criteria (Turban, 1995).
– Dialogue Generation and Management System (DGMS): the interface between the decision-maker and the different components of the software (the graphic interface, the three-dimensional (3-D) or bi-dimensional graphic visualization and so on) (Crossland et al., 1995).

The Spatial Decision-Support System may be developed as a general-purpose tool for a decision-making, problem-solving environment. There, the problem can be explored, understood and redefined, and trade-offs between conflicting objectives investigated and prioritized actions set.

The following are some criteria for a useful SDSS:

– The system must be easy to use;
– The user must be able to develop interactive scenes;
– A model that supports the specific application should be interactively developed;
– Environmental assessments are spatially defined;
– The system should be able to link other programs and should be capable of normal management functions.

It is then possible to add to the list other distinguishing capabilities and functions of SDSS. It is necessary to 1) provide mechanisms for the input of spatial data; 2) allow representation of the spatial relations and structures; 3) include the analytical techniques of spatial and geographical analysis, and 4) provide output in a variety of spatial forms, including maps. The result would be a Spatial Decision-Support System with extended capabilities and flexibility.

However, the real power of the DSS has not yet affected GIS technology. The missing link between sophisticated spatial data systems (GIS) and refined choice models, pattern-seeking systems and integrated decision support tools (DSS) seems to be a planning analysis methodology based on a planning analysis theory. While Spatial Decision-Support System generally include pattern-seeking models of geographical analysis, they would benefit by the integration of choice models in a flexible, decision research approach. There is a definite trend to investigate the use of GIS to facilitate group decision-making. Due to the complex nature of spatial problems, a multi-disciplinarily team normally needs to be involved in the decision-making. The users of a Spatial Decision-Support System are typically professional operators and managers. They can perform 'what-if' analyses, identify trends, or perform mathematical/statistical analyses of data to solve unstructured problems (Conger, 1994). They can facilitate the creation of alternative choices as well as report comparisons between these alternatives.

Already in 1960, Simon defined the three principal phases in a decisional process:

- *Intelligence*: localization of the decisional problem to be solved;
- *Project*: collection of every possible alternative solution to the problem;
- Choice: *evaluation of the best choice by means of the concept of multi-criteria analysis (typical economic concept).*

The first phase uses the geographical database of the GIS to analyse the environment, localize the problem and study the possible solutions. The 3-D maps and ortho-images help in the making of good and comprehensible future decisions.

In the second phase, GIS can handle the collection of a set of possible decisions with the purpose of solving the next identified problem. Note, however, that a generator of alternatives causes the decision-maker to be unable to express his abilities by intuition and professionalism.

The last phase rarely uses GIS, because it is difficult to insert in the system the preferences and the criteria of evaluation of the users. Nevertheless, GIS can be used to visualize aspects of every decision. In this perspective, integrating GIS, DSS and spatial analysis into SDSS constitutes a step towards converting knowledge into intelligence. A good SDSS seems to be able to deal with the capabilities of humans as problem solvers, with short-term and long-term limitations, associative memory structures, conservative biases, and decision-making illusions. And it must leave the possibility to be accepted by the user.

3   A REAL WORLD PROBLEM: THE WATER MANAGEMENT

Since antiquity, rivers have been a focus of human activity. The benefits obtained from rivers are important to humanity; therefore protecting against floods and other river disasters is an important task. Thus, it is necessary for environmental enhancement to understand the behaviour of rivers.

In this research, the river under analysis is located around the Ancona hills. It crosses some little towns and then flows into a channel along the Ancona (Italy) Airport, turning at last in a bigger river: the Esino River (Figure 1). The stream starts from an elevation of 110.7 meters a.s.l. and reaches the coast with a mean slope of 0.82% in 12.03 kilometres. Using 3-D data in GIS we calculated the middle elevation of the basin: 106.57 meters a.s.l. We analysed only the final stream of the river where it crosses an urban and industrial area.

No matter what the purpose is, the success of any river study largely depends on the volume of data available and on the tools available for the storage, retrieval and analysis of data. The data collected in this research include the cartography at different scales, cross-sections of the stream, the hydraulic parameters and the historical data on several rain gauges. The spatial data are fundamentally of different types, scales and accuracies. They include:

- Many points surveyed by tachemetry in the country along transversal cross-sections of the stream of water (with distance among the sections of about 250 meters);
- A 3-D Vector Map of the urban area at a scale of 1:2000 and the Regional Technical Map at a scale of 1:10.000 (substantial contour lines and quoted points), both realized in the year 2000.

Given the 3-D Regional Technical Map of the area, the limits of the basin, it was possible to get the Triangular Irregular Network (TIN), and to extract information on the extension of the plain to the basin and on the inclusive volume (Figure 2).

In general, the realization of the Digital Terrain Model (DTM) introduces different problems related to the available data: if the data are regularly distributed, it is easy to get a regular DTM; otherwise a preliminary interpolation is necessary.

In this application the principal problem occurred in the generation of DTM. Inside the river, DTM went according to the accuracy of the surveyed points along the sections, while outside it followed the only available cartographic data (Figure 3).

Unfortunately, the surveyed data were not as abundant as desired, and had to be interpolated to obtain a more accurate DTM within the river. After that, it was possible to extract many other

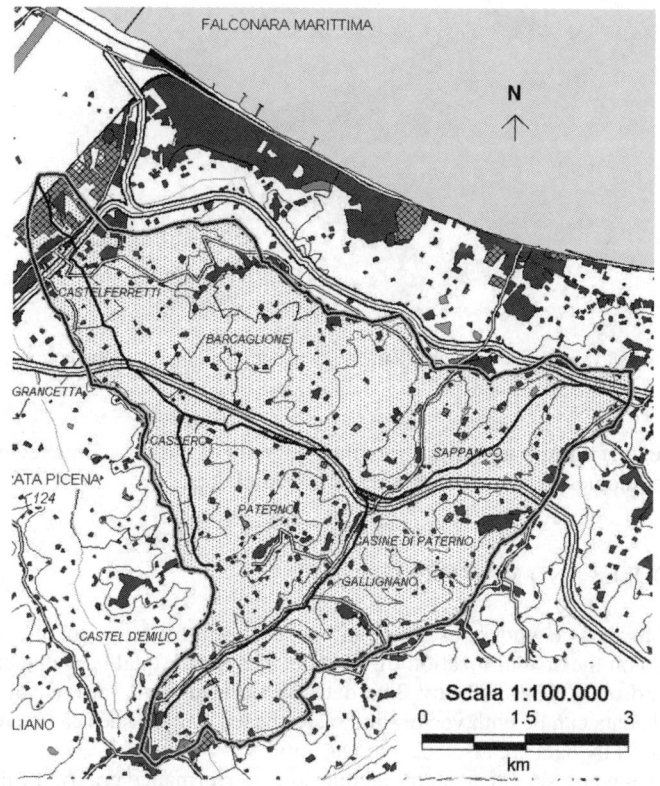

Figure 1.   Delimitation of the river basin.

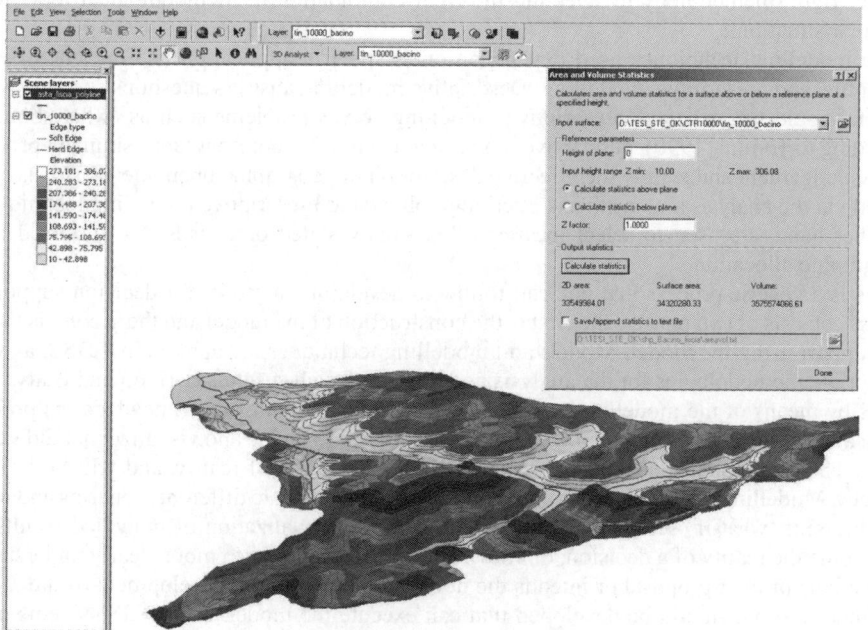

Figure 2.   3-D representation in the GIS of the basin area.

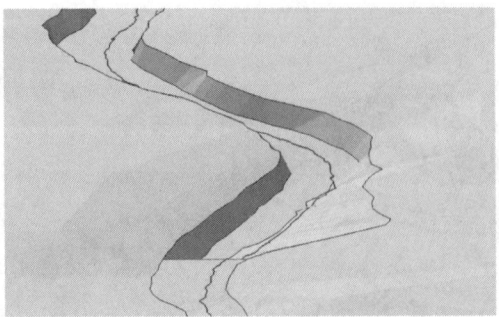

Figure 3.   The TIN by available cartographic data is not accurate along the river section.

sections.In the next paragraph we write about the necessity of the modelling techniques and illustrate some methodologies to correctly organize the spatial data.

4   ABOUT MODELS AND METHODOLOGIES

The term spatial analysis is widely used, but in the strict sense it refers to an operation in which the result depends on the spatial location of the data. The detail, quality and accuracy of the collected data depend on their availability. The data are derived from various sources of different types. Geographic data can be both vector- and raster-based. Most issues can be mapped as vector data, but raster and TIN data are also accepted as input.

The initial stage in any scientific study includes the performance criteria of the model, model parameters and the identification and definition of any variable, which will be used. Consequently the analysis needs to choose a methodology to follow. The emphasis should move away from finding the 'right' methodology to selecting the right combination of methodological tools for any particular situation.

The modelling techniques used depend on one's specific application and are classified into descriptive and prescriptive modelling. Descriptive modelling answers questions such as 'what is' or 'what could be', whereas prescriptive modelling solves problems such as 'what should be'. According to Tomlin (1990), descriptive modelling techniques represent facts, simulate processes, express judgement and supply the effective description of geographic phenomena. This is accomplished via the analysis and synthesis of cartographic data. Prescriptive modelling techniques, on the other hand, are used to select locations that satisfy stated objectives, also referred to as a cartographic allocation.

This is a generic process that one can follow in developing a model for decision support. The process consists of two phases: the first is the construction of the model and the second is the decision support using the model. Models and modelling techniques are applied in a DSS, as well as appropriate methodologies for the analysis and design of such systems (Britton and Batty, 1992).

It is by means of the modelling process that new information or new knowledge is produced. This feature is the added value of a GIS; that is, the ability to create and visualize that did not exist before. The model is an idealized or simplified representation of reality, and will be limited in accuracy. Modelling allows planners and decision-makers to create different scenarios and to compare different 'what-if' scenarios with one another. The visualization of modelled results often bring home the reality of a decision, or allows the decision-makers to more clearly understand the implications of their proposal or intends the development. After the development of the model, a computer program should be developed that can execute the model. Some GIS systems provide some form of development tools. Any problem relating to spatial location can be addressed through some form of spatial analysis. A spatial analysis solution is therefore central to any SDSS. The way in which these solutions are incorporated into a specific SDSS determines the flexibility of these systems for development and future modifications.

5   CARTOGRAPHIC MODELLING PROCESS

The first step in the development of a floodplain analysis is to collect highly detailed information grouped into three categories: hydrologic, hydraulic and topographic. Hydrologic and hydraulic data consist of measurements of rain and discharge hydrographs, stage historical records, rating curves relating stage and discharge, etc. (Da Deppo et al., 2002). Topographic data describe the geometry of the simulated river system. The topographic data suitable for building river hydraulic models may be divided into two basic categories:

– Qualitative data are a reconnaissance type of description of the river, and may be obtained by field investigations, inquiries, satellite and aerial photographs, newspaper reports, etc.
– Quantitative topographic data are needed for the model representation of the river and its flooded plains. The three essential types of quantitative topographic information required in river hydraulic models are: longitudinal profiles along banks, dykes and roads; cross-sections or transverse profiles across the water stream; and contour imagery of the inundated area.

Producing some cross-sections of a river channel is fundamental to all river studies. Traditionally, this information is collected in the field using surveying techniques and taking elevations along the cross-sections of the river. Collecting data in this manner is time-consuming and costly. Luckily, there is now the possibility of extracting cross-sections from DTM. The ground elevations are stored using formalized data-structured rules such as TIN or GRID. Each model has advantages and limitations. A GRID is a simpler model of a surface, and is a widely accessible format. TIN can produce a more accurate representations of surfaces and features, but usually requires a more extensive data collection effort using aerial photogrammetry or other remote-sensing techniques, such as Ligh Detection and Ranging (LIDAR). In a TIN the elevations are spatially irregularly sampled. This data structure allows a variable point density in areas where the terrain changes sharply, yielding an efficient and accurate surface model. In addition, it is necessary to have some break lines, linear features that represent natural features such as streams and ridges or man-made features such as roadways. While the surface is always continuous, its slope may not be.

We then derived many cross-sections to accurately simulate the hydraulic characteristics and to delineate the floodplains. The cross-sections are localized where there is a representative channel reach section, and where there are significant changes in hydraulic and/or hydrologic characteristics over short distances, such as changes in geometry, slope, roughness or discharge.

Using the cartography at a scale of 1:2000 we traced the stream of the river along two consecutive sections. We subsequently adapted it in altimetry, connecting the two points of the Talweg. Such a procedure called for particular attention to be paid to avoid possible discordances between the two interpolated sections and the qualitative data of the cartography (contour lines). Tracing the connections among the notable points of the sections (Talweg, banks, extreme sting) we produced a 3-D surface that clearly described the stream of water. This was turned into a TIN, creating a regular grid of 2×2 meters. The result was a DTM that inside the river has the same order of altimetric accuracy of the surveyed data, and outside of it has the same order of altimetric accuracy of the cartography (Figure 4).

The accuracy of the simulated water levels, and eventually the accuracy of the delineation of the floodplain, largely depends on the shape as well as the extent of these cross-sections. In a flood model it is important to specify a detailed cross-section geometry that not only extends over the floodplain, but is also truly capable of carrying the total flood discharge through it.

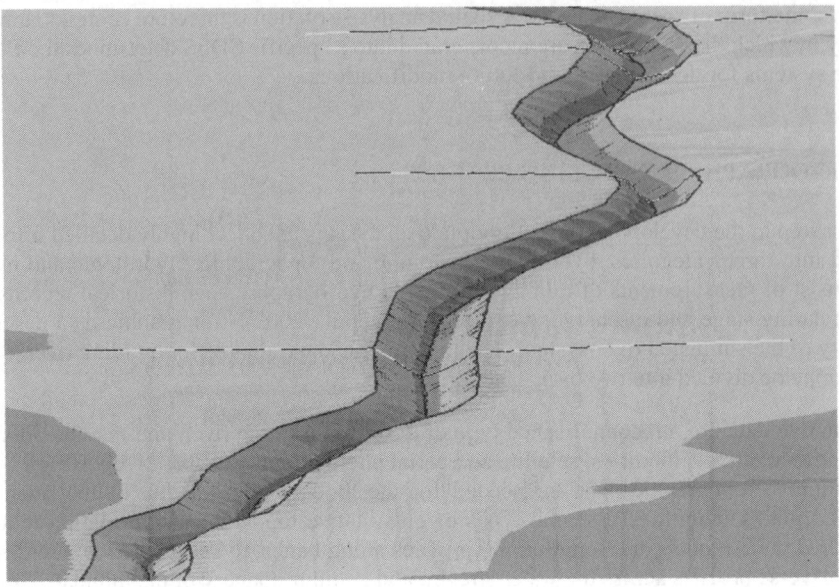

Figure 4.   A new interpolation using more accurate altimetry by cross-sections.

6   SOFT-SYSTEM ANALYSIS AND DESIGN METHODOLOGIES:
AN INTEGRATION BETWEEN GIS AND HYDRAULIC MODELS

The aim of the hydraulic model is to develop accurate floodplain maps based on flows from the hydrologic model using all available data including GIS data sources (Maidment and Djokic, 2000).

A floodplain is the normally dry land area adjoining rivers that is inundated during floods. An automated floodplain delineation process determines the extent of inundation by comparing simulated water levels from a river hydraulic model with ground surface elevations.

The HEC-RAS, Version 2.2 is a piece of public domain software used in countless hydraulic studies that has a greater capability to import and export data in GIS format than other types of software. HEC-RAS was developed by the US Army Corps of Engineers – Hydrologic Engineering Center (HEC) of the United States. It produces one-dimensional hydraulic models for a full network of river channels. One of the major steps in developing a hydraulic model with HEC-RAS is to enter the necessary geometric data, which consist of the connectivity information for the stream system, cross-section data, and hydraulic structure data. The physical and hydrologic parameters used to develop input for the HEC-RAS model were derived from the best available geographic and technical information. The primary sources of GIS data include elevation data by TIN and contour lines.

Historical storms and rain data from all available gauges within or near the watershed are used to develop the precipitation data for the HEC-RAS model. In this case study, we developed a simple cinematic hydrological model to deduce the discharge from the historical records of the rain data. We obtained the following discharges: 54.93, 67.16, 78.89, 94.06 m³/s for recurrence intervals of 5, 10, 20, 50 years, respectively.

As a tool to link HEC-RAS with GIS, HEC-GeoRAS was used, another piece of public domain software developed by the HEC. This software is specifically written to extract HEC-RAS geometry data from *ArcView* for input in HEC-RAS and then to delineate the boundaries of the floodplain from the HEC-RAS output. The HEC-GeoRAS with *ArcView* is used to create the base HEC-RAS model and to produce the spatially referenced floodplain boundaries in a highly

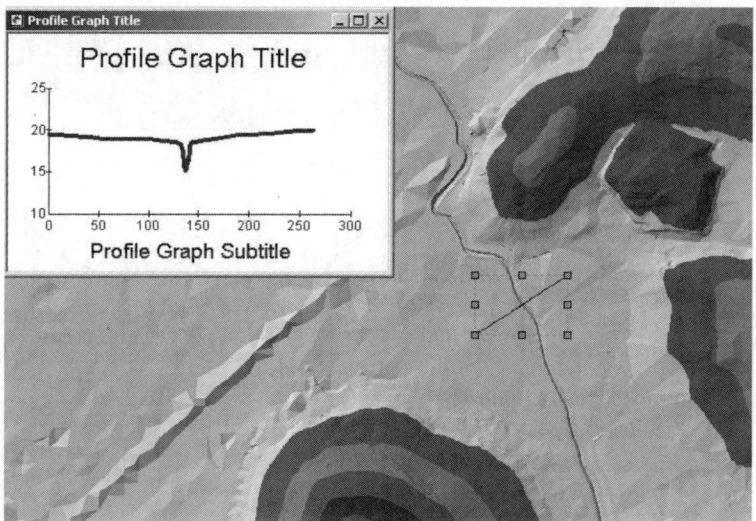

Figure 5.   Cross-section perpendicular to the flow in the channel by TIN in *ArcView*.

accurate and geo-referenced coverage that can be used with other GIS watershed data. The result-ing floodplain delineation polygons were then used to identify buildings in a floodplain, locate street flooding, and to provide an effective planning tool. Floodplain polygons are then overlaid onto the existing GIS coverage in *ArcView* to provide an effective planning and decision-making tool (Ding et al., 1992). In addition, *ArcView* can be effectively used to display, quarry and export all of the processed data and to produce useful thematic maps.

An example of this integration between GIS and hydraulic modelling is performed in this research by three phases, each one producing different problems.

In the first phase the geometric data are built in the GIS. Such a phase is called the PRE-RAS phase. Here, TIN defines the elevations, the geographical references and the cross-sections that are to be exported. The cross-sections developed in *ArcView* are arranged perpendicularly to the flow in the channel and floodplain (Figure 5). HEC-GeoRAS creates a file from the GIS coverage that was directly imported into the HEC-RAS model containing the cross-section connectivity, geometry and flow lengths (Figure 6). Three flow lengths were inputted into the model: channel centreline, left over-bank, and right over-bank (Figure 7). HEC-GeoRAS calculates the distances between an arbitrary section and the following one using such lines of flow.

In the second phase the hydraulic model of flood is determined, adding some parameters not derived from the GIS: the stream value, the flow typology (slow, fast), the possible presence of bridges, bridles, etc. This is the crucial phase of the integration because HEC-RAS really becomes the motor of DSS that allows decisions to be made. In this procedure the already imported data related to the geometry, need to be modified: by, for example, adding or removing some cross-sections. Another task of this phase is the completion of the geometry with the addition of possible levees, bridges, culvert, bridles, and so forth.

Nevertheless, the software is essentially a one-way model. It is not able to manage calculations of bi-dimensional information. Thanks to the informative levels produced in *ArcView* the imported geometry is geometrically correct and useful for the input model.

Such an aspect can already be seen as a decisional aspect that, however, still happens in a spa-tial reference that has not been specified, or preferably in the 'computational model' of the envi-ronment. At this point the model that has been produced needs to be exported into the GIS.

The third and last phase consists of recovering in the GIS the profile of flood determined in the previous phase, such as the polygon of the flooded area (Figure 8).

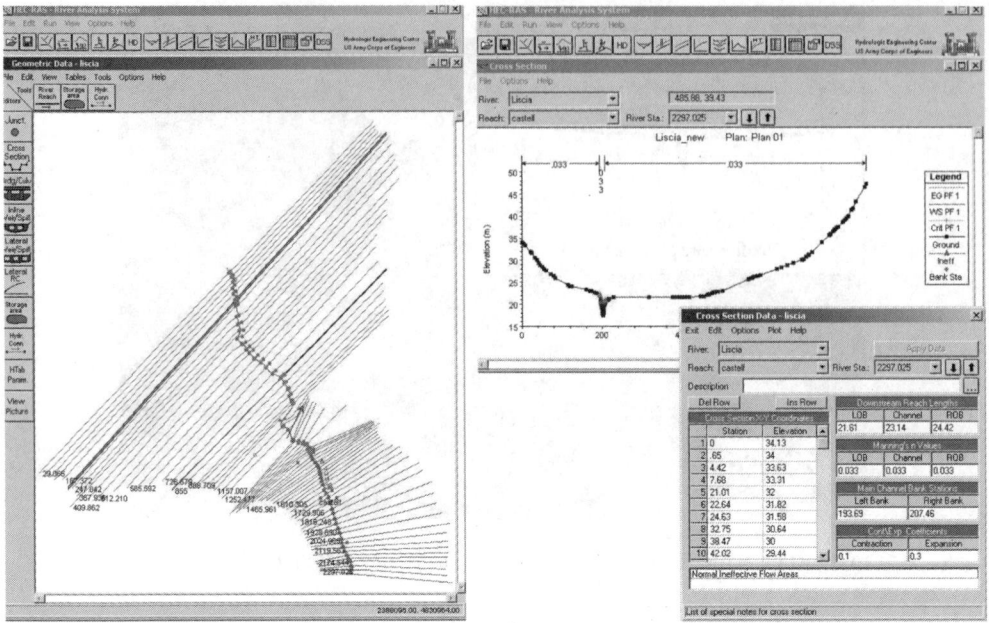

Figure 6.   Cross-section geometry imported in HEC-RAS.

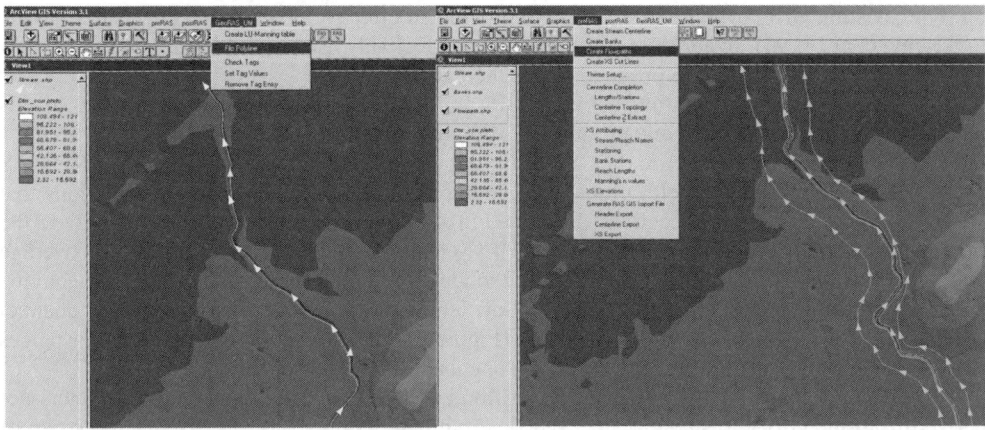

Figure 7.   The stream of the river and the three flow lengths.

In the case study, for example, we observed that the problems of flood were localized in the sections of valley in the right over-bank of the channel that did not support a maximum flood of $50\ \text{m}^3/\text{s}$ of water.

The accurate analysis of the cartography of the zone inundated by the flood shows industrial or urban buildings not well described by the TIN. More accurate data in altimetry would have led to better results. A punctual analysis was made of the floodplain, observing the altimetric data in the right over-bank of the channel compared with those in the left.

We determined a critical analysis of the polygon to fill some gaps, first caused by the impossibility of realizing a DTM of equal accuracy to the surveyed section data and in agreement to the mono-dimensional use of the hydraulic modelling (Figure 9).

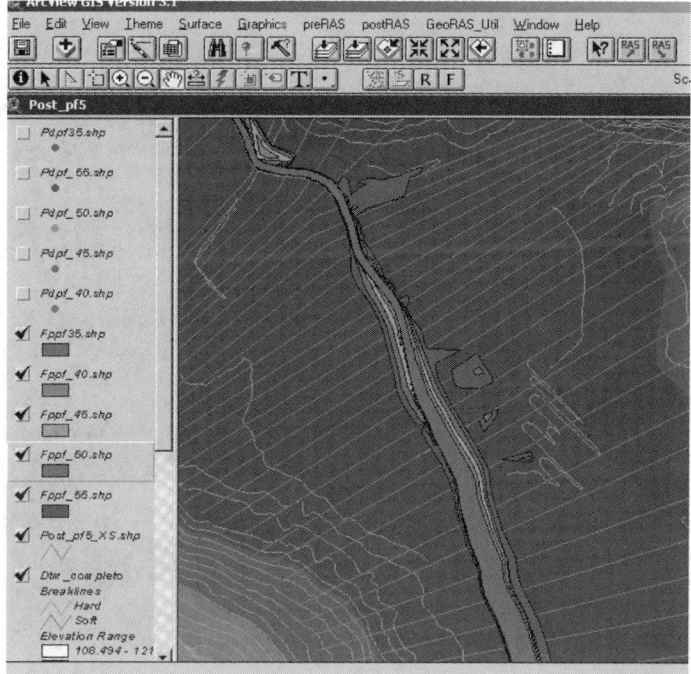

Figure 8. The profile of the flood generated by hydraulic modelling and imported in *ArcView* to determine the floodplain.

Figure 9. Critical analysis of the floodplain in the GIS.

Furthermore, the analysis of the profiles of flood has been repeated under conditions of permanent motion, with the result being a flood of 58 m^3/s supportable among the valley sections.

The analysis was conducted under steady conditions because reliable data on flood hydrographs were not available. In future, however, it would be desirable to conduct an unsteady analysis to enhance the quality of the delineation of the floodplain.

After the critical analysis of the polygons created of the floodplain map, a series of other analyses were then possible; for instance, the analysis of risk and other queries in the GIS (Figure 10).

239

Figure 10.   The final representation of the flood area overlaid on the 3-D ortho-image.

In fact, the flooded area, a function of the set-up of the water stream, produces a risk that really shows where the interventions are needed (this is the moment of decision).

7   CONCLUSIONS

The SDSS that generally includes models of geographical analysis could benefit the integration of choice models in a flexible, decision-research approach. Batty (in Goodchild and Densham, 1990) considered that in comparison with the development of GIS, which was easy enough, the development of SDSS is much more problematic. Furthermore, the SDSS should incorporate great flexibility in the design to the point where the user might have to adapt the system to the specific constraints imposed by the problem.

The approach to SDSS should measure the impact of environmental factors according to the environmental risk that can be displayed in the GIS. The analysis should rely upon existing data, and risks should be estimated by a product of exposure and hazard. From that the necessity comes, for the realization of a SDSS, to interact with the typical parameters of the risk analysis. The risk analysis should be a multi-criteria analysis taking into account other important elements (such as how the ground is being used) to sift the most opportune decisions.

We were limited to visualizing the floodplain analysis in the GIS because suitable methodologies for the definition of the hydraulic risk were not available. However, we observed that the analysis is a function of the return time of the event. Therefore, the statistic-probabilistic concept of return time is always introduced as a parameter in multi-criteria analysis to establish the hydraulic risk. One of the future developments of the research could be a multi-criteria analysis of the hydraulic risk. In fact, in the philosophy of the SDSS it is fundamental for the study of the risk to establish the weights to attribute the different elements involved in the phenomenon. Will such weights depend on the sensibility of the subjects called to take some decisions in worth naturally: for instance the loss of human lives or to monetary damage. The approach can be similar to an economic cost-benefits analysis where a value is given (for instance, in terms of score) to the elements involved.

The problem is to want to properly build a SDSS that foresees the possibility of rapidly taking a decision. In this case study, the resulting advantage is the possibility to check and, therefore, to manage and to interpret, the various phases of the process of taking the decision.

However, if the decision is made to intervene on the stream of water, there can be advantages under certain aspects and disadvantages under others; therefore the choice always depends on the sensibility of the decision-maker.

ACKNOWLEDGEMENTS

We give special thanks to the engineer, Stefano Leti, for the fruitful collaboration during his degree thesis.

REFERENCES

Armstrong, M. P. 1992. GIS and Group Decision-Making: Problems and Prospects, GIS/LIS '92, vol. 1: 20–29.

Batty, M. Densham, P. J. 1996. Decision Support, GIS, and Urban Planning, http://www.geog.ucl.ac.uk/~pdensham/s_t_paper.html ESRI, (1996): Introducing GIS http://www.esri.com/resources/gis.html

Britton, H. Batty, M. 1992. Locational Models, Geographic Information, and Planning Support Systems. Technical Paper 92-1. National Center for Geographic Information and Analysis.

Conger, S. A. 1994. The New Software Engineering. Wadsworth Publishing, Belmont, California.

Craig, W. J. Moyer, D. D. 1991. Progress on the Research Agenda: URISA '90, *URISA Journal*, 3 (1).

Crossland, M. D. Wynne, B. E. Perkins, W. C. 1995. Spatial Decision Support Systems: An overview of technology and a test of efficacy. *Decision Support Systems*, 14: 219–235.

Da Deppo, L. Datei, C. Salandin, P. 2002. *Sistemazione dei corsi d'acqua* 4a ed., Libreria Cortina, Padova.

Densham, P. J. 1991. Spatial decision support systems, In: D. J. Maguire, M. S. Goodchild and D. W. Rhind (eds) Geographical information systems: principles and applications, London: Longman.

Ding, Y. Fotheringham, A. S. 1992. The Integration of Spatial Analysis and GIS, Computers, Environment and Urban Systems, 16: 3–19.

Goodchild, M. F. Haining, R. Wise, S. 1992. Integrating GIS and Spatial Data Analysis: Problems and Possibilities, *International Journal of Geographic Information Systems*, 6(5): 407–423.

Hargrove, W. W. 1995. Perspective on Future Directions in GIS. GIS World, vol. 9, Iss. 3, Fort Collins, Colorado.

Maidment, D. Djokic, D. 2000. Hydrologic and Hydraulic Modeling Support with Geographic Information Systems. *Environmental Systems Research Institute*, Inc., Redlands California.

Moon, G. 1992. Capabilities Needed in Spatial Decision Support Systems, GIS/LIS '92, vol.2: 594–600.

NCGIA, 1996. Report from the specialist meeting on collaborative spatial decision making, *Initiative 17, National Center for Geographic Information Analysis*, UC Santa Barbara, September 17–21, 1995.

Petch, J. 1993. Concepts for Spatial Thinking. UNIGIS Course Notes, Module 2, 3rd ed., Manchester, UK.

Simon, H. A. 1960. *The new science of management decision*, New York: Harper & Row.

Sprague, Jr. R. H. Watson, H. J. (1996): *Decision support for management*, Upper Saddle River, N.J.: Prentice Hall.

Tomlin, C. D. 1990. *Geographic Information Systems and Cartographic Modelling*, Prentice Hall, Englewood Cliffs, N.J.

Turban, E. 1995. *Decision Support and Expert Systems 4th ed.*, Prentice-Hall International, 241–242, N.J.

Van der Merwe, S. E. 1997. Spatial Modelling and Decision Support for New Linear Developments, PhD Thesis, Manchester Metropolitan University.

Design and implementation of a prototype temporal geospatial information system (TGIS)

Mahdi Talebi & Mahmoud Reza Delavar
Dept of Surveying and Geomatic, Engineering Faculty, University of Tehran, Tehran, Iran

ABSTRACT: New users of geospatial information systems (GIS) are often surprised to discover that their software has great difficulty in responding to requests such as, 'Display the land parcel changes between 2002/10/05 and 2002/10/15'. In our world any feature has three basic components including time, space and attribute. Therefore, geospatial data should be described by spatial, temporal and attribute components. Missing temporal component means that GIS analyses of past events and future trends are difficult or impossible. The primary objective of this paper is to design and develop a prototype temporal geospatial information system (TGIS) based upon an existing static GIS. The prototype is able to create, store and update TGIS data sets and perform display and analysis operations. Ideally, the working system would have a menu-driven user interface for a set of temporal data management, editing and query functions. The TGIS data model would be selected and designed to minimize the volume of data required to retain the necessary information for temporal analysis. The spatial data sets used in this paper are a digital cadastral map of Qom city in Iran, and the prototype TGIS functions include feature split, feature merge, topology queries, etc. The prototype TGIS was developed based on existing ArcView GIS software. Some macro programs were written with Avenue language to provide menu-driven temporal data management, editing and display as well as query functions. The prototype TGIS can trace, display and query versions of features using an extensive set of criteria such as 'Where was feature X at time T?', Snapshots of trends can be created and played to show changes over time.

1 INTRODUCTION

The main subject of this paper is the modelling of spatio-temporal information, where both the position and history of the information are of interest to the users. Time has to be regarded as multidimensional (Raper, 2000; Roshannejad, 1995). However, two orthogonal time dimensions are emphasized in temporal database research: valid time, which is the time when a fact is true in the real world and transaction time which is the time when the fact is current in the modelled reality; i.e., in the database. The valid time dimension is considered to start in the infinite past (the beginning of time), and progress into some infinite future (the end of time), whereas the transaction time dimension is considered to stretch from the time of the creation of the database up to the present time (Langarn, 1992; Al-Taha, 1990). Thus, for the vector updating model used in this paper, this two types of time should be used for the time stamping of the database.

The main advantage of this model is that feature objects are traceable over time. The main disadvantage is that there is no easy way to store changes in spatial topology over time. In spite of these formidable processing requirements, this model can at least use the feature and attribute time stamps to answer spatio-temporal queries and perform analysis. In practice, this model can best be implemented using current temporal geospatial information system (GIS) software technology. Feature updates are added to the primary layer, with the old versions of any modified features being moved into an archive

243

layer. The advantages of this method are: (i) The current data set is usable for normal GIS functions, and (ii) the deleted features (features that have ceased to exist) and any previous versions of the deleted features can be found and traced using the archive layer. The primitive objects of this research are:

Storing multiple versions of feature and feature attributes with minimum rendundany of data.

Using a data model for the support of queries, representation and analysis of data with time and space constraints.

Be sure that the existing static GIS functionality performs properly once used without the TGIS functions.

Run temporal geospatial information system (TGIS) in a static GIS environment without using external software.

Providing a menu-based user interface with a set of functions for input, management, display and query.

The secondary objective was to examine the functional limitations of TGIS that resulted from the decision to extend an existing vector GIS data model. Particular areas that were studied include the design of the TGIS data model and the user interface, system performance, spatio-temporal topology and data visualization.

2   DESIGNING A TGIS DATA MODEL

Previous researchers have suggested five data models for TGIS (Spaccapietra, 2001). The snapshot model cannot perform feature-based temporal queries or analyses, although can be used to store and view complete sets of historical data. The update and space-time composite model can analyze temporal data, as each feature and attribute is time-stamped with birth and death dates (Langran, 1992; Armenakis, 1991).

The update model has the advantage that it can be implemented within existing GIS software, and the space-time composite has the advantage of storing spatial topology over time. TGIS software will evolve over the next few years and the 4D model is probably a strong contender for a TGIS data structure if the GIS industry can fund the massive amount of software development required. The integrated TGIS data model makes use of raster and vector data structures to help solve TGIS topology problems but suffers from data redundancy. The prototype TGIS will use the vector update model (Burrough et al., 1988).

In this paper a vector updating model has been used. The model uses multiple versions of features and attributes of features, and for this reason the time stamping of the data base can be used. The initial design was driven by a requirement that the static GIS operate as usual, but has additional TGIS functions (Bishop, 1994). ArcView was selected because the product is the most popular fully functioning GIS software in geospatial information (GI) community. It is well documented and provides a macro programming language (Avenue) which makes rapid prototyping possible, is 'toolbox orientated' (in that the user can select functions in any sequence to make an application), and supports both vector and raster data models. The prototype TGIS has been developed using the vector data model. The second objective was to examine the functional limitations of TGIS that resulted from the decision to extend an existing vector GIS data model. Particular areas that were studied include the design of the TGIS data model and the user interface, system performance, spatio-temporal topology and data visualization (Burrough, 1998).

The prototype TGIS was successfully developed and is based upon ArcView GIS software. Macro programs were written with the private macro language of ArcView GIS (Avenue) to provide menu-driven temporal data management, data editing and data display and query functions. In the next sections the operation of TGIS functions used in the prototype TGIS implementation is described.

2.1   *The TGIS function to add temporal fields*

In this research on the implementation of a prototype TGIS, seven temporal fields have been added to the feature attribute table (FAT), consisting of: the tgis_id, Fdb_in, Fdb_out, Prev_id,

Table 1.   Complete list of items be added to FAT for the TGIS design.

Item name	Purpose of item
Tgis_id	Tracks all feature versions. (TGIS keeps this unique for each feature version; however, each feature version can have multiple instances due to line feature splitting)
Fdb_in	Stores the feature version database time at the start time
Fdb_out	Stores the feature version database time at the end time
Fev_in	Stores feature version event times at the start time
Fev_out	Stores feature version event times at the end time
Prev_id	Stores the 'tgis-id' of the previous feature version
Next_id	Stores the 'tgis-id' of the next feature version

Next_id and Lver. The list and the purpose for using these seven temporal fields are represented in Table 1. For any operation in TGIS, it is necessary to construct infrastructure consisting of views, themes, feature attributes, tables and the relationships between them. In some cases this process is performed by examining and rectifying existing elements.

2.2   The TGIS function to add new temporal features

This function is used for entering new spatio-temporal features into an active theme. When applying this function, a dialogue appears on the user interface. After entering the start event time, a user can draw the new temporal feature.

2.3   The TGIS function for feature selection

This function is used to select features in TGIS. Such tools in TGIS should be able to recognize current features from other feature versions. For this purpose the Lver (last version) field is used. Value one represents the last version of the feature and value zero indicates that the current feature version is not the last feature version. Using this function, the last version of the feature is selected so that other functions can be applied to the feature.

2.4   The TGIS function for modifying and updating feature event times

In some cases, there are some errors in temporal values, or if we want to change the values of temporal fields, this function applied to selected features, and the feature time is then entered. This function is used in cases where once one map time is stamped, it is necessary that the start event time of the features be changed (Peuquet, 1995; Beller, 1991).

2.5   The TGIS function for spatial modification, feature splitting and merging

For spatial modification or change the features, select the required feature. The spatial updating function is applied and the start time event entered in the dialogue.To split one feature, select the required feature and apply the splitting function. Then, with split tools the split line is lined on the feature and the start event time entered in the dialogue. For the multiple splitting of one object the method should be repeated. To merge some objects, select the required features and apply this function. Then, after entering the start event time in the dialogue by applying the union function, the selected features are merged. In their operations, these functions use prev_id and next_id fields so that the prev_id points to the previous versions and the next_id points to the next version of the features.

2.6   The TGIS functions for movement on multiple feature versions

These functions are used for movement on multiple versions of features on active themes. One of the applications of these functions is to examine trends in changes to features (Dorling, 1992).

2.7   *The TGIS function to export feature versions to separate files*

This function is used for saving selected levels of feature versions in a separate file with a shape file format. This function is used for examining trends in changes to spatio-temporal features. For any level of changes, a snapshot can be produced.

3   IMPLEMENTATION OF PRACTICAL SAMPLES OF TGIS

After describing the previous functions for examining the abilities and effectiveness of the proto-type TGIS, one digital cadastre map sheet of the city of Qom in Iran at a scale of 1:500, along with a feature attribute text file is used. In Figure 1, the customized user interface of ArcView is repre-sented. It can be seen that the TGIS functions have added to the user interface of ArcView GIS and that the non-temporal GIS function is also active.

   The spatial data set used is shown in Figure 2. The land parcels are a portion of the digital cadas-tre map of the city of Qom in Iran. These land parcels were used for specific tests of TGIS func-tions such as feature split, feature merge, topology queries, and so forth.

3.1   *Applying the TGIS function to add temporal fields*

With applying this function, a set of temporal fields are added to the active theme so that spatio-temporal operations can be done. The result of applying this function is illustrated in Figure 3.

3.2   *Applying the TGIS function to add new temporal features*

In applying this function, a new temporal feature with a start event time can be added to existing feature sets. A sample of the application of this function is represented in Figure 4, where a new feature with tgis_id = 482 has Fdb_in = 2002/11/28 and Fev_in = 2002/12/05.

Figure 1.   Customized user interface of ArcView GIS.

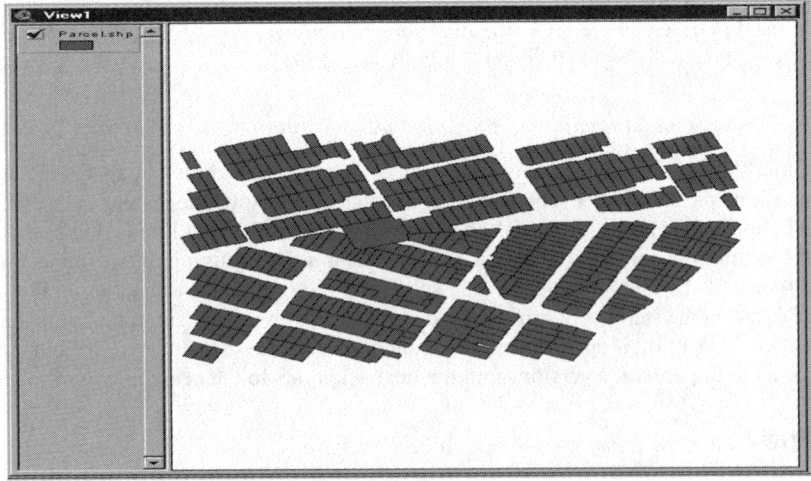

Figure 2.   Digital cadastre map of the city of Qom in Iran.

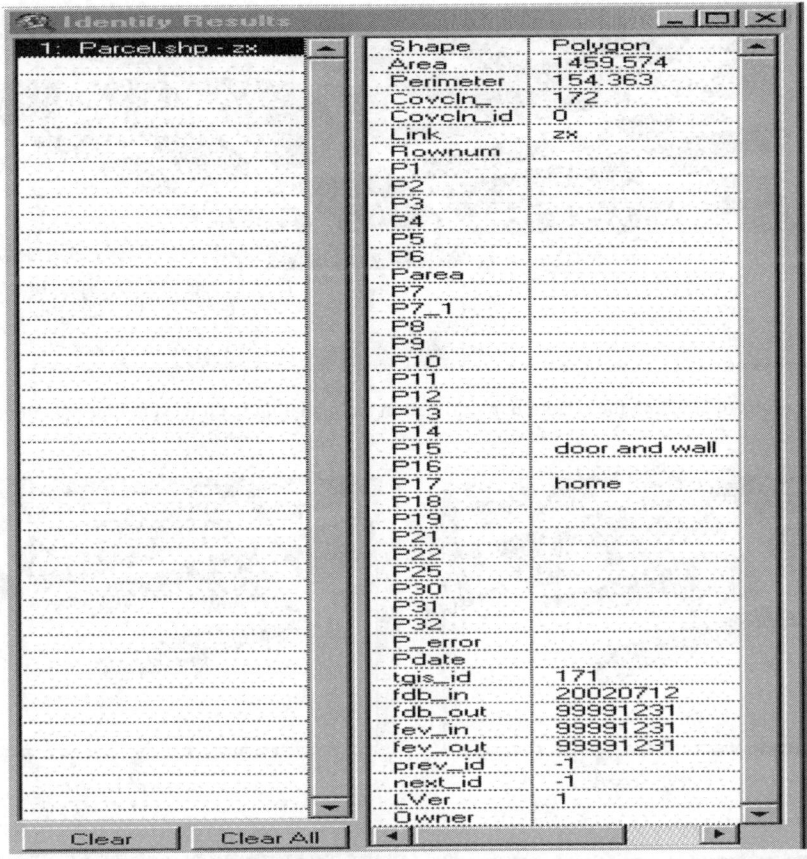

Figure 3. Results of applying the function of adding temporal fields.

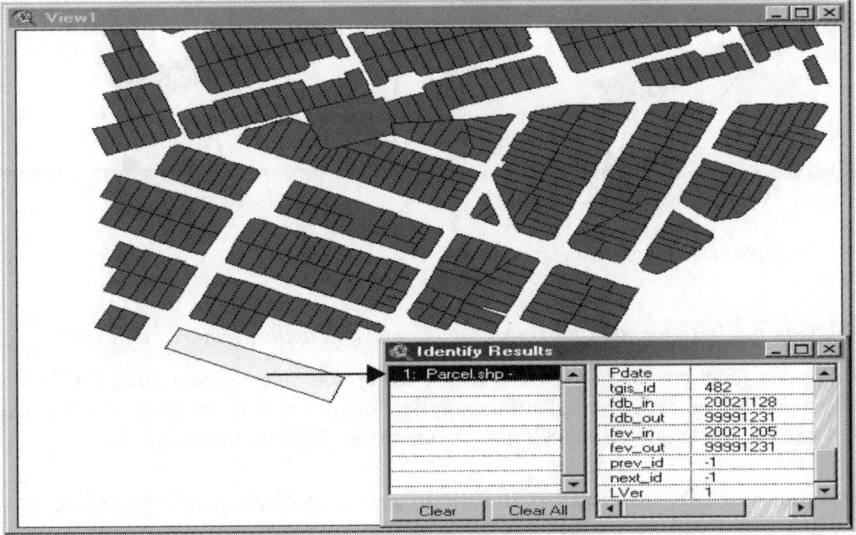

Figure 4. Results of applying the function of adding new temporal features.

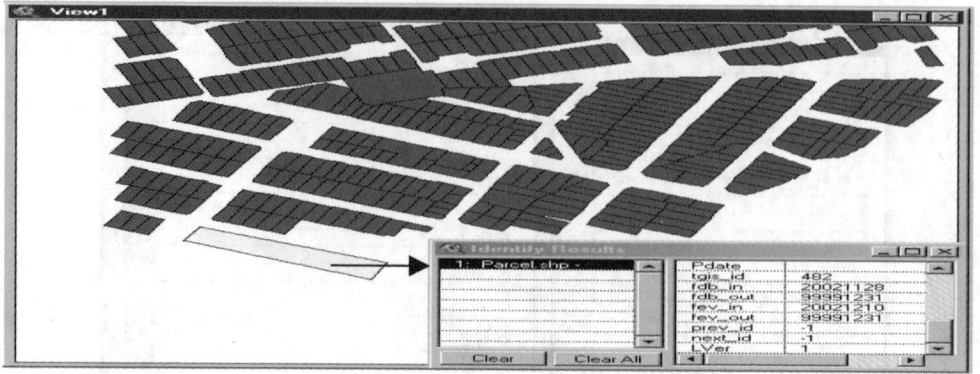

Figure 5.   Results of applying the selection, modification and updating feature event time function.

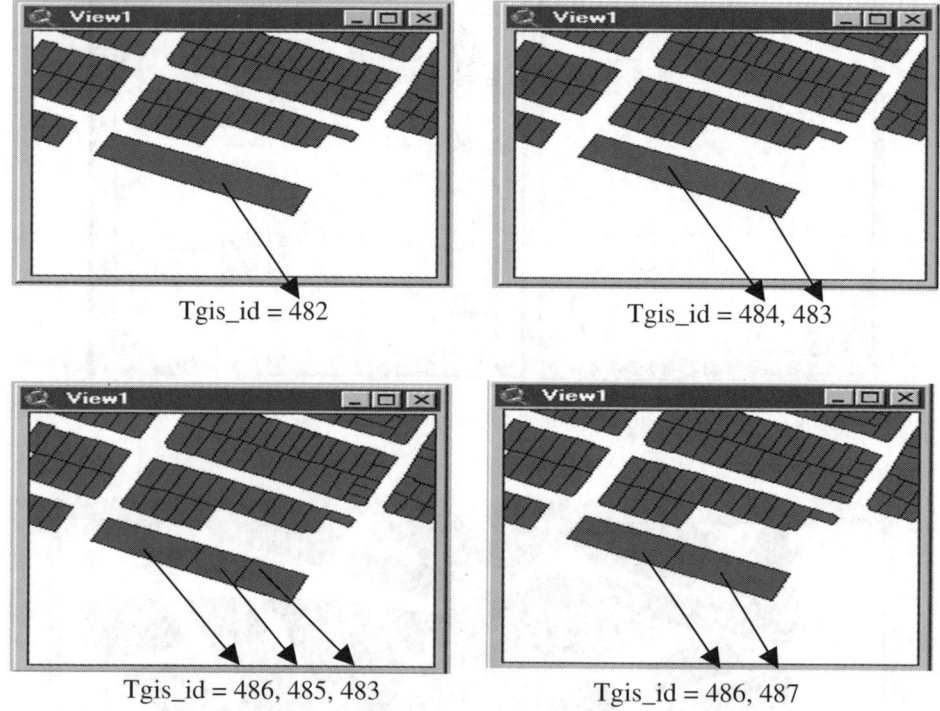

Tgis_id = 482

Tgis_id = 484, 483

Tgis_id = 486, 485, 483

Tgis_id = 486, 487

Figure 6.   Example of splitting and merging.

3.3   *Applying the TGIS function for selection, change and update feature event time*

In applying this function, the last version of the features is selected and other functions can be applied to them. In Figure 5, a sample of the application of this function is represented, where the Fev_in of a feature with tgis_id = 482, has changed from 2002/12/05 to 2002/12/10.

3.4   *Applying the TGIS function for spatial modification, features splitting and merging*

When applying these functions, selected features can be spatially modified, split or merged. In Figures 6 and 7 a sample of the application of these functions can be seen.

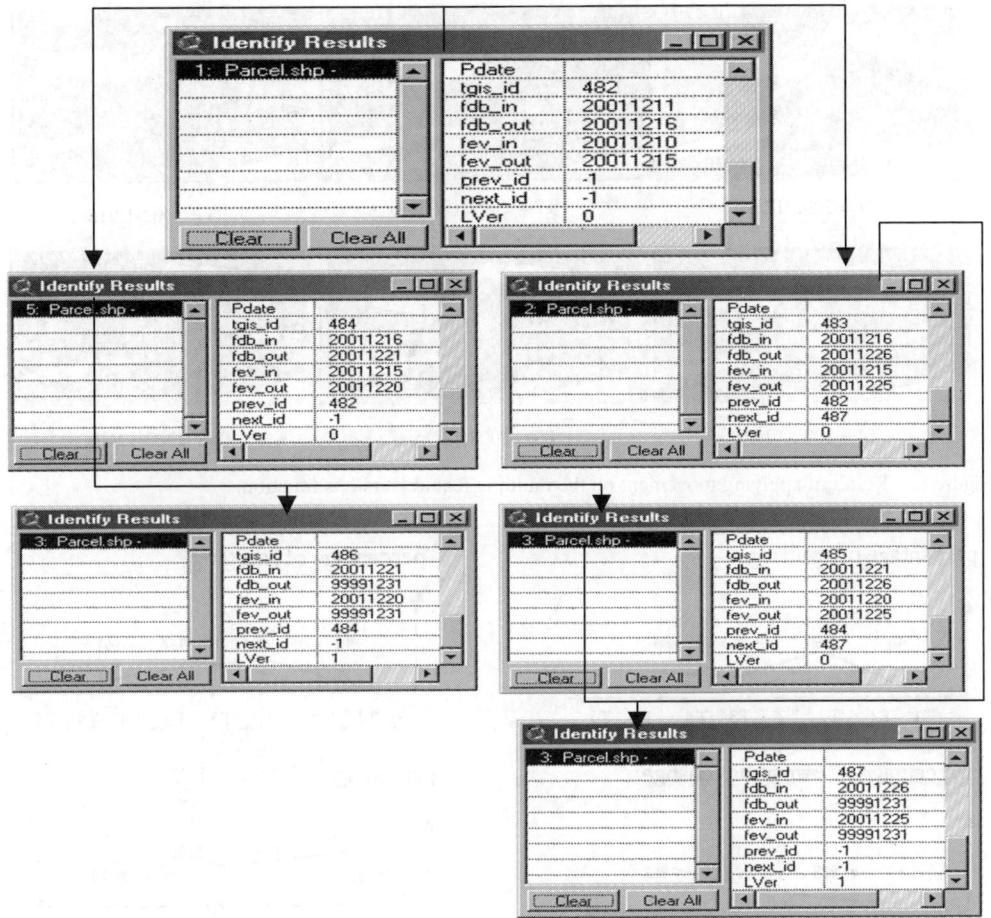

Figure 7.   Feature attribute table of the split and merged polygons.

In Figure 6, it can be seen that the polygon with tgis_id = 482 and Fev_in = 2001/12/10 can be selected and, in time, Fev_in = 2001/12/15 be split into two polygons with tgis_id = 483, 484. Then, in time, Fev_in = 2001/12/20 and the polygon with tgis_id = 484 has been split into two polygons with tgis_id = 485, 486. Again, in time, Fev_in = 2001/12/25, and the two polygons with tgis_id 485, 483 have been merged to the polygon with tgis_id = 487.

In Figure 7, a feature attribute table of each parcel can be seen, in which temporal fields and their change have been highlighted.

3.5   *Applying the TGIS function for movement on multiple feature versions*

When applying these functions, it is possible to move on multiple feature versions and change levels. In Figure 8, an example of the application of this function is represented (DiBiase, 1992).

3.6   *Applying the TGIS function for exporting feature versions to a separate file*

When applying this function, multiple change levels in separate files in the shape file format can be saved. For example, in Figure 8, for any times T1, T2, T3, T4, T5 and T6, a shape file can be drawn.

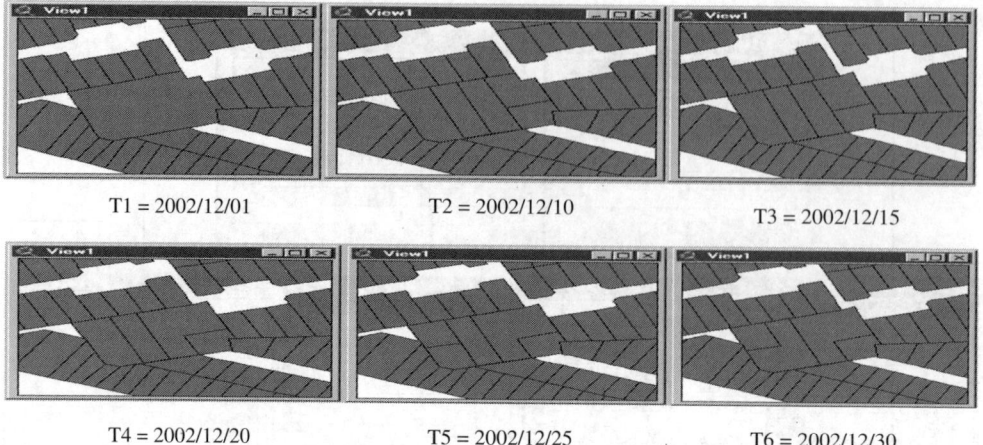

T1 = 2002/12/01   T2 = 2002/12/10

T3 = 2002/12/15

T4 = 2002/12/20   T5 = 2002/12/25

T6 = 2002/12/30

Figure 8.   Result of applying movement on the multiple feature versions function.

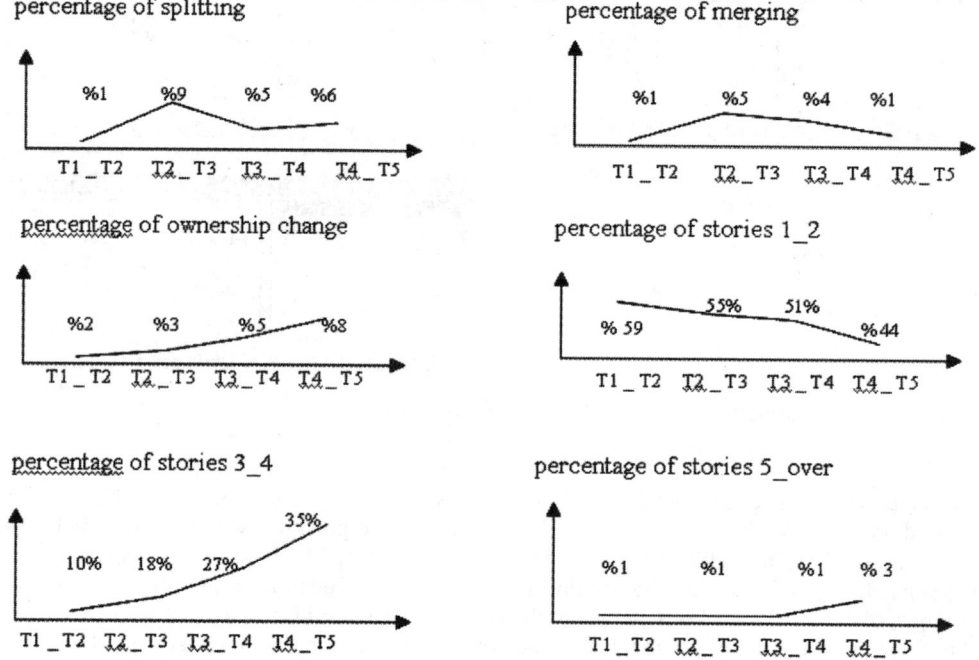

Figure 9.   Trends in spatial and attribute changes in the prototype TGIS.

3.7 *An analytical example of the prototype TGIS developed*

For examination of the prototype TGIS, a primitive map (a cadastral map sheet of the city of Qom in Iran at a scale of 1:500) has been used as a base map with an arbitary production date such as 2001/01/01. Some changes and modifications have been applied to the attributes and spatial elements of the parcels on the map. With a concentration on these changed features, some analytical results can be achieved. The diagramme of these changes for time intervals can be illustrated. These analyses are based upon the start and end event times fields, the prev_id field and next_id field, so that the changes can be traced (Pequet, 1994, 2001).

Table 2. Trends in changes in time intervals.

Operation	T1_T2	T2_T3	T3_T4	T4_T5
Percentage of splittings	3	9	5	6
Percentage of mergings	1	5	4	3
Percentage of ownership change	2	3	5	8
Percentage of buildings with 1 to 2 stories	59	55	51	44
Percentage of buildings with 3 to 4 stories	10	18	27	35
Percentage of buildings with 5 or more stories	1	1	1	3

In this example, from spatial changes only split and merge where selected, and from attribute changes only the name of the owner (owner field), number of stories 1 and 2 (T1_2 field), number of stories 3 and 4 (T3_4 field), and number of stories greater than 5 (T5_over field) were selected, and so five epochs of time such as T1 = 2001/01/01, T2 = 2001/04/01, T3 = 2001/07/01, T4 = 2001/10/01 and T5 = 2002/01/01 were selected. The diagrammes in Figure 9 and Table 2 represent trends in the changes in the four time intervals. The percentages of change from dividing the number of changed polygons of any type over the total number of polygons can be achieved.

4 CONCLUSION AND RECOMMENDATIONS

A prototype TGIS has been developed that has been successfully tested as an addition to the ArcView GIS. The system has functions for the development, editing, display and query of TGIS data sets, with the usual 2D GIS functions still available. The macro programs can easily be modified or new functions can be added. In the pratical example on test data (a digital cadastral map of the city of Qom in Iran), many analytical results can be represented such as:

- The maximum statistic on the change in ownership occurring in the time interval T4_T5 is 8% and the minimum statistic occurring in the time interval T1_T2 is 2%.
- In the time interval T1_T2, 59%, in the time interval T2_T3, 55%, in the time interval T3_T4, 51%, in the time interval T4_T5, 44% of buildings had one to two stories.

The prototype TGIS is limited by several factors. First, TGIS requires advanced hardware due to increased volumes of data and heavy demands on processors for the modelling and visualizing of graphics, although the prototype software could have been be made to be more efficient. Second, the vector update data model does not maintain feature versions of spatial topology. The prototype data model stores information on the history of the topology to help improve the performance of temporal queries. However, temporal analysis requires that the topology be built before the data is processed. Yet a better data model for polygons should be investigated (Worboys, 1994, 1995). The vector update data model also makes the detection of errors difficult, and further work should include the addition of comprehensive error checking and retroactive update functions. The ambitious research objectives were met: a prototype TGIS based upon ArcView workstation software was developed and the functional limitations resulting from the underlying 2D data model were examined. The research results show that a simple TGIS can be developed using current GIS science and technology. The prototype TGIS can give users the opportunity to experiment with spatio-temporal data. It is hoped that once the users become enthusiastic about the prototype product, GIS software vendors will make a serious attempt to develop a commercial TGIS product. TGIS is an exciting area for GIS researchers and many new ideas and software programs can be expected to be generated in the near future.

TGIS software will be developed by GIS and database system vendors will be driven by the demands of users with complex applications and large quantities of low-cost data. A TGIS must support a time coordinate for spatial data and link geographic features to multiple feature attribute records. The time value may represent database or event time. The TGIS database must be flexible enough to be able to deal with temporal resolutions varying between minutes and millions of years. Compared to the specialized temporal modelling software that is available at present, a TGIS will consist of a generic spatio-temporal database with a set of temporal data editing and analysis tools that the user can apply as required.

REFERENCES

Al-Taha, K. K. and Barrera, R. 1990. Temporal Data and GIS: An Overview. In Proceedings of GIS/LIS '90, (Bethesda, MD: American Congress on Surveying and Mapping: 244–254. Anaheim, CA., USA.

Armenakis, C. 1991. The Temporal Dimension of Geographical Data, Technical Report, Applied Research and Technology Service, Geographical services division, EMR, Canada.

Barrera, R., Frank, A. and Al-Taha, K. K. 1991. Temporal Relations in Geographic Information Systems: A Workshop at the University of Maine. ACM SIGMOD Record, 20: 85–91.

Beller, A. 1991, Spatial/Temporal Events in a GIS. In Proceedings of GIS/LIS '91, (Bethesda, MD: American Congress on Surveying and Mapping): 766–775. Atlanta, USA.

Bishop, I. 1994. The Role of Visual Realism in Communicating and Understanding Spatial Change and Process. In Visualization in GIS, edited by H. M. Hearnshaw and D. J. Unwin, Chichester: 60–64. UK: John Wiley & sons Ltd.

Burrough, P. and McDonnell, R. 1998. Principles of Geographic Information Systems. Oxford, OUP.

Burrough, P. A., van Deursen, W. and Heuelink, G.1988. Linking spatial process models and GIS: a marriage of convenience or a blossoming partnership. In Proceedings of GIS/LIS '88, (Bethesda, MD: American Congress on surveying and Mapping): 598–608. San Antonio, Texas, USA.

DiBiase, D., MacEachren, A. M., Krygier, J. B. and Reeves, C. 1992. Animation and the Role of Map Design in Scientific Visualization, Cartography and Geographic Information Systems, 19: 201–214.

Dorling, D. 1992. Stretching Space and Splicing Time: From Cartographic Animation to Interactive Visualization. Cartography and Geographic Information Systems, 19: 215–227.

Langran, G. 1992. Time in Geographic Information Systems. Technical Issues in Geographic Information Systems. Taylor & Francis, London.

Peuquet, D. and Wentz, E. 1994. An approach for time-based analysis of spatio-temporal data. In Advances in GIS Research. Proceedings, 6th International Symposium on Spatial Data Handling, Edinburgh, UK, edited by T.C. Waugh and R.G. Healey: 489–504. Taylor and Francis, London.

Peuquet, D. and Duan, N. 1995. An event-based spatio-temporal data model (ESTDM) for temporal analysis of geographical data. International Journal of Geographical Information Systems, 9: 7–24.

Pequet, D. J. 2001. Making Space for Time : Issues in Space–Time Data Representation. Geoinformatica :11–32.

Raper, J. 2000. Multidimensional Geographic Information Science: 1–121. Taylor and Francis, London.

Roshannejad, A. A. and Kainz, W. 1995. Handling identities in spatio-temporal databases Proc. Auto carto 12.

Spaccapietra, S. 2001. Spatio-Temporal Data Models and Languages. Geoinformatica : 5–9.

Worboys, M. F. 1994. Unifying the spatial and temporal components of geographical information. In Advances in GIS Research. Proceedings, 6th International Symposium on Spatial Data Handling, Edinburgh, UK, edited by T.C. Waugh and R.G. Healey : 505–517. Taylor and Francis, London.

Worboys, M. F. 1995. GIS: A Computing Perspective :1–120. Taylor and Francis, London.

An integrated Web-GIS approach for indicating risky places for children: a case of Mid-town Toronto

Cun Wang, Zhizhong Xu, Yonggang Hu, Alice Croitoru & Vincent C. Tao
Geospatial Information and Communication Technology (GeoICT) Lab, Department of Earth and Space Science, York University, Canada

ABSTRACT: Approximately one third of deaths among children between the ages of 1 and 14 years are related to unintentional injuries. Studies are commonly aimed at examining the health (physical and mental) and family factors that place children of this age group at risk; other risk factors that relate to the spatial environment in which children play tend to be overlooked. In view of this, this paper addresses four critical spatial risk factors: traffic, slope, water bodies and crime. The relationship of these four risk factors is studied and an index model is derived. Based on this model a GIS-based risk analysis of a study area is carried out and analysed. In addition, as it is argued that the potential impact of such a model could be realized by making risk-related information available to the public, the emphasis is put on providing parents with this information via the Internet. Using an advanced Web GIS system, this information could be displayed using a simple web browser; thus, parents can assess the risks children are likely to face in their playgrounds. Consequently, such information can help people to avoid hazardous areas and lead to a reduction in the rate of unintentional injuries.

1 INTRODUCTION

A variety of factors influence the overall safety of children. Suicide and incidents of family violence (Children's Safety Network, 2002), drowning, near drowning and other water-related injuries (Population, 2000), abduction (The Lost Children's Network, 2002), child abuse or neglect (Family Services Act of Ontario, 2002), traffic accidents, accidental falls (Statistics Canada, 1996) and other self-inflicted injuries (Baker, 1994) all contribute to the overall safety of children (CICM, 1994). Although considerable attention is commonly put on various factors, such as mental factors, disease factors, drug abuse, family violence, firearm accidents or HIV infection, fundamental environmental factors have often been overlooked in spite their importance.

The high impact of environmentally related factors is evident in a simple review of the leading causes for death among children aged 1–14 years: approximately one-third of deaths in this age group are associated with unintentional injuries (Statistics Canada, 1996). Consequently, unintentional injuries, including traffic accidents, drowning and falls, are ranked as the number one cause of death. A similar situation may be found in the United States. According to statistics for North America (American Academy of Pediatrics, 2002), within the category of unintentional injuries, traffic accidents rank first, followed by drowning and falls. Over half of these falls are directly related to terrain slopes. One should note that another key factor, which appears as 'Other causes' in Table 1, includes crime. It is evident that if parents become more aware to these two risk factors (crime and unintentional injuries) a considerable percentage of child injuries and deaths could be prevented.

In light of this, this paper focuses on two key issues in child risk management. The first issue relates to risk factors that are commonly overlooked; namely, *crime, traffic, water* and *terrain slope*. Another

Table 1. Leading causes of death (deaths per 100,000 of the population), (Statistics Canada, 1996).

Rank	Cause of death, ages 14 and under	Males		Females	
		Deaths	%	Deaths	%
1	Unintentional injuries	228	36.7	140	29.4
2	Cancer	95	15.3	84	17.6
3	Congenital anomalies	59	9.5	61	12.8
4	Suicides	32	5.2	9	1.9
5	Heart diseases	18	2.9	14	2.9
6	Hereditary and degenerative diseases of the central nervous system	15	2.4	12	2.5
7	Pneumonia and influenza	13	2.1	6	1.3
	Other causes	161	25.9	150	31.5
	Total	621	100	476	100

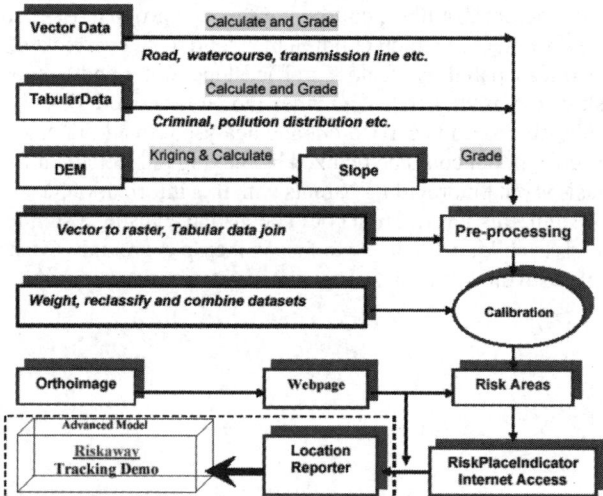

Figure 1. Risk assessment model.

focus is on exploring new technologies that enable parents and other stakeholders to review spatially enabled analyses of child risk factors. This risk assessment model is derived from the Cartographic model (Michael, 2000) (Figure 1). For this purpose a web-based Geographic Information System (GIS), with which parents may review various risk factors in their immediate neighbourhood, is applied. At the basis of this approach stands the assumption that all parents are interested in the well-being and safety of their children. Consequently, by providing risk-related spatial data, parents may manage the risks to which their children are exposed.

The implementation of the above approach is carried out using real-world data. For this purpose a study area of approximately 51 sq. km (ranging from the area of Steeles Ave. north to Lawrence Ave. south, and Jane St. west to Bathurst St. east in mid-town Toronto, Canada) is identified and analysed for a special group; namely, children aged 14 and under.

2   THE MODEL

The model implemented in this paper is derived from an index model with four risk factors. Each risk factor has a weight that is obtained from census data and a risk place survey. In order to review this model, a more formal definition of the various components of the model will be given. This will be followed by a review of the model formalization.

2.1 Definitions

- *Risk*: Factors that may cause injuries, disabilities or even death for children;
- *Risk factors*: crime, traffic, water and slope, used in the risk area indicator model;
- *Traffic*: Injuries related to motor vehicles, other road vehicles and pedestrian injuries;
- *Slope* (fall): Injuries related to falling from slopes and high elevations;
- *Water* (drowning): Injuries caused by drowning, near drowning and other water-related injuries;
- *Crime*: All offences that may cause injuries to children;
- *Model weights* (W_i): Coefficients of each risk factor – W_1: for Traffic (T); W_2: for Slope (S); W_3: for Water (W); W_4: for Crime (C).

2.2 Weights of an index model

In order to generate risk areas, an index model is used in this paper to evaluate risk factors. From these risk factors index values are computed, and a raster map based on the index values (Lo, 2002) is generated. This index model can be expressed as a linear equation:

$$R = (W_1F_1+W_2 F_2+W_3 F_3+W_4 F_4) / \Sigma W_i \qquad (1)$$

or

$$R = \Sigma f (W_1F_1+W_2 F_2+W_3 F_3+W_4 F_4) \qquad (2)$$

where R is value of risk, F_i are risk factors with levels 0–9. The F_1, F_2, F_3 and F_4 represent traffic, slope, water and crime, respectively,

$$F_i: \text{(traffic, slope, water, crime)} \qquad (3)$$

and W_i are risk coefficients, where:

$$\Sigma W_i = (W_1+W_2+W_3+W_4) = 1 \qquad (4)$$

By evaluating each risk coefficient W_i and implementing the above model, a map of risk place indicator can be derived. Such a map could then be used for risk analysis and management. Thus, in order to implement this model the weights should be evaluated.

2.2.1 W_1 – traffic related coefficient

According to the Vital Statistics Compendium, 1996 (Statistics Canada, 1996), the recent five-year average of the mortality rate for motor vehicle accidents is 23.0 (males and females) per 100,000 of the population of Canada (see Figure 2).

As Toronto is the fourth-largest city in North America, it is also possible to use some updated American data for reference. According to statistics for the year 2000 from the National Highway* Traffic Safety Administration (NTHSA, 2002), more than 41,000 people were killed in the US in traffic accidents, an average of 115 deaths a day. The traffic-related mortality rate is approximately 18 per 100,000 of the population. The NTHSA also indicates that young people lead the categories of victims of fatal motor vehicle accidents [* highway refers to main traffic roads, and excludes local roads].

In the case of children, the total traffic mortality rate of the age group 0–14 is 19.5 (6.31 + 6.08 + 7.07) per 100,000 of the population (statistics of Maternal and Children Health Bureau of

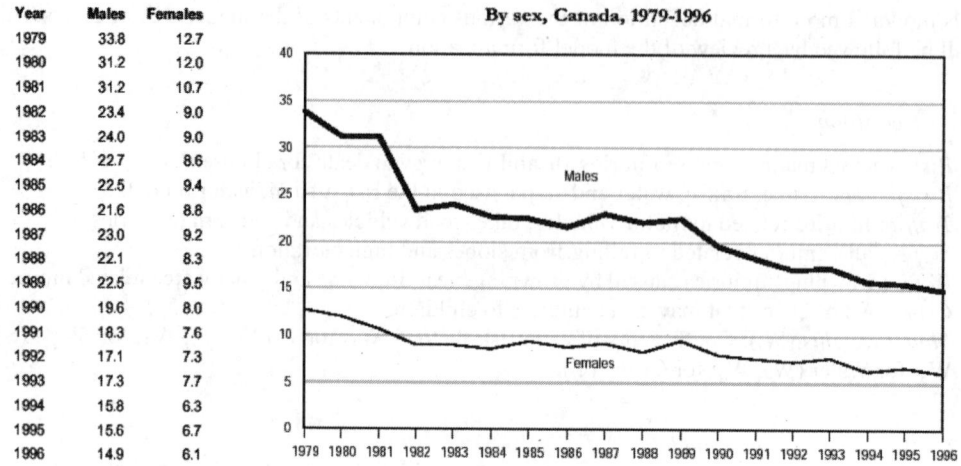

Year	Males	Females
1979	33.8	12.7
1980	31.2	12.0
1981	31.2	10.7
1982	23.4	9.0
1983	24.0	9.0
1984	22.7	8.6
1985	22.5	9.4
1986	21.6	8.8
1987	23.0	9.2
1988	22.1	8.3
1989	22.5	9.5
1990	19.6	8.0
1991	18.3	7.6
1992	17.1	7.3
1993	17.3	7.7
1994	15.8	6.3
1995	15.6	6.7
1996	14.9	6.1

Figure 2. Motor vehicle accidents (deaths per 100,000 populations), (Statistics Canada, 1996).

Table 2. Traffic mortality rate by age, 1999 (deaths per 100,000 of the population), (Health, 2002).

	Age group (years)			
	0–4	5–9	10–14	15–19
All transport	4.79	4.50	5.53	27.61
Pedal cyclist	0.03	0.37	0.47	0.42
Pedestrian	1.49	1.21	1.07	1.58
Total	6.31	6.08	7.07	29.61

HRSA (Health Resources and Services Administration, 2002)). More details are listed on Table 2. Consequently, W_1 is set to 19.5.

2.2.2 W_2 – slope related coefficient

As indicated earlier, over half of fall-related injuries are caused by slopes. According to the Vital Statistics Compendium, 1996, (Statistics Canada, 1996), the average mortality rate for falls in the recent five years is 15.6 (males and females) per 100,000 of the population in Canada (Figure 3). Consequently, W_2 is about 7.8 per 100,000 of the population. It should be noted that this factor should be further adjusted, as it includes falls by infants indoors, falls by elderly people and suicides.

What is the rate for the US? The National SAFE KIDS Campaign (Safekids) states that falls continue to be the leading cause of nonfatal unintentional injuries among children aged 14 and under. In 2001, more than 2.5 million children in this age group required treatment in hospital emergency rooms. The falls resulting for children aged 14 and under are most often due to slipping or tripping, including in sports or recreational activities (21 per cent), and playground injuries (18 per cent). One out of every fifty injured children suffers from severe, permanent disability, or even death. Compared to US population of 285 million (U.S. Census Bureau, 2001), the Canadian rate can be calculated by:

$$W_2 = 2.5\,M \div 285\,M \times 100{,}000 \times (21\% + 18\%) \times 1/50 \approx 6.8 \text{ per } 100{,}000 \text{ population, hence,}$$

W_2 is set as 6.8.

2.2.3 W_3 – water related coefficient

The preliminary adapt of child deaths due to external causes, 2000 by the HRSA (Health Resources and Services Administration, 2002), shows that the figure of death by drowning is half that of death from motor vehicle accidents; therefore, the W_3, water-related coefficient is set as 9.8.

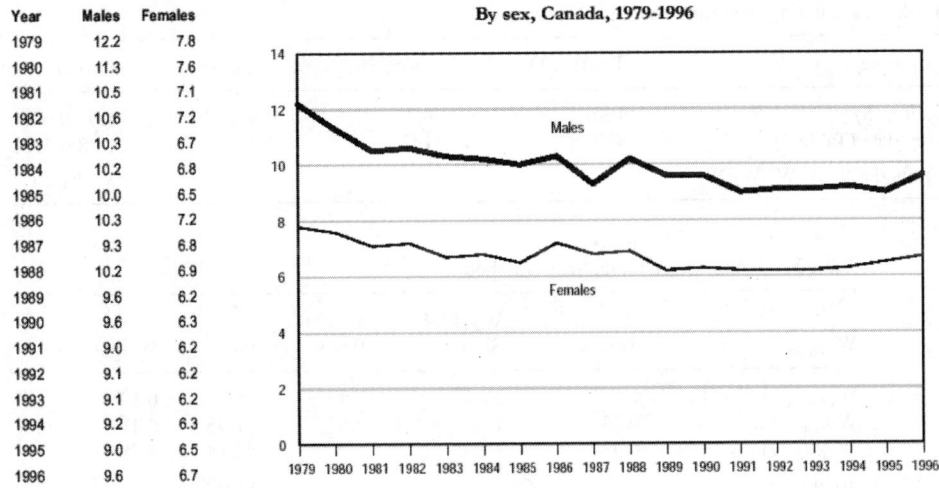

Year	Males	Females
1979	12.2	7.8
1980	11.3	7.6
1981	10.5	7.1
1982	10.6	7.2
1983	10.3	6.7
1984	10.2	6.8
1985	10.0	6.5
1986	10.3	7.2
1987	9.3	6.8
1988	10.2	6.9
1989	9.6	6.2
1990	9.6	6.3
1991	9.0	6.2
1992	9.1	6.2
1993	9.1	6.2
1994	9.2	6.3
1995	9.0	6.5
1996	9.6	6.7

Figure 3.   Accidental falls (deaths per 100,000 of the population), (Statistics Canada, 1996).

Table 3.   Scores (sums of the grade) for risk factors.

Risk factors	Crime	Slope	Traffic	Water	Pollution	H.voltage	Noise	Disco & Bar	Sum
Total score	1080	143	462	260`	130	15	0	110	2200

2.2.4 W_4 – crime related coefficient

The study area is mainly located in Division 31 of the Area Command of the Toronto Police (Toronto Police, 2002). According to Toronto Police statistics (2000), Division 31 has a violent crime rate of 1637 per 100,000 of the population (3242/198,073). Combining these four coefficients into Equation (2) gives:

$$R = 19.5T + 6.8S + 9.8W + 1637C \qquad (5)$$

It appears that $R \approx f(C)$, since $W_4 \gg W_1 + W_2 + W_3$. This is caused by the way in which this weight is derived: W_4 is a combination of mortality rate and the rates for other types of violent crime; while W_1, W_2 and W_3 include only the mortality rate. Consequently, W_4, the crime-related coefficient, should be further refined for children.

2.3   Tuning of the risk model

Because the risk model will be used by parents who live in the study area, a random risk place survey was carried out in order to adjust and tune the model. The survey was carried out in 10 different shopping locations in mid-town Toronto. Thirty-three forms were filled by parents who were accompanied by their children. From the survey data the score ratio of [traffic:slope:water] was derived. Parents were asked to choose four factors from eight risk factors: Crime, Slope, Traffic, Water, Pollution, High Voltage, Noise and Disco & Bar, as shown in Table 3, and to give a grade of between 1 (lowest risk) to 10 (highest risk) to each of their four selected factors. The results obtained are summarized in Table 3. The score in each column is the sum of the grade given by parents to each relevant factor. These are then normalized to a percentage score of weight for the various risk factors.

In order to derive the final W_1:W_2:W_3 ratio, the results obtained from the survey are combined with the ratios that were derived earlier (Equation 5). This process is described in Table 4. In this

Table 4. Calculating the ratio of traffic:slope:water.

Data source	Traffic (T)	Slope (S)	Water (W)	Ratio of T:S:W
Equation (5)	19.5	6.8	9.8	1.99:0.69:1
Survey data (Table 3)	462	143	260	1.78:0.55:1
Combined ratio of $W_1:W_2:W_3$				3.77:1.24:2

Table 5. Comparison table, Seety method.

W_{column} \ W_{row}	W_1 (3.77) Traffic	W_2 (1.24) Slope	W_3 (2) Water	Sum	Weight
$W_{1Traffic}$ (3.77) (T)	1	3.04	1.89	5.93	0.538
W_{2Slope} (1.24) (S)	0.33	1	0.62	1.95	0.177
W_{3Water} (2) (W)	0.53	1.61	1	3.14	0.285
Total sum				11.02	

table, the number of first lines are obtained from Equation 5, then divided by the number for water (9.8). The first ratio of T:S:W is derived as 1.99:0.69:1. Using the same method, the second ratio is derived as 1.78:0.55:1. The combined ratio is added from the above two lines.

In Table 5, values are calculated from each W_{column} (the number in the first column) divided by each W_{row} (the number in the first line). For example: $W_{1Traffic}(3.77) \div W_2(1.24) = 3.04$. The weights are obtained from the Sum divided by Total Sum (11.02). Based on these weights it is found that:

$$W_1:W_2:W_3 = 5: 2: 3 \quad \text{(Traffic: Slope: Water)} \tag{6}$$

Furthermore, the results of the survey indicate that the score given by parents to the crime factor is approximately half of the total score (1080 for crime vs a total of 2200). As a result, W_4 can be set to 0.5. As the sum of all weights should be 1 (Equation 4), the resulting weights W_1 through W_4 can be set as follows:

$$W_1+W_2+W_3+W_4 = 0.25 + 0.1 + 0.15 + 0.5 = 1 \tag{7}$$

Consequently, the final adjusted model is:

$$R = 0.25T + 0.1S + 0.15W + 0.5C \tag{8}$$

Since the values for traffic, slope, water and crime are reclassified as being between 0 to 9, (Bernhardsen, 1999), the overall range of risk should be 0 to 9 as well, according to Equation 8. Another justification for setting W_4 to 0.5 can be found in the results of the field survey: most parents indicated that they would prevent their children from using areas that have a criminal risk rank of more than 5. This preference of the parents was also implemented in the model by a transformation of the risk ranks according to the crime risk factor, using a logical AND operator. The implementation of this operator was carried out using the key described in Table 6.

The above transformation is implemented by the following mathematical model:

$$R = \overline{C} \cap (W_1T + W_2S + W_3W) + W_4C \tag{9}$$

$$R_{new} = \{\text{high if } C \geq 5, R\} \tag{10}$$

Table 6. Criminal rate.

Crime factor	0	1	2	3	4	5	6	7	8	9	Total
Total Mark	0	0	1	1	2	10	7	0	1	0	22

Thus, risk values of crime over 5 are reclassified as 9 (maximum value), and values of risk over 5 are defined as high risk. Consequently, all of the crime risk areas ranked over 5 will be shown as high-risk areas. Finally, we define range 0–2 as safe, 2–3 as low risk, 3–5 as medium. The output of our model was:

$$R_{output} = \text{Reclassify} (\text{Safe}_{0-2}, \cup \text{Low}_{2-3}, \cup \text{medium}_{3-5}, \cup \text{high}_{5-9}) \tag{11}$$

3 SPATIALLY ENABLED RISK ANALYSIS

In order to provide the ability to assess the risk model within its spatial domain, several spatial data layers were collected. These included the road network (traffic by hour), the hydrology network (rivers), and a digital terrain model (DEM). In addition, population data and boundaries were collected and integrated. This allowed a more detailed modelling of the crime rate since the crime rate of the Toronto Police is defined as crime offence to population ratio. Consequently, a product of population density and crime rate is used as the criminal variable in the model (Stan, 1995).

3.1 *Indices of spatial safety indices*

In addition to the various risk factors that are quantified and calibrated, spatial indices should be derived. These indices should describe and define the boundaries of the environment that is considered safe for children. In the spatial domain two indices are considered, namely safe distance and slope.

Safe distance: The safe distance is considered as the minimal distance between a child and a potential risk factor a parent would allow and would consider as safe. The survey that was carried out indicates that parents indicated a distance of 50 to 100 m as a safe distance from rivers and main roads (Table 7).

As a result, a distance of 100 m is set as a maximum buffer distance to roads and rivers, and the risk values are set as the following: 0–10 m as risk value 9, 10–20 m as risk value 8 … 90–100 m as risk value 0. Therefore, the layers of traffic and water are reclassified with values of 0 to 9.

Safe slope: A safe slope is defined as the maximal terrain slope that parents consider as safe for their children. The survey results indicate that a slope of over 30° is considered by parents as dangerous (Table 8).

Therefore, when reclassifying the slope, all values of a slope of less than 30° are set to zero. As it is impossible for children to play on a slope of over 60°, the value of a slope of over 60° is also set to zero. The slope layer is derived from DEM, and 30° to 60° was reclassified as a value of 1 to 9.

Since the distance and slope degrees are obtained from parents who have children aged 14 and under, the above analyses are suitable to the group described in the model.

3.2 *Criminal analysis*

The enumerated population data, downloaded from Census Canada, shows the population density for each unit of area. Thus, the criminal variable can be obtained from Equation 12:

$$\text{Criminal Variable} = (\text{population density}) \times (\text{criminal rate}) \tag{12}$$

The criminal variable layer is then reclassified from 0 to 9, and all values of over 5 to 9 are changed (according to Equation 10).

Table 7. Safe distance.

Distance from	<20 m	50 m	100 m	150 m	>200 m	Total
River & pond	0	6	11	1	4	22
Main road	2	8	10	0	2	22

Table 8. Analysis of slope degree.

Dangerous angle of slope	0°–20°	20°–30°	30°–40°	40°–50°	50°–60°	60°–70°	Total
Total marked	0	2	12	7	1	0	22

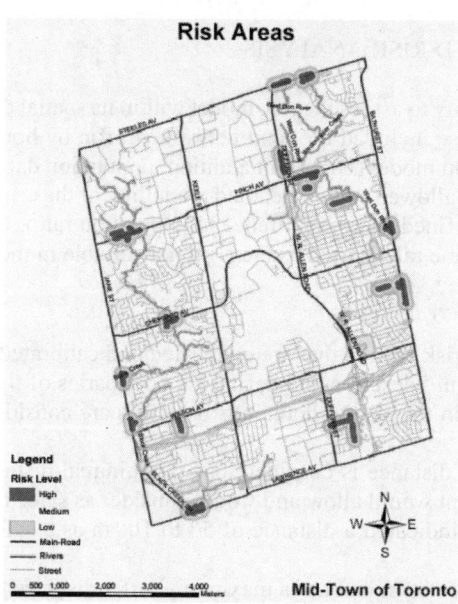

Figure 4. Risk areas for children.

All four layers (traffic, slope, water and crime) are now normalized with a unique value range. By doing a raster layer calculation with weights in Equation 8, a final layer is generated to show Risk areas for children.

4   RESULTS

Using the risk model, the spatial indices and the spatial data, a reclassification of the total risk into three levels (low, medium and high) is carried out. Based on the results obtained, a map showing all risk areas for mid-town Toronto is compiled (Figure 4). In the map, the green areas indicate low risk, the yellow areas with a green circle are the medium-risk areas, and red areas are high-risk areas. It should be emphasized that most of high-risk areas are in the vicinity of traffic intersections, rivers and regions with high population densities, and along the slopes of river banks or the edges of roads. Hence, the model successfully indicated areas of risk in terms of traffic, water, and terrain. It should be noted that in the centre of the study area there is a small airport that was not accessible for security reasons. As a result this area was not ranked.

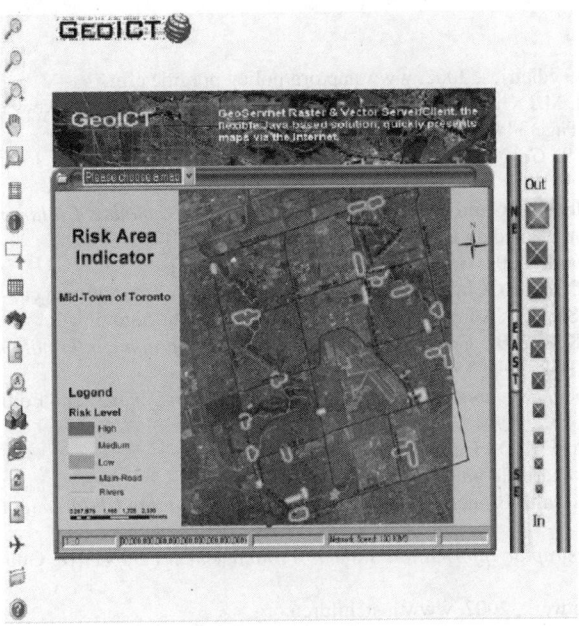

Figure 5.   Risk place indicator.

In order to allow parents to utilize the results of this study and to further explore its application for the benefit of parents, the resulting map is also made available through a web based GIS system – *GeoServNet* (GSN) that was developed at GeoICT (Figure 5). In this system the results of the analysis may be viewed with a background of an ortho-photo of the area through the Internet. By logging to the GSN webs page using an ordinary web browser parents can then easily review the locations of the risky areas. In addition, the GSN interface can provide parents with specific information (for example, by typing a postal code into the system, the map will show certain areas of interest). GSN can also provide 3-D visualizations of slopes, thus allowing parents an easier review of the terrain.

5   CONCLUSIONS AND FUTURE WORK

Allocating safe playgrounds and determining potentially high-risk areas for children are essential if unintentional injuries among children under 14 are to be reduced. Yet, the ability of such information to reduce injuries depends highly on its availability to the public. In this paper we have demonstrated the benefits that could result by combining a spatially based risk analysis model with advanced web GIS capabilities. This combination provides not only the means for an efficient implementation of the risk model, but also the means to provide such information on demand to parents. If, through this technology, only half of unintentional child injuries could be prevented, the total mortality rate of children under the age of 14 will drop by 33%. Hence, the potential impact of this technology on child safety within communities is likely to be high.

Further research and development of this technology is still required in order to develop it into a full child risk management system. Combined with an automatic child-tracking system, and linked with a spatial database (such as the one shown in Figure 5) an alert system that will send an instant message via the Internet and a short message through cellular phone to parents could be devised.

REFERENCES

American Academy of Pediatrics, 2002. www.aap.org/policy/pprgtoc.cfm.

Baker, S.P., B. O'Neill, M.J. Ginsburg and G. Li. 1992. *The Injury Fact Book.* New York: Oxford University Press, 2nd edition. Page 344.

Bernhardsen, Tor, 1999. *Geographic Information Systems – An Introduction.* John Wiley and Sons Inc., New York. Chapter 14.

CICM, Canadian Institute of Child Health. 1994. *The Health of Canada's Children: A CICH Profile.* 2nd edition. Ottawa, Ont.: Canadian Institute of Child Health. Page 175.

Children's Safety Network, 2002. www.childrenssafetynetwork.org.

Family Services Act of Ontario, Children Protection, 2002. www.ccas.toronto.on.ca.

Health Resources and Services Administration, 2002. www.mchb.hrsa.gov.

Lo, C.P. and Yeung, A.K.W. 2002. *Concepts and Techniques of Geographic Information System.* Upper Saddle River, NJ: Prentice Hall. Page 152.

Michael N. DeMers, 2000. *Fundamentals of Geographic Information Systems.* 2nd Edition. John Wiley & Sons, Inc. Chapter 13.

Safe Kids. National SAFE KIDS Campaign at www.safekids.org.

NTHSA, 2002. www.samarins.com/driving.

Population and Public Health Branch, 2000. *For the Safety of Canadian Children and Youth.* Health Canada: Chapter 11.

Stan Arnoff, 1995. *Geographic Information System: A Management Perspective.* Ottawa, WDL Publications. Page 115.

The Lost Children's Network, 2002. www.lostchildren.org.

Toronto Police, 2002. www.torontopolice.on.ca.

Statistics Canada, 1996: *Vital Statistics Compendium*, Chapter 7, page 99.

U.S. Census Bureau, 2001. http://quicfacts.census.gov/gfd/states/00000.html.

Advances in Spatial Analysis and Decision Making, Li, Zhou & Kainz (eds)
© 2004 Swets & Zeitlinger, Lisse, ISBN 90 5809 652 1

Knowledge-based land use dynamic monitoring supported by the '3S'-integration technique

Changsheng Xue, Qingquan Li, Deren Li & Bijun Li
National Laboratory for Information Engineer in Surveying, Mapping and Remote Sensing,
Wuhan University, Wuhan, China

ABSTRACT: The New National Land and Resources Investigation Project in China was launched in 1999, and anticipated to last 12 years. Land use dynamic monitoring is an important part of this grand, cross-century project. In this paper, the general structure and technical approach of knowledge-based land use dynamic monitoring supported by the '3S'-integration technique is presented. The emphasis is on the implementation of this method, which is involved in satellite imagery processing, land cover classifications based on the spectral repository and land use classifications based on knowledge, field investigations (DGPS surveys), and the establishment of a land use database and land use dynamic monitoring system. To test the approach, we selected a site for the case study in Wenling city in Zhejiang province, China. The experiments show that this approach greatly improves the accuracy of land use classifications and work efficiency, compared with the traditional technical approach of land use dynamic monitoring.

1   INTRODUCTION

Land resources are an important resource, because they are related to almost all matters involving people. In China, it is a fact that the amount of cultivated land has been shrinking even as the economy has been growing. Such a development makes it necessary for the land management division of the government to obtain detailed information about the land and changes in land use. In 1988, the detailed information on land was gathered all over the country. Fourteen years later, the situation has changed greatly, and a second detailed investigation on land resources needs to be carried out. Thus, the Chinese government launched the New National Land and Resources Investigation Project in 1999, which is planned to last 12 years. For this project, new thoughts and new methods have been advocated. Land use dynamic monitoring is an important part of this grand and cross-century project. In section 1, this paper describes the general structure of knowledge-based land use dynamic monitoring supported by the '3S'-integration technique. Section 2 gives a detailed description of its implementation. Section 3 tests the thought by the test. Finally, Section 4 concludes with some summary remarks.

2   THE GENERAL STRUCTURE OF KNOWLEDGE-BASED LAND USE DYNAMIC MONITORING SUPPORTED BY THE '3S'-INTEGRATION TECHNIQUE

Knowledge-based land use dynamic monitoring supported by the '3S'-integration technique is composed of three parts: Remote Sensing imagery processing, analysis and classification, field surveys and a land use database. The general structure is shown in Figure 1.

In the integrated system, the technique of Remote Sensing (RS) has been used to supply real-time information on the target and its surroundings. By means of this technique, all kinds of information

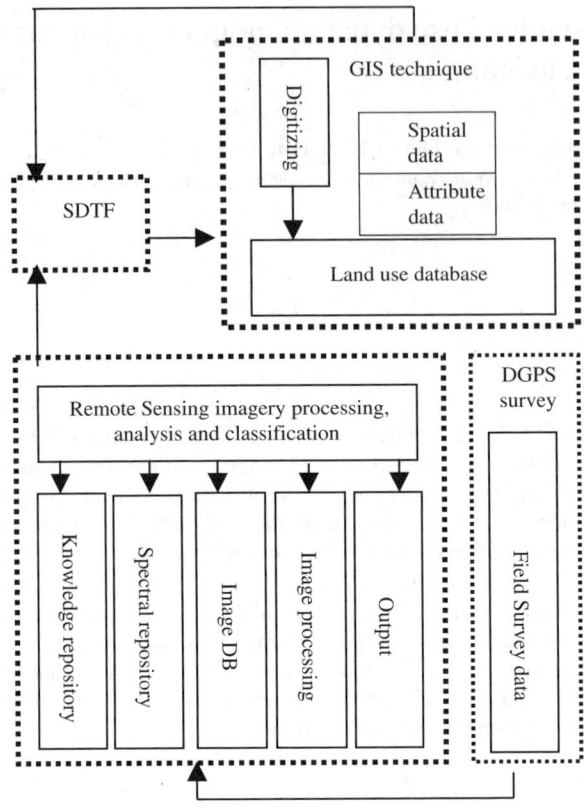

Figure 1. The general structure of knowledge-based land use dynamic monitoring.

can be found on changes to the land on the surface of the earth, and land use data can be updated. The Global Positioning System (GPS) can acquire the instant coordinates and attitude data using various kinds of sensors and carrier platforms (such as vehicles, ships, aircrafts, satellites, etc.) in real time. The Geographical Information System (GIS) provides a database management platform for the integration, display, and analysis of the data collected from GPS and RS images. Moreover, it can supply geographic knowledge for the acquisition of intellectual data.

In order to integrate the system, an internal contact, or a common frame of reference must be created among the RS, GPS receiver and GIS data management systems. That is to say, first, the coordinate system must be unanimous among the RS, GPS and GIS data. Second, different data file formats should be conveniently transferable between satellite raster imagery and GIS vector data, or between GPS survey data and GIS data. With a clear matching relation in space and time, such multi-sensor data can be inputted, managed and analysed in GIS, and provide detailed background information for land use dynamic monitoring.

3 THE IMPLEMENTATION OF THE THOUGHT

The technical approach of knowledge-based land use dynamic monitoring supported by '3S' integrated technique is shown in Figure 2. The technical approach will be described in detail, and will involve the processing of satellite images, land cover/land use classifications, field investigations (DGPS surveys), updating the land use database and the establishment of a land use dynamic monitoring system.

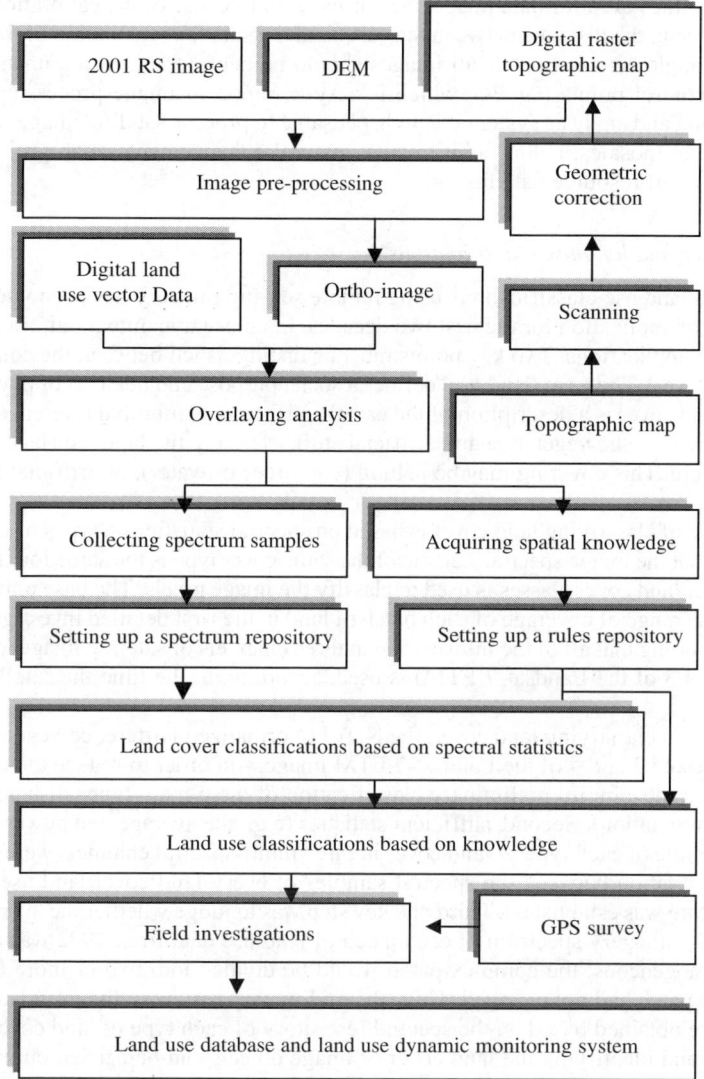

Figure 2. The technical approach of knowledge-based land use dynamic monitoring supported by the '3S' technique.

3.1 *Satellite imagery processing*

Usually, a panchromatic band covers a broad wavelength range of the visible and near infrared spectrum, while a multi-spectral band covers only a narrow spectral range. Most sensors of earth resource satellites provide panchromatic and multi-spectral images at different spatial resolutions to effectively utilize such images. The ability to effectively combine high-resolution panchromatic and low-resolution multi-spectral images into one colour image is demanded. The spectral and texture information for fused images comes separately from low-resolution multi-spectral images and high-resolution panchromatic images. The purpose of fusing images is to make use of the advantages of multi-resource data, improve the spatial and spectral resolution of fused images, and advance the accuracy of image interpretation. Before the images are fused, it is necessary that the brightness values of multi-data images be normalized. Because of the calibration of detectors, the angle of the sun, the distance between the earth and the sun, the attenuation of the atmospheric and the phase angle between dates, all images should be georeferenced to a unanimous system using ground control points (GCPs), which is very common in image processing. Thus, in an image processing and analysis system it is indispensable to process satellite images for image rectifications, image mosaics, colour adjustments, textural enhancements, map projections, image matching, and multi-resource data fusion.

3.2 *Land cover and land use classifications*

Land cover and land use classifications using remote sensing imagery has been widely employed in many fields of application for the past two decades. Land use dynamic monitoring is one of the most important applications. Two key points must be distinguished between the concepts of 'land use' and 'land cover'. The term 'land use' defines a social purpose and not a set of physical qualities. By contrast, land cover is a description of the earth's surface with minimal reference to social purpose, referring to '… the vegetative and artificial stuff' covering the land's surface in terms of its physical structure. This covering may be natural (e.g., trees or water), or artificial (e.g., concrete or tarmac).

The principle of classifying land cover is based on spectral statistics on types of land cover. The assumption is that the image spectral statistic of the same cover type is the same for all of the pixels. The classifier of land cover classes is used to classify the image pixels. The base unit in classifying land cover is the range of coverage of each patch of land in the first detailed investigation into land use in 1992. In doing that all of the information in three channels of satellite imagery (for example, band 7, 5 or 1/4/3 of the Landsat-7 ETM) is used, according to the time the satellite image was received.

The first step in classifying land cover, the NDVI (Normalized Difference Vegetation) was created based on Band 3 and 4 of the Landsat-7 ETM imagery in order to reduce the amount of data and to achieve a meaningful preliminary classification of three main types of land cover (water, buildings and vegetation). Second, sufficient statistics (e.g., the average and power difference) on the spectral sample of each type of land cover in three multi-spectral channels were collected. The corresponding relation between the spectral samples of every land cover/land use classification and image feature was established. Third, the key step was to judge whether the average and power difference of the imagery spectrum of each patch of land use data from 1992 was homogeneous. If it was inhomogeneous, the complex patch would be divided into two or more than two ones. Otherwise, the patch had not changed. Thus, the preliminary results of the exercise in classifying land cover were obtained based on the spectral repository of each type of land cover.

Classifying and identifying the land cover of image objects can be carried out mainly according to their spectral statistics. By contrast, the classifying of land use at a higher level needs to be determined not only on the basis of land cover objects and but also according to spatial information. Therefore, knowledge on the spatial distribution of land use types and on ecological conditions must be accumulated. According to the data and to field investigations, some useful spatial knowledge was collected: The distribution of land use depends upon altitude. For example, cultivated land, water, urban areas and garden plots exist, respectively, in different altitudes. Data on slopes can also distinguish certain types of land use from others. The same object presents a different brightness because of different terrains. In other words, 'the same object presents a different spectrum'. Concretely, the spectral features of woodland in the north side of a hill differ from those on the south of a hill. The land use database of 1992 provides a significant amount of information for reference to use in classifying land use when it is difficult to separate a garden plot from woodland. The above knowledge will be regarded as the rules of land use classification.

Base Rules: IF (the Condition), THEN (the Conclusion), CF (the Conclusion Faith Operator).

The value range of CF is [0, 1]. When its value is equal to 0, the faith of the conclusion is minimal; when it is equal to 1.0, the faith of the conclusion is maximal. The detailed rules are described in the case study. For the land cover and land use classification flow, see Figure 3.

266

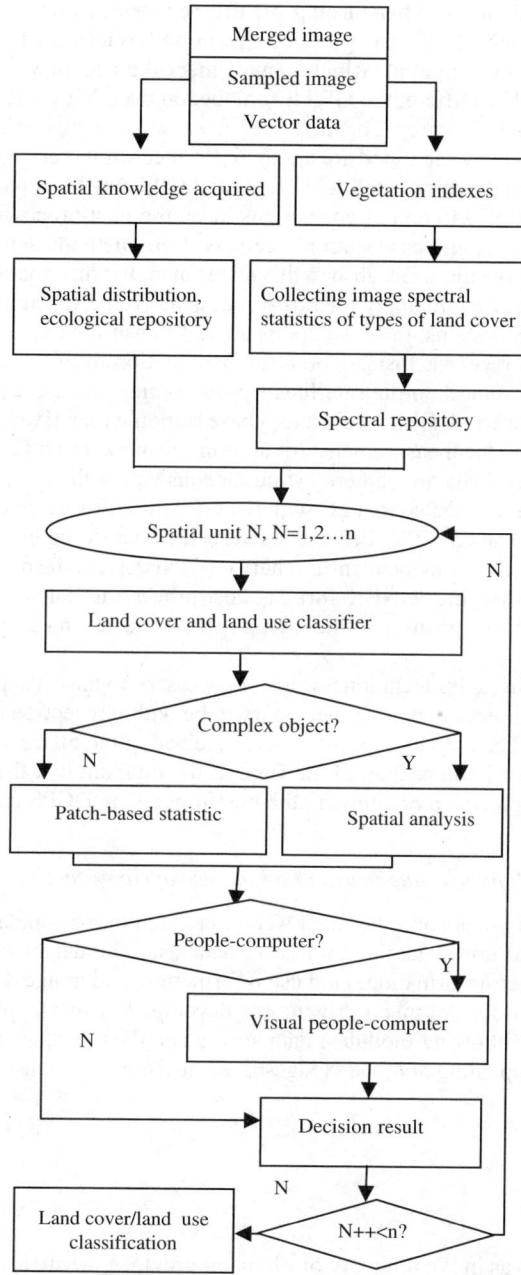

Figure 3. Land cover/use classification flow.

3.3 *Field investigation*

The main goal of this field investigation was to check the ground truths and to acquire the spatial coordinates of patches. The former has not been described in detail. Generally, up-to-date data on small areas can be obtained by field surveys or GPS surveys. Obviously, time and cost are two important factors that often prevent us from making field measurements. By contrast, the GPS survey method is fast and cheap. After turning off the Selective Availability, the GPS system's Standard Positioning Service ('SPS') can produce a position fix accurate to about 20 metres in a stand-alone mode with no differential, which cannot meet the accuracy requirements for land use dynamic monitoring. The Differential GPS technique, on the other hand, greatly improves the accuracy of positioning to 3–5 metres. This improved accuracy depends on the fact that two GPS receivers that are not too widely separated are apt to experience similar errors, as they measure the 'pseudo-range' distances to the same satellite. Differential GPS may be implemented in two ways: as the real-time reception of differential corrections or as the post-processing of data collected simultaneously from the rover and base station receivers. Both methods depend on having a base station located at a known position. Because in this experiment the information obtained must not be transmitted in real time, we will apply the post-processing DGPS technique.

In post-processing GPS systems, the corrections are not transmitted in real time from the base station to the roving GPS receiver. Instead, both the base station to the rover store the measurements they make of the distances to the satellites ('pseudo-range' measurements), for later post-processing by special software. A GPS receiver as a base station with a fixed antenna is used at the office of the person making the measurements (or at a survey marker), and another GPS is held by hand as the roving receiver. Data are gathered simultaneously at both receivers. The rover is then brought back to the office, connected to a PC with the GPS PC software, and the data is extracted from the rover to the PC. The GPS PC software converts the data stored in the GPS receivers into an industry-standard 'Receiver Independent Exchange (RINEX)' file format. Data from the base station are also converted to the RINEX format. Post-processing software then processes the RINEX files and computes the position of the roving receiver made a measurement of the satellite pseudo-ranges.

With this post-processing GPS technique, it is not necessary to have a separate DGPS receiver or to subscribe to a DGPS service, nor is it necessary to be within reception range of the transmitter of the DGPS service. The corrected positions are received in the office after the data has been gathered and post-processed, rather than in the field at the moment that the measurements were made. The measurement precision of applying the post-processing DGPS can reach 3–5 metres.

3.4 *Establishment of a land use database and a land use dynamic monitoring system*

In this research, surveying data and other data were converted into a standard digital format and formed an integrated land use dynamic monitoring database. The database consisted of at least three data sets: the base geo-information, land use information, and image data. This management system, all functions of which are menu-driven, was developed in an MS Windows environment. The system includes the following modules: data access and data management, map production, statistics summary, data updating and knowledge-based decision functions.

4 THE CASE STUDY

4.1 *The introduction of the study area*

The study area selected was in Wenling city of Zhejiang province, where a 1:10000 land use database of 1992 in 1995 was built. DEM data were captured from 1:50000 topographic maps by obtaining the contour values. The situation had changed greatly in the past 10 or so years, making it very necessary that land use dynamic monitoring be implemented and the land use database updated.

4.2 *The implementation of knowledge-based land use dynamic monitoring supported by the '3S'-integration technique*

In this research, the image data resource consists of two scenes of high-resolution panchromatic CBERS-2 satellite imagery (with 3-metre spatial resolution) taken in March 2001, and a scene of low-resolution multi-spectral Landsat-7 ETM remote imagery (with 30-metre spatial resolution) taken in April 2001. The two scenes had been resamped to a 3-metre grid image to match the resolution of CBERS-2 panchromatic imagery. By extracting the coordinates of the predominant terrain features from the corresponding 1/10000 topographic map, 63 points with good distribution in the area under investigation were used. A second-order polynomial was used for the purpose of geo-referencing to a BJ-54 coordinate system. The average RMS error for the registration and rectification is 2.4 meters and the maximum does not go beyond 3.2 metres in mountain areas. Digital Elevation Model (DEM) data of the 1/50000 scale was used to produce ortho-rectified images of the 1/10000 scale.

The experiment tests whether band 4, 3 and 1 of Landsat-7 ETM satellite images, which were received in April, are able to distinguish among types of land cover (see Figure 4). We collected spectral statistics for eight types of land cover (see Table 1).

Spatial distribution knowledge and ecological conditions relating to types of land use were collected. More than 60 rules based on this knowledge were summarized according to the local conditions of the study area. After repeated debugging, only 31 rules, which are in accord with local conditions, were finally fixed upon. The detailed rules are described as follows:

IF DEM < 10 THEN landuse = woodland CF 0.1;
IF DEM >= 10 & DEM < 40 THEN land use = woodland CF 0.8;

Figure 4.   Sampling the spectral samples.

Table 1.   Spectral samples statistic table of eight types of land cover/land use.

No	Land cover type	Number	Band4_A	Band4_V	Band3_A	Band3_V	Band1_A	Band1_V
1	Paddyfield	59	135	4.41	123	10.82	222	9.61
2	Garden plot	27	130	7.31	120	14.35	190	13.34
3	Woodland	20	124	4.58	108	9.10	182	23.82
4	Grass	30	127	5.82	115	14.48	185	22.44
5	Residence	21	158	9.11	173	19.18	182	12.99
6	Road and railway	20	146	8.81	168	17.62	177	11.33
7	River	20	162	9.14	181	16.33	166	21.17
8	Unused land	28	132	4.74	122	9.17	212	28.03

Note: Band 4, 3 and 1 represent TM bands, A-Average, V-Power difference. Every type of land cover/use is made up of at least 20 parcels.

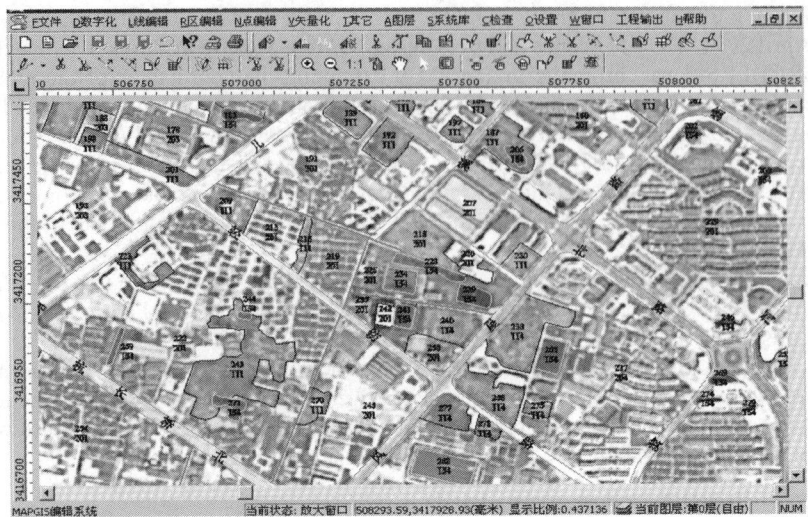

Figure 5.   Land use database and management system.

IF slope >= 10° THEN landuse = garden plots CF 0.1;
IF slope >= 3° & slope < 6° THEN landuse = garden plots CF 0.6;
IF landuse = woodland THEN garden plots CF 0.3;
IF landuse = garden plots THEN woodland CF 0.2;
IF landuse = paddyfield THEN residence CF 0.5;
IF DEM < 10 & slope < 2° THEN Woodland CF 0.1.

There can be no doubt that image analysis based on spectrum and spatial knowledge can significantly contribute to our understanding of land cover/land use. At the same time, automatic imagery recognition as a means of classifying land cover/land use is not a panacea. In spite of a high degree of automation, a significant amount of work is still required post-process and edit the extracted information. For example, the automatic classification must be checked by interactive interpretation between people and computers and field investigations. This will improve the accuracy of the land cover/land use classification and reduce the ratio of the omitted patches. The ultimate target of land use dynamic monitoring is to establish a land use database that utilized repeatedly and a land use dynamic monitoring system for integrating, displaying and analysing all kinds of vector and raster data (see Figure 5).

5   ANALYSING THE RESULTS OF LAND USE DYNAMIC MONITORING

A land use database with a scale of 1/10000 was updated, supported by the '3S'-integration technique. It is necessary that the cause of the errors be analysed in the course of land use dynamic monitoring and that the accuracy of the updated result be evaluated. We will describe these from three aspects as follows: The small changed patches are omitted mainly because of the limitation in the resolution of the satellite images. In this study, a large number of changed patches under 200 square metres using 3-metre spatial resolution images had been omitted. Fortunately, the percentage of errors in area caused by the image resolution was only 3%. Errors in the classification of land cover/land use were caused when 'the same image spectrum represents a different object, while the same object reflects a different image spectrum'. The percentage of the accuracy of the automatic land cover/land use classification, which was tested on small areas in the field investigation, was 88.2%. In other words, the correctness of the land cover classification based on the

270

spectral repository and the correctness of the land use classification based on spatial knowledge indoors was 88.2%. On the basis of the result, the classification accuracy using the people-computer interactive interpretation method was 93.7%. The geometric accuracy of changed patches is mainly determined to the spatial resolution of the image and the precision of the GPS survey. The former has been discussed. Differential GPS acquisition data with a precision of 3–5 metres meets the requirements of land use dynamic monitoring and of updating the land use database at the scale of 1/10000.

6   CONCLUSION

Knowledge-based land use dynamic monitoring supported by the '3S'-integration technique was accomplished. The test proved that classifying land cover based on the spectral repository of land cover types and classifying land use based on knowledge advanced work efficiency and the accuracy of image classification as compared with the traditional technical approach of land use dynamic monitoring.

REFERENCES

Jie, Y. 2002, Investigation of Land Database Updating Based on High-Resolution Airborne Images, *Symposium on Geospatial Theory, Processing and Applications*: 10–14.
Yang, J. H., Study on Land Use Map Revising in County Scale Using High-Resolution Digital Satellite Remote Sensing Image. *China Land Science*, 2002. 3: 41–45.
Tseng, Y. H. 1997, A Region-Based Classification by Integrating Geographical Data with Multitemporal Satellite Imagery for Rice-Corp Inventory. *The Proceedings of GIS AM/FM ASIA '97 & Geoinformatiics '97 Mapping the Future of the Asia-Pacific.*

Spatial representation

Spatial data warehouses: a methodological framework

Omar Boussaïd

Laboratoire ERIC, Université Lyon 2, Bron Cedex, France

Marie-Aude Aufaure

Laboratoire d'Ingénierie des Systèmes d'Information, Université Lyon 1, Villeurbanne Cedex, France

ABSTRACT: In this paper, we propose a methodological framework dedicated to spatial data warehouses. Due to the specificities of such data, we have to take into account spatial measures and dimensions during the construction of a multidimensional data cube. OLAP tools can be used during the exploratory step and can be seen as a preprocessing phase before data mining. Generalization and specialization should be used during the execution of these methods. A few studies have already been conducted to define a methodology for designing multidimensional models for spatial data. This paper deals with the design of a spatial multidimensional model and also with spatial data analysis for decisional applications.

1   INTRODUCTION

With the increasing digitalization of spatial information, we need to solve the problem of how different categories of users can access and exploit such data for decision-making purposes. The major characteristics of spatial data relate to their volume, complex representation, time variability, and multi-source and multi-scale representation.

A geographic object (Rigaux et al., 2002) is composed by: (1) a thematic description (a set of alphanumerical attributes like the name and population of a town) and (2) a spatial representation in terms of geometry (localization, shape, etc.) and topology (spatial relationships).

One of the main characteristics of spatial data comes from the great importance of spatial relations between objects. These spatial relations are related to the interactions of objects in space and can be topological (intersection, etc.), metrical (distance, etc.) or directional (north of, behind, etc.). These relations are crucial during the phases of query and analysis. Tobler (1979) defined the 'first law in geography' as follows: 'Everything is related to everything else, but nearer things are more related to each other than distant things.' Another important characteristic that we have to take into account for spatial analysis is spatial heterogeneity. The results of spatial analysis depend on different parameters like local conditions. Global parameters do not describe a geographic phenomenon in a particular place well. Local characteristics must be combined with global ones.

In Geographic Information Systems (GIS), geography is expressed by successive map layers such as is done with a road network. These layers are strongly correlated. Consider, for example, a map of rainfall and a map of density of population: the density of the population may depend on agricultural production and, consequently, on the amount of rainfall in a region. Thus, these two maps are closely related.

Spatial analysis can be performed in several ways. The results will be less or more precise, depending on the tools used by the analyst. The user can query geographic data using a visual language or by exploring the contents (Aufaure & Trepied, 2001). But, these tools are not powerful enough to perform an analysis; the user can only query or visualize and is in charge of the interpretation of the results. The huge volume of data now exceeds human capabilities for analysis. To face this problem, new approaches to extract knowledge (Han & Kamber, 2001) from spatial data have been developed (Ester et al., 1997; Han et al., 1997; Zeitouni et al., 2000; Zeitouni, 2002). The mining of spatial data can be performed with a descriptive or a predictive objective. Consider as an example the field of road accidents management. For such an application, it is logical to explore the properties and relationships of the neighbourhood. Data warehouses and OLAP tools in particular can be efficient for use in data exploration.

Because very little work has been conducted on the methodological aspects of conceptual multi-dimensional modelling, we propose a methodological framework in order to help the analyst to build his/her multidimensional model of spatial data and then to perform a decisional analysis taking spatial criteria into account. This process is illustrated with a road accident application.

This paper is organized as follows: a spatial data warehouse and spatial dimensions and measures are described in section 2, while section 3 contains a description of a methodological process for decision-making applications. This section is mainly dedicated to conceptual modelling. The different analytical steps that we propose are presented in section 4. Finally, section 5 concludes with some ideas of our perspectives and future work.

2 PRINCIPLES OF SPATIAL DATA WAREHOUSES

As we said previously, spatial data represent a huge volume of data and the decision-making component is prevalent. Therefore, data warehouses and OLAP tools seem to be well-suited to manage such data. A data warehouse (Immon, 1996) is a set of subject-oriented, integrated, chronological and persistent data organized to support a decision-making process. Data from heterogeneous sources are stored in a warehouse. In the case of spatial data, the main difficulty comes from the choice of spatial and non-spatial dimensions of interest and from the need to build a model dedicated to multidimensional analysis. This model should integrate different dimensions: (1) non spatio-temporal ones, (2) spatio-temporal dimensions at a low level that are transformed into non spatio-temporal ones at a higher level (using concept hierarchies for generalization/specialization that are a priori knowledge (Widom, 1995) and (3) pure spatio-temporal dimensions. These dimensions represent the perspectives of analysis so their choice is a crucial step. Measures can be classical ones or pointers on spatio-temporal objects. Visualization is then performed using the data hypercube. Visualization should be useful in helping the user choose the dimensions of interest, and the degree of generalization/specialization necessary to producing a useful summary of the data. Multidimensional analysis constitutes a first step in data exploration.

Few studies have focused on the field of spatial data warehouses. We note the work of two Canadian teams: DBMiner and GeoMiner (Han et al., 1997) at Simon Fraser University and the work carried out at Laval University (Rivest et al., 2001).

SOLAP concept (*Spatial On-Line Analytical Processing*) (Rivest et al., 2001) can be defined as a visual platform designed for the multidimensional analysis and exploration of spatio-temporal date. Spatial dimensions can be: (1) non-geometrical, as in the name of a town, where no geometrical representation is associated with the dimension; (2) geometrical, where the dimension remains geometric whatever the level of generalization; and (3) combined, where the dimension is geometric at a low level of generalization and non-geometric at a high level of generalization; for example, a town is represented by a polygon and can be generalized by a nominal attribute like 'town of Paris'. A spatial data warehouse has to be able to manipulate numeric as well as spatial measures (Bédart et al., 2001). Three kinds of measures have been proposed in (Rivest et al., 2001). The first is represented by a set of geometric shapes obtained by geometric spatial dimensions combined with each other. In this case, a spatial operation like an intersection or merge is performed. The second kind of spatial measure is computed by topological operators or spatial metrics. The last is a set of pointers on geographical shapes. The two first cases imply the need to pre-compute spatial measures to build the multidimensional data cube, which can be very time-consuming.

3 TOWARDS A METHODOLOGICAL FRAMEWORK

Spatio-temporal data are complex. Spatial databases allow for the modelling and physical storage of spatio-temporal data. SQL language can access and manipulate such data but is not suitable for use in analysing spatial data that takes into account geographic and topologic properties and spatial relationships.

We need more powerful tools, especially a methodology, to exploit such data. Here, a spatial data analysis is developed using different steps such as data collection, modelling, loading and information return.

3.1 *Data collection*

In the following, we present our application concerning road accidents. The aim is to extract relevant information about road risks from spatial and alphanumerical data on accidents, road networks, environment and circulatory flows.

For this application, we use different data sources from a great French urban centre (Chelghoum et al., 2002). A geo-referenced database contains data collected from the accidents that occur in this area. Other information describes different geographic layers such as roads, buildings, schools, etc. Another database is dedicated to the description of accidents that occur in this urban centre at a precise period in time. Many studies have been conducted about road risks in this area. We only make use of these data to illustrate our methodology for modelling and analysis.

Spatial data comes from various and heterogeneous sources. The problem of integrating data has been well identified. In data warehousing, data integration is performed by the ETL (*Extracting, Transforming and Loading*) process, which deals with the tasks of extracting, transforming and loading data. The final model, obtained by a synthesis of all of the original models, is not always a 'good' model in terms of a user's needs in analysis. We need a method to build a spatial data warehouse model.

3.2 *Conceptual modelling*

A data warehouse is a database oriented towards decision-making projects. Queries are complex, often involving a huge volume of data and requiring a good response time. This database must be built on a model that corresponds to these requirements. Star models (Immon, 1996; Chaudhuri & Dayal, 1997) (Figure 1) are used, but a method to build such a model should be very useful. The process used to build a model is based upon the designer's knowledge and is rather empirical.

At present, there are few methodologies dealing with the designing of data warehouses. They are not popular and tend to be linked to applications in a particular field, so there is little generalization. In spite of this drawback, we note that the approach proposed by Kimball (2002) seems to have interesting implications for the process of multidimensional modelling. It permits the identification of, first, the objectives of analysis; second, the dimensions; third, the granularity of the data; and, last, the facts. However, the modeller must be familiar with the fields to be studied in order to be able to efficiently choose the data representing indicators (*measures*) that can be observed according to axes of analysis (*dimensions*).

Thus, we propose a method to emphasize Kimball's approach. After fixing the objectives of analysis to reflect the needs of the user, the choice of dimensions and measures is not trivial. We advocate pre-processing the available data. In our running example, a descriptive statistical study allowed us to discover interesting elements to describe road accidents, such as places where accidents happen and the time when they happen, meteorological conditions and implied responsibilities. The designer unfamiliar with road accidents can exploit the results of the statistical study to easily choose the dimensions. In our case, we have defined four dimensions: *Localization, Conditions, Responsibilities and Dating* (Figure 1).

In determining the measures, selecting the number of accidents as a measure would be relevant. However, the lack of additional information about weather, traffic, and so forth make this indicator

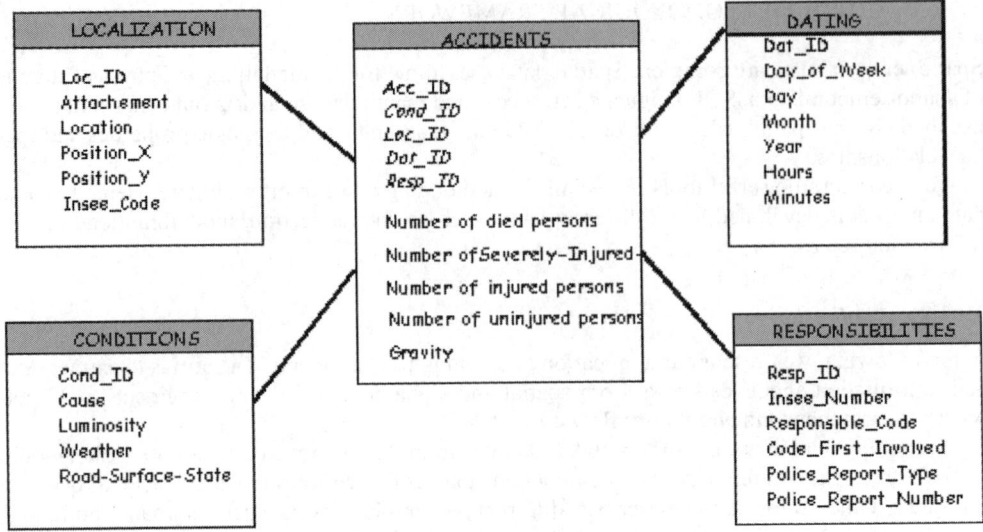

Figure 1.   Star model for the road accident application.

inefficient. We prefer using other measures such as the *Number of people who have been killed, Severely injured, Injured, Uninjured* and *Gravity*. With available data, we think that these measures are more attractive for use in comparing different cases of accidents.

Our aim is to build a multidimensional model that will help us to analyse our spatio-temporal data. We want to study road accidents, so we have to determine what parameters would be useful in predicting road risks. In a geographic model, measures can be either numerical attributes or spatial ones. In the first case, measures can be divided into three categories according to the kind of aggregate function used: distributive (sum, min, etc.), algebraic (average, etc.) or holistic (median, rank, etc.). In the second case, the measures contain one or more pointers on spatial objects. These objects can be parts of roads composed of sections and crossroads, for example, or geographic areas grouped on the same meteorological conditions. Dimension tables contain alphanumerical descriptors that explain the measures and represent the axis from which the indicators should be observed.

Hierarchies are built for some tables of dimension to offer new analytical perspectives to the user (Jagadish et al., 1999) (Figure 2). To obtain these hierarchies the data need to be aggregated either by *month-year* or by *week-quarter-year*. These hierarchies allow the data to be explored from a microscopic or a macroscopic point of view. The user can, for example, present results per year or focus on the number of dead or severely injured persons, etc., during a week-end in spring.

Among all the dimension tables of our model, two are purely spatio-temporal dimensions (Localization and Dating tables). Data still remain spatio-temporal whatever the hierarchy level. The two other tables (Conditions and Responsibilities) contain data on non-spatial attributes. Some dimensions, spatio-temporal or not, are homogeneous; this implies that hierarchies can be easily built. It is more difficult for non-homogeneous dimensions but these ones represent a pertinent and realistic axis for analysis. For example, we can note the *Dating* table is homogeneous whereas the *Conditions* table is not.

The starting point of our methodological process to build a conceptual model is a set of analysis objectives and a data dictionary. After that, we build a multidimensional model based upon a star model (Immon, 1996; Chaudhuri & Dayal, 1997). The facts – some aspects of an activity – that we want to analyse are then identified and expressed by means of measures (*indicators*), dimensions (*analysis axes*) and hierarchies (*coarse dimensions*). The central point of the model is the fact table, named *Accidents*, and all of the dimension tables are connected to this table in order to facilitate the joins. Most data are stored in the fact table. This particularity of the star model leads to a speeding

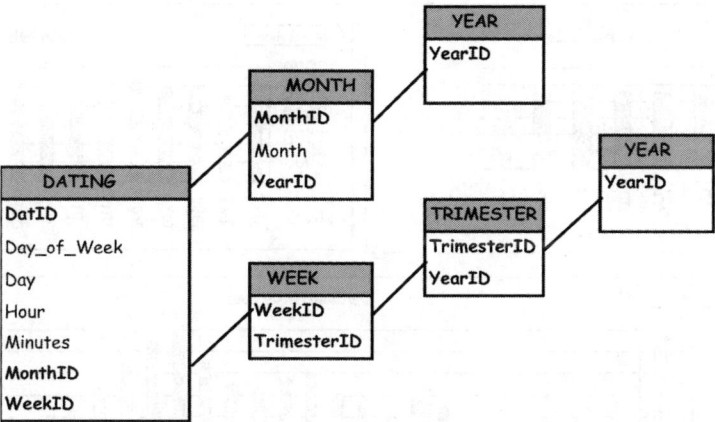

Figure 2.   Hierarchy on dating table.

up of the time needed to respond to queries. In our application, the choice of accident risk as a fact, justifies the creation of only one fact table. Taking into account other facts, they can be performed with other fact tables. The model will be then transformed into a fact constellation model. In the same way, the creation of a hierarchy produces *snowflaked* models.

3.3   *Spatial data warehouse loading*

Once the multidimensional model is defined, data loading is generally performed using ETL tools. In our application, we have written scripts in Visual Basic to load data in our spatial data mart. We have only considered road accidents with respect to their geographic localization, dating, responsibilities and conditions in which they occur. To simplify our model, data on road networks, road traffic and environment have been intentionally neglected.

4   SPATIAL DATA ANALYSIS

The phase of analysis can be divided into different steps. These steps constitute a methodology to help the user explore the data. We now describe these steps.

4.1   *Data description and summary*

First, the user progressively discovers information about his data using basic descriptive techniques such as those provided by popular tools of descriptive statistics. The aim is to point out the relevance of some data. With this information, the user elaborates on a hypothesis that may be confirmed or not by more sophisticated methods of analysis such as spatial data mining techniques. In Figure 3, we note that more accidents occur between 4 and 7 p.m. and during June or October, for example. The user will then try to explain the reasons for this increase in the next steps, described below.

4.2   *Data exploration and navigation*

The multidimensional view of data is close to the cognitive model of the analyst. Spatial OLAP tools are very efficient and user-friendly for use in analysing data. OLAP systems pre-aggregate the data and offer data models that are optimized to perform analyses, are easy to use, have a friendly user interface and give preformatted results. OLAP datacube operations materialize the

279

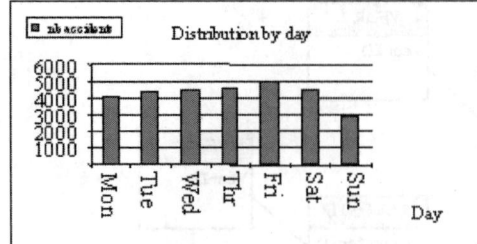

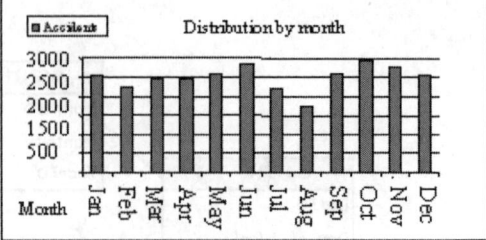

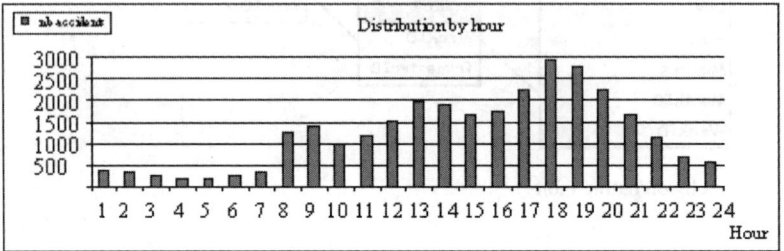

Figure 3.   Descriptive results.

different views and allow interactive analysis. In our application, the characterization of the areas of risk taking into account neighbourhood properties can reveal useful information. OLAP tools can be seen as a method of preprocessing: the user can interactively search for relevant patterns on which he/she will then focus on using data mining techniques.

The OLAP analysis allows data to be navigated along dimensions and facts for better visualization. Suppose that in our application on road accidents we want to verify if the security around schools is correctly managed. We applied a slice-and-dice operation to isolate data referring to schools. We then observed the nature of the related accidents. For instance, we can state if the accidents are more dangerous when pupils arrive or leave school. The geographical objects of our multi-dimensional model can emphasize the extracted information through the results of the analysis.

We can also elaborate on other hypotheses by exploring the geographical repartitioning of accidents. We have seen that there are slightly more accidents inside agglomerations than outside, whereas accidents in rural locations tend to result in more fatalities.

4.3   Data mining

During the exploratory step the user selects relevant objects or geographic areas on which he wants to build a predictive model. Decision trees (Han & Kamber, 2001) are used in this predictive step to classify, predict or diagnose, and are applied when attributes are either nominal or numeric. The results are visually expressed by logical rules or SQL queries. The main interest for our application is to predict the risk of accidents using predictive attributes. The model is defined using attributes to explain such things as the number of persons killed, the locality, weather, etc. The tree is built using the most discriminating attribute in a set of possible segmentation attributes (using, for example, the entropy). It is interesting to use this technique starting from the multi-dimensional data cube because we can generalize/specialize some parts of the decision tree according to dimensions and measures of interest. Many studies have been conducted in the field of spatial data mining (Ester et al., 1997; Han et al., 1997; Zeitouni et al., 2000), but most cases do not take multidimensional aspects into account.

The steps described above show the process of analysis that we propose. Each step has a specific objective. The first one consists of the task of describing the data. The next one allows us to closely observe the data and to elaborate on some of the hypotheses we have been studying using the help of data mining techniques.

5   CONCLUSION

Decision-making applications have become more and more prevalent for spatial data. The user has to find relevant information in a huge volume of data. This volume exceeds human capacities for analysis. Consequently, new tools need to be developed to help the analyst. Spatial data warehouses combined with spatial data mining constitute a challenging solution in a decision-making context. Thus, we proposed in this paper a methodological framework to help the user when he/she plans to use the data warehousing approach, especially in the case of spatio-temporal data. Our approach allows various techniques of analysis to be combined, such as statistical, data mining and OLAP techniques.

One of our aims is to develop data mining techniques that can be applied in a multidimensional space, the spatial datacube (or hypercube). Functions of generalization and specialization will be exploited during the phase of analysis. We plan to build decision trees. This technique can be easily interpreted and exploited by the end-user. Another perspective is to introduce spatial and temporal fuzziness in the multi-dimensional space and the decision tree. The starting point of the analysis could then be expressed in a fuzzy or imprecise way. Some studies have been defined in this way for alphanumerical data (Laurent et al., 2001), and it would be interesting to extend this approach to spatio-temporal data.

REFERENCES

Aufaure, M.A. and Trépied, C. 2001.What Approach for Searching Spatial Information, *Journal of Visual Language and Computing*, 12(4), pp. 351–374.

Bedart, Y., Merett, T. and Han, J. 2001. Fundamentals of spatial data warehousing for geographic knowledge discovery, In: *Geographic Data Mining and Knowledge Discovery*, H. Miller and J. Han eds., Taylor and Francis.

Chaudhuri, S. and Dayal, U. 1997. An overview of data warehousing and olap, technology. *Sigmod Record*, 26(1): 65–74.

Chelghoum, N., Zeitouni K. and Boulmakoul, A. 2002. A Decision Tree for Multi-layered Spatial Data, *10th International Symposium on Spatial Data Handling*, SDH, Ottawa, Canada, July 2002.

Cood, E.F. 1993. Providing OLAP to user-analysts: an IT mandate, *Technical Report*, E.F. Cood and Associates.

Ester, M., Kriegel, H.P. and Sander, J. 1997. Spatial Data Mining: A Database Approach, *Proceedings of the 5th Symposium on Spatial Databases*, Berlin, Germany, 1997.

Han, J., Koperski, K. and Stefanovic, N. 1997. GeoMiner: A System Prototype for Spatial Data Mining, *Proc. ACM-SIGMOD Int'l Conf. on Management of Data (SIGMOD '97)*, Tucson, Arizona (System prototype demonstration).

Han, J. and Kamber, M. 2001. *Data Mining: Concepts and Techniques*, Morgan-Kaufmann.

Han, J. and Kamber, M. 2001. *Data Mining: Concepts and Techniques*, Morgan Kaufmann.

Jagadish, H.V., Lakshmanan, V.S. and Srivastava, D. 1999. What can hierarchies do for data warehouses?, In: *Proc. of the 25th VLDB Conference*, Edinburgh, Scotland.

Immon, W.H. 1996. *Building the DataWarehouse*, John Wiley and Sons.

Kimball, R. and Ross, M. 2002. *The Data Warehouse Toolkit*, Second Edition, Wiley Computer Publishing Editor.

Laurent, A., Bouchon-Meunier, B. and Doucet, A. 2001. Towards Fuzzy-OLAP Mining, In: *Proc. Workshop PKDD 'Database Support for KDD'*, Freiburg, Sept. 2001.

Rivest, S., Bédart, Y. and Marchand P. 2001. Towards better support for spatial decision-making: Defining the characteristics of Spatial On-Line Analytical Processing (SOLAP), *Geomatica, the Journal of the Canadian Institute of Geomatics*.

Rigaux, P., Scholl, M and Voisard, A. 2002. *Spatial Databases with Application to GIS*, Morgan Kaufmann.

Stefanovic, N. 1997. Design and Implementation of On-Line Analytical Processing (OLAP) of Spatial Data, *M.Sc. Thesis, Computing Science*, Simon Fraser University, September 1997.

Tobler, W.R. 1979. Cellular geography, In: Gale S. Olsson G. (eds) *Phylosophy in Geography*, Dortrecht, Reidel, pp. 379–86.

Widom, J. 1995. Research Problems in Data Warehousing, *Proceedings of CIKM*, pp. 25–30.

Zeitouni, K. 2002. A Survey on Spatial Data Mining Methods Databases and Statistics Point of Views, Chapter in Shirley Becker Editor, IRM Press, pp. 229–242.

Zeitouni, K., Yeh, L. and Aufaure, M.A. 2000. Join Indices as a Tool for Spatial Data Mining, *International Workshop on Temporal, Spatial and Spatio-temporal Data Mining*, TSDM 2000 Lecture Notes in Artificial Intelligence no. 2007, Roddick J.F. and Hornsby K., Eds., Springer, pp. 102–114, Lyon, France.

Advances in Spatial Analysis and Decision Making, Li, Zhou & Kainz (eds)
© 2004 Swets & Zeitlinger, Lisse, ISBN 90 5809 652 1

Rich media and enhanced GIS: theoretical background and profile of an exploration into innovative ways to (geo)informate

William E. Cartwright & Chris Pettit
Department of Geospatial Science, RMIT University, Melbourne, Victoria, Australia

Bob Williams
Command and Control Division, Defence Science and Technology Organisation, Edinburgh, SA, Australia

ABSTRACT: A current research and development program is addressing the problem of finding appropriate geographical information by developing a (geo)information realisation resource based on the concepts of the GeoExploratorium. It has as its main goal to provide tools for geographical knowledge building and exploring. This research examines the formulation of an initial GeoExploratorium prototype component to assist in military intelligence and strategic planning exercises in Townsville. The prototype enables a number of spatial and two-dimensional and three-dimensional spatial datasets to be combined and analysed in order to formulate and explore 'what-if' intelligence strategies. Geographical Information System (GIS) and multimedia technologies underlie the GeoExploratorium prototype. Using both these technologies enables spatial analysis and data visualization to be undertaken, which subsequently enhances the exploration of geographical knowledge. The research examines new ways to prospect for, discover and disseminate geospatial knowledge within a military context. The paper reports on the concepts behind the design of the GeoExploratorium and findings from the evaluation of the initial prototype.

1 INTRODUCTION

The increased access to sophisticated computers by the general public has led to an awareness that resources like discrete multimedia products and their distributed counterparts on the Internet, and particularly through the use of the World Wide Web, has revolutionised the way in which information is both accessed and used. Cartographers have embraced the use of interactive multimedia, delivered via discrete or distributed means, as a method of providing products that are easily useable with 'everyday' skills, using modest computer platforms and accessible communications resources like the Web.

This paper outlines a research project that has been implemented to apply the ideas developed for a GeoExploratorium, an electronic discrete/distributed 'space' that would allow inexpert users to explore geographical information using metaphorical approaches that they felt most comfortable with. The underlying foundation theory will be discussed, the design of a prototype described and the research and development methodologies explained. Finally, early evaluation findings will be outlined and its impact on the direction of the prototype development explained.

2 OVERVIEW

System requirements for the visualisation of geographic phenomena are interaction, displays which allow for area-based data to be depicted (allowing the user to undertake pattern comparisons, to discover relationships, to 'see' geographic movement and to process and display the temporal characteristics of data), a good user interface and it should be based on efficient algorithms. Routines used in interacting with a GIS can be divided into three categories: basic drawing tasks; data retrieval, manipulation and creation of data displays; and the visualization modules. The latter should allow for interactive viewing, 'brushing' (providing operations for geographic correlation analysis), 'alternagraphic display' (to compare spatial patterns), area masking (for examining geographic movement associated with individual regions), and temporal browsing (for investigation of temporal patterns in an animated fashion or in a browsing mode) (Tang 1992). Multimedia can be used to enhance the temporal browsing aspects of data access through graphical interfaces.

Most GIS complete these tasks through the use of map interfaces and mapping metaphors. However, there has been a trend to incorporate interactive multimedia and hypermedia with GIS, delivered discretely on hard disk or CD-ROM or via distributed means through the Internet and the World Wide Web.

There has been much interest in harnessing multimedia and Geographic Information Systems to provide a better and more useable depiction of geography. GIS and multimedia have been incorporated to produce packages in the areas of GIS education and training, improving human-computer interaction using hypermedia systems, access to new sorts of data, using hypermaps to help classify multimedia geographical information and developing new visualization techniques (Lewis 1991). It is some years now since Lewis proposed such enhancements provided by a GIS/multimedia package, and it is now timely to explore how such a 'liaison' might function.

3   THE GEOEXPLORATORIUM

Metaphor models form a pivotal link between learning and memory through the abstraction of relevant properties of a situation into a simplified and convenient form. In doing so they are usually dynamic and their development is effected by situational factors. Users interact with artifacts and then form mental models. Therefore interface metaphors attempt to map knowledge already held by a user group to a normal problem area (Smyth & Knott 1994). An area or resource is needed to fully exploit the use of a set of metaphors that provide a different means of access to spatial information.

What is being explored is the development of an application based on the concepts of the GeoExploratorium, a virtual space that would enable users of the Map Shop to explore geographic information using different metaphors. Cartwright (1997) developed the conceptual ideas of supplying information as part of a GeoExploratorium: a multimedia enhanced package that combines tactile, discrete and distributed multimedia components. The GeoExploratorium combines a number of metaphors (Cartwright & Hunter 1999) to allow users to choose the resources and the relevant delivery method that they are most comfortable with for their specific 'discovery' and 'exploration' methodologies. The multi-metaphor resource provided by the GeoExploratorium enables users to choose the package or parts of a package that is most appropriate for individual tasks of geographical information exploration. As well as providing access to multimedia, hypermedia and interactive maps, the GeoExploratorium would provide links to other 'world wide' resources.

The metaphor set to be used in the application being developed was proposed by Cartwright (1999) and its purpose is to provide component parts of the GeoExploratorium. It provides complementary ways to understand and comprehend geographic information using multimedia. The metaphors were originally conceived and developed with discrete multimedia products as 'targets', but the concept can be extended to distributed multimedia. Figure 1 shows this concept.

The metaphor set includes the Storyteller, the Navigator, the Guide, the Sage, the Data Store, the Fact Book, the Gameplayer, the Theatre and the Toolbox. A brief summary of each of the nine metaphors is as follows:

3.1   *The Storyteller*

Interactive storytelling can be used to enhance the information to which users gain access. Some users may wish to be 'told' a story while they view map or graphical depictions on a screen.

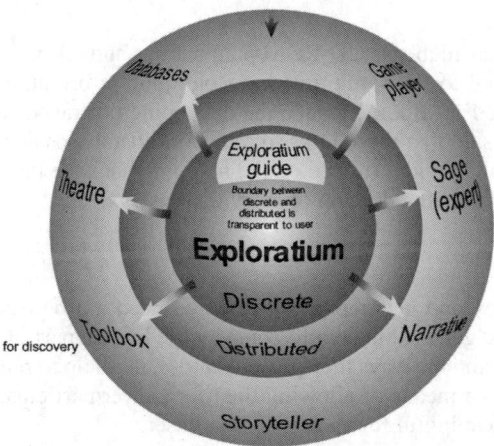

Figure 1. GeoExploratorium – both an enclosed and an open spatial information virtual 'space', linked to databases and resources that have integrity, currency, presentation flexibility and historical lineage. It relies on both the diverse nature of data and information available, and the innovative nature of new presentation methods.

The story can be told using digital sound or by allowing the users to read text from the screen. Storytelling, whilst not expected to be used vigorously needs to be made available for those users interested in finding out more about particular phenomena by being painted 'word pictures'.

3.2 The Navigator

For some 'power users' the use of navigation tools will be unwanted, but for novice or naive users The Navigator would allow them to move through the package in the most efficient manner. Final navigation strategies would be developed on the basis of user needs, modified according to actual usage patterns.

3.3 The Guide

This metaphor can be used to find the 'exact' type of information and the 'exact' portrayal device for a particular user for a particular use. The Guide, experienced in the subject matter being mapped and the efficient use of the delivery package, can help users gain access to appropriate information more quickly and precisely.

3.4 The Sage

The Sage could be seen as being very close to The Storyteller. The main difference is that The Sage injects experience and expert knowledge into the telling. Multimedia can contain elements of oral, print and electronic expert storytelling. Multimedia mapping packages enable The Sage to provide audio descriptions of static or dynamic maps and to appear in videos that elaborate on the underlying facts behind the phenomena being depicted.

3.5 The Data Store

The Data Store metaphor uses, and is closely associated with, the Information Superhighway/ Infobahn metaphor. Multimedia, provided in discrete form using CD-ROM or in distributed form on the Internet, offers the ability to make available large amounts of support data that can be used to complement the main mapping package. A multimedia mapping package with eons of support data (produced specifically for mapping applications or for general release) provides the means to enhance mapping packages with a 'Virtual Knowledge Library'.

3.6 The Fact Book

Collections of 'facts' are available as CD-ROM publications and on-line services. These resources and Web sites can be used to 'link' multimedia geographic information to the real world. Facts available via links to The Fact Book enhance the data being portrayed, and give credibility to the information. It could be said that these links to real world databases and publications allow users to convert the information being portrayed on the mapping package into knowledge (The gaining or the imparting of wisdom is the domain of individual users).

3.7 The Gameplayer

Multimedia and gameplay and geographic information need not be based on the typical puzzle solving/shoot-em up type games, but should consider the human part of the interpretation of 'the game'. Computer games and the ways they present landscapes help to reinforce certain ideologies. Gameplay can be used as a means of allowing the user/viewer/participant to discover patterns of phenomena which are meaningful for each individual user.

3.8 The Theatre

The use of the Theatre metaphor involves functionality and thus allows the user to be engaged/ pleased with the experience. This means that the user must understand the activity well enough to do something. Some users may prefer this type of 'discovery' activity to traditional human-computer interface methods. This metaphor enables everyday activities, life in general and items specific to certain problems to be depicted.

3.9 The Toolbox

The Toolbox can provide many tools for individual users' needs. The most appropriate toolbox should be furnished through user browsing and tool choice. Toolbox items will vary from user to user and the package merely needs to offer the user the means of finding and retrieving appropriate tools.

It is argued that the combination of these metaphors, when used with the map metaphor would provide the means to deliver the contents of the GeoExploratorium using multimedia components.

4 APPLICATION – GEOGRAPHIC INTELLIGENCE

A Technological advances in computer systems over the past two decades have provided mapmakers and land resource managers with capabilities to perform increasingly sophisticated mapping and geographic analysis functions. For much of this period development has followed two distinct streams; one concerned with automating the map-making process and favoured by the traditional mapping organisations, and one focusing on environmental analysis and land planning functions generally used by regional planners, asset and facilities managers, etc.

We have now reached the time when we now have a range of technologies and these technologies are enabling the fusion of the two trends. However, the World Wide Web, the *e*-phenomena and society's desire to be better informed now mean that the simple fusion of the existing capabilities from those two traditional trends is no longer adequate. We now need to know far more about our environment than ever before and we need to make decisions far more quickly than ever before. We need to embrace the concept of a geospatial information infrastructure and geographic intelligence.

4.1 Specialist adviser on geography and the environment – a SAGE

The availability and access to geographic and environmental information is growing at a staggering pace; a pace which far outstrips our ability to use it credibly for, not only the planning and conduct

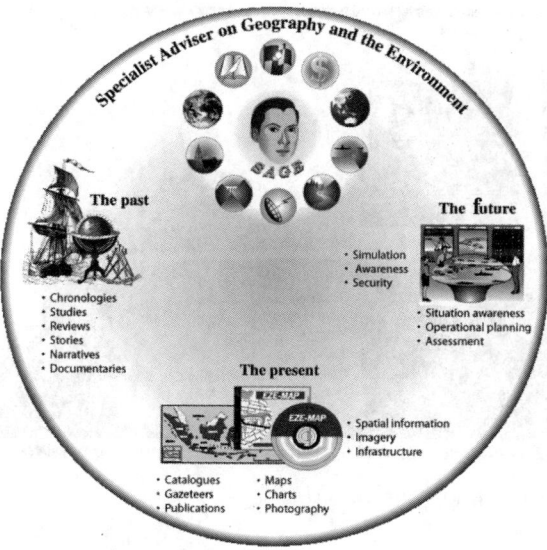

Figure 2.   William's SAGE concept (Williams 2002).

of operations and routine activities, but for strategically more important purposes such as strategic national development, homeland defence, security and so on. We now need to consider the concept of designing and creating *virtual advisers* or artificial agents. One such concept might be that of a Specialist Adviser on geography and the Environment – a SAGE. However, if we embrace the concept of an artificial agent then we need to investigate the scope and context of the *world* in which the SAGE would function.

It seems intuitive that the SAGE would need to know about the past. The SAGE would need to have available a comprehensive knowledge of the sciences, technologies and disciplines of the past and the implications to present databases and archives. The SAGE would also need to know about previous studies and reviews, previous initiatives and activities; and the SAGE should have access to documentaries, narratives, histories, and so on. In addition, it seems that the SAGE would need access to a plethora of policy guidance documents, agreements, memoranda, etc.

The SAGE needs access to the nation's geospatial information infrastructure (the ASDI) via accredited portals, etc. The knowledge base needs also to include white papers, and a comprehensive range of scientific and technical sources suitable to both naïve and expert users. Furthermore, it seems that the SAGE needs a visionary component. The knowledge and vision of domain experts needs to be elicited and this information organised into capability development strategies. Such strategies need to be cognisant of acquisition processes and scope the near, mid and long terms acquisition programs of an organisation. Overall, the concept for the SAGE can be shown in the illustration in Figure 2.

5   PROTOTYPE – TOWNSVILLE GEOKNOWLEDGE PROJECT

The Townsville prototype is being developed to incorporate both multimedia and GIS elements, accessed through the appropriate use of metaphors, in the first instance incorporating Cartwright's Sage metaphor with William's SAGE concept. The prototype has been designed to be delivered via a World Wide Web browser and requiring minimal plug-ins – for *Flash* and *QuickTime* movies. As this is being used as a testbed for the GeoExploratorium two access interfaces have been provided as part of the initial interface, a 'map' of information resources and a 3D *Information Landscape* that the user can move through. The remaining section of this paper concentrates specifically on the 3D *Information Landscape* interface.

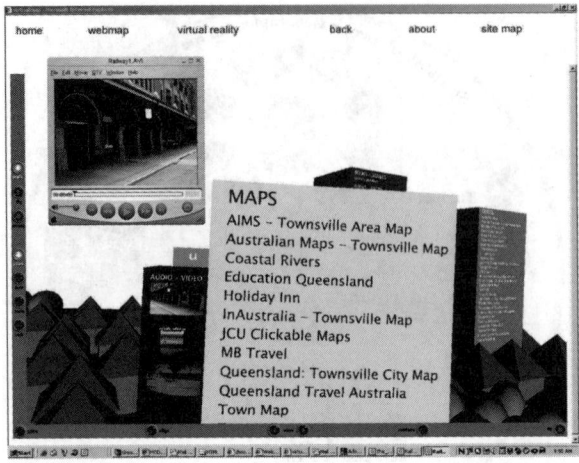

Figure 3.   Video links (interactive).

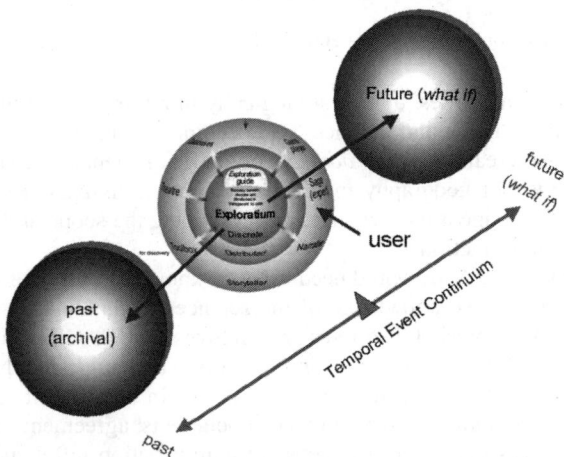

Figure 4.   GeoExploratorium and Temporal Event Continuum.

The VRML components of the *Information Landscape* are designed to provide a World where users can browse and 'walk' through the information access 'space', that is a virtual terrain built on information availability, rather that physical or human geographical information. Similar concepts have been applied for accessing news stories (Sparciano 1997) and building DTMs of 'likeness' of news stories that can be analysed through standard GIS applications (Fabrikant 2003). Testing will evaluate whether users prefer the traditional 2D map interface or the *Information Landscape*.

Using the *Information Landscape* users can move about the 'terrain' and move into areas of (information) interest to 'prospect' for geographical information related to pre-determined areas of need, or merely browse through the space. The VRML World contains links to information that includes videos and audio that are able to be activated by clicking on the image (which is an active 'running' video) (Figure 3) or audio logo that appear on the 'walls' of areas of information availability 'buildings'. The appropriateness of this method and whether users access this information in an intuitive way will be evaluated as part of this research program. This information access method provides access to current information. But, to best empower users to make better decisions a hybrid product will be developed which will allow users to 'slide' between current, archival and projected information using a '*Temporal Event Continuum*' slider control. This information 'dial-up' method will be developed parallel to the population of the *Information Landscape* using *what if* scenarios. This will be realised by users accessing information through the *GeoExploratorium* 'portal', and in the initial stage of prototype development only the Sage metaphor will be employed. Users will then be able to slide along the *Temporal Event Continuum* vector and choose information related to where it occurs at a certain time, present, past or future. This concept is illustrated in Figure 4.

What if scenarios will add GIS functionality to the *GeoExploratorium*, and the planned method for achieving this is outlined in the following section.

6   ADDING GIS FUNCTIONALITY TO THE GEOEXPLORATORIUM – WHAT IF SCENARIOS

GIS and multimedia technologies underlying the GeoExploratorium prototype enable both spatial analysis of data visualisation to be undertaken. Specifically, the GeoExploratorium prototype component conceptualised in this paper focuses on datasets acquired for Townsville to ultimately explore *what if* intelligence strategies. In recent time *what if* scenarios have been developed and tested in the context of urban and regional strategic planning both nationally (Pettit, Pullar & Stimson 2002) and internationally (Landis 1994, 1995; Klosterman 1999).

If *what if* scenario modelling can assist in making better urban and regional planning decisions then inductively we explore the possibility of *what if* scenario modelling assisting in making better decision in a military intelligence and strategic planning context. Modelling of '*what if*' scenarios involves the formulation and evaluation of a number scenarios based upon a number of input parameters, and specified constraints and opportunities. A control scenario is typically developed first and is commonly referred to as a *business as usual* scenario (Landis 2001; Pettit & Pullar 2001). The remaining scenarios can be used to measure the deviation from the *business as usual* scenario in order to visualize and analyse the effects particular military intelligence or strategic planning decisions may have. This enables likely future outcomes of decisions to be more fully explored in order to assist in making critical strategic decisions.

Packages used to develop such *what if* scenarios are commonly referred to as planning support systems (PSS) (Harris 1989; Harris & Batty 1993; Klosterman 1997) and are developed using GIS technology. This paper provides a conceptual model for developing a *what if* scenario PSS tool as a component to a GeoExploratorium prototype. By coupling the PSS within the GeoExploratium it is envisaged that access to such geospatial decision support system technology will be available to military personnel who do not have particular expertise in driving traditional GIS tools. This is because the interface to the PSS component can be developed with a soft multimedia front end which is easily navigatible by the military personnel who are not geospatial scientists.

7   CONCLUSIONS

The use of a Geographical Information System incorporating multimedia devices as a tool for the visualisation of geographical relationships can be seen as perhaps one of the ways in which spatial decision support (where decisions are based on the evaluation and consideration of data which is spatially unique and geographically referenced) can be effectively made available. The hardware and software developments of both multimedia and Geographical Information Systems have now reached a stage where the technological issues have been resolved. The linking of such powerful systems allow for the presentation of spatial data, which is spatially accurate and timely and presented in such a way that it supports the decision-making process.

The addition of multimedia elements to a Geographical Information System can improve the user's visualisation of reality when it is displayed graphically as three-space data and time. As spatial data about natural and cultural objects change over time, in terms of position, weighting and dominance, it is important that the display of the quantitative information (the information, when given an accurate four-dimensional position can be termed geographical information) correctly gives the user a narrative of space and time which captures the essence of what is being depicted and hence aids visualisation.

This paper discussed the use of multimedia with several GIS packages. Although initially developed as stand-alone products these types of GIS are now being delivered using the World Wide Web. As GIS vendors strive to make their software more competitive and geographical information providers move towards the delivery of their data and services using contemporary communications systems there has been much interest in making GIS work on-line. However, whatever the delivery mechanism, the essential components of multimedia working with GIS make for a powerful visualization tool for geographical information users.

ACKNOWLEDGEMENTS

This research is supported through a research grant from the Department of Defence, Australia, and particularly the Command and Control Division, Defence Science and Technology Organisation, Edinburgh, South Australia.

REFERENCES

Cartwright, W. E. 1999. Extending the map metaphor using web delivered multimedia. *International Journal of Geographical Information Science*, 3(4): 335–353.

Cartwright, W. E. 1997. The Application of a New Metaphor Set to Depict Geographic Information and Associations, *proceedings of the 18th International Cartographic Conference,* Stockholm, Sweden: International Cartographic Association, June: 654–662.

Cartwright, W. E. & Hunter, G. J. 1999. Enhancing the map metaphor with multimedia cartography, *Multimedia Cartography*, Cartwright, W. E., Peterson, M. P. & Gartner, G. (eds), Heidelberg: Springer-Verlag: 257–270.

Fabrikant, S. I. 2003. Spatialization: Charting the sea of information, *GepCart'2003, Taupo, N Z: N Z Cartographic Society.*

Harris, B. 1989. Beyond Geographical Information Systems: Computers and the planning professional. *Jrnl of Amen Plg Ass* 55: 85–92.

Klosterman, R. E. 1997. Planning Support Systems: A New Perspective on Computer-Aided Planning. *Jrnl of Planning Education & Research* 17: 45–54.

Klosterman, R. E. 1999. The What if? Collaborative Planning Support System. *Environment and Planning B: Planning and Design* 26: 393–408.

Landis, J. 1994. The California Urban Future Models: A New Generation of Metropolitan Simulation Models. *Environment and Planning B: Planning and Design* 21: 399–420.

Landis, J. 2001. CUF, CUFII, and CURBA: A family of spatially explicit urban growth and land-use policy simulation models. In *Planning Support Systems: Integrating Geographic Information Systems, Models, and Visualisation Tools.* (ed.) Brail, R. K. & Klosterman, R. E. Redlands, California, ESRI Press: 157–200.

Lewis, S. 1991. Hypermedia Geographical Information Systems, *Procs of EGIS '91*, Brussels: EGIS Foundation, 1: 637–645.

Pettit, C. & Pullar, D. 2001. Planning Scenarios for the Growth of Hervey Bay. In *6th Int Conf on Geocomputation*, Sept, Brisbane.

Pettit, C. Pullar, D. & Stimson, R. 2002. An Integrated Multi-Scaled Decision Support Framework used in the Formulation and Evaluation of Land-Use Planning Scenarios. In *The 1st International Environmental Modelling and Software Society Conference*, 24–27th June 2002, Lugano, Switzerland, IEMSS.

Smyth, M. & Knott, R. 1994. The Role of Metaphor at the Human Computer interface, *procs OZCHI94*: 287–291.

Sparciano *et al* 1997. City of News, http://flavia.www.media.mit.edu/~flavia/CityOfNews.html

Tang, Q. 1992. From Description to Analysis: An Electronic Atlas for Spatial Data Exploration, *ASPRS/ACSM/ RT 92 Technical Papers*, Washington, D. C.: ASPRS-ACSM, vol. 3: 455–463.

Williams, R. J. 2002. Geographic Intelligence: The Key to Information Superiority, Conference, Adelaide, South Australia.

Advances in Spatial Analysis and Decision Making, Li, Zhou & Kainz (eds)
© 2004 Swets & Zeitlinger, Lisse, ISBN 90 5809 652 1

Sphere digital space based on manifold: definition, properties and applications

Miao-le Hou & Xue-sheng Zhao
China University of Mining and Technology, Beijing, China

Jun Chen
National Geometrics Centre of China, Beijing, China

ABSTRACT: SGDM (Sphere Grid Data Model) is an efficient method of dealing with global data because of its advantages of multiple resolution and hierarchy. Several studies have been examined many aspects of the model. However, SGDM has no distinct descriptions and lack of round mathematical basis for various applications. To overcome this deficiency, this paper constructs a regular mesh structure for sphere; i.e., the digitization of the spherical surface as a common spatial framework, such as a planar. First, the definition of sphere digital space based on manifold is given. The characteristics of sphere digital space are then presented in detail, and compared with planar digital space in the aspects of topology, metric and order relationships. In the end, some potential applications and possible future studies are discussed briefly. Computers are digital, and most efforts at image acquisition and communication are currently directed towards digital approaches. Therefore, it is convenient to analyse the global data and make decisions in sphere digital space.

1   INTRODUCTION

Up to now, studies on global data have mostly been based on map projections; i.e., the real world (three-dimensional) is transformed into a two-dimension planarity. As a result, the global data inevitably produces overlaps and gaps. The map projection transforms the sphere manifold into planar Euclidean space; therefore, the distances, orientations and areas in a large field are not accurate at all (Hu, 2001). In addition, people simply use a planar map after a map projection to browse and make an analysis, and then come to conclusions and make decisions. If large quantities of global information can be directly dealt with on a spherical surface in a computer, the map projection, which involves complex calculations, can be avoided. With the development of computer techniques, it has become possible to directly store, manage and analyse large quantities of global information. Making the global data directly on the sphere and constructing a sphere dynamic data model is one of the key problems for the Digital Earth (Zhao *et al.*, 2002).

Recently, SGDMs have been developed to represent the surface of the earth, offering the advantages of hierarchical organization, continuous ordering, and equivalent subdivision (Nulty, 1993; Bartholdi and Goldsman, 2001). Studies of these SGDMs include the following: Dutton (1989, 1997) discussed hierarchical tessellations in the context of the generalization of GIS data. White *et al.* (1992) designed a sampling data model to monitor the global environment. Lee and Samet (2000) deal with the global navigation with it. Bartholdi and Goldsman (2001) set up the continuous indexing of the globe based on hierarchical subdivisions. Zhao *et al.* (2002) developed the QTM-based Voronoi sphere data model. However, there have been no distinct descriptions with mathematical language in SGDM, and it is lack of round mathematical basis for various applications, such as the definition, properties, the topological structure and basic topology model and so on.

Therefore, in order to analyse the global data conveniently, a regular mesh structure will be presented in this paper as a common spatial framework, just as planar. Sphere regular mesh system, i.e. the digitization of the spherical surface can be called as sphere digital space, which is the base of the basic topology model and spatial relationship calculation on spherical surface. Computers are digital, and most image acquisition and communication efforts are currently directed towards digital approaches. Therefore, it is convenient to analyse the global data and make decisions in sphere digital space.

The discussion is organized as follows. The next section presents some basic concepts, which are needed in the paper. Section 3, describes the common sphere digital space based on manifold. The prosperities of sphere digital space are then discussed in Section 4. Some potential applications are investigated in Section 5. Conclusions are drawn in the final section and possible avenues for future studies are discussed.

2   PRELIMINARY CONCEPTS

The discussions here depend on several fundamental mathematical concepts, as follows. The details can be obtained from (Bishop, 1980; Boothbay, 2000).

Definition 1.   The n-ball is usually defined to be the set

$$B^n = \{(x_1,..., x_n) \ , \ \ R^n \colon x_1^2 ... + x_n^2 \leq 1\} \tag{1}$$

The frontier of this set is the $(n-1)$-sphere

$$S^{n-1} = \{(x_1,..., x_n) \ , \ \ R^n \colon x_1^{2+} ... + x_n^2 = 1\} \tag{2}$$

Definition 2.   A Hausdorff space is a topological space where every two points lie in a disjoint open set.

Definition 3.   Set M is a Hausdorff space, where any $x \in M$. It has a neighbourhood homeomorphic to the interior of a B^n (an open set) for some fixed. Thus, set M is called an n-dimensional manifold.

The homeomorphism mentioned above is expressed as $\varphi \colon U \rightarrow \varphi(U) \subset R^n$, if $\varphi(U)$ is an open set in R^n. (U, φ) is called a local coordinates system of manifold M. Intuitively, a manifold is a topological space that is locally Euclidean. A two-dimensional manifold is locally flat.

Definition 4.   Set $R^n = (x_1, x_2, ..., x_n)$ and $x_i \in R$. The distance $\rho(x, y)$ of any two points $X = (X_1, ..., X_n)$ and $Y = (Y_1, ..., Y_n)$ can then be defined as:

$$\rho(x, y) = \sqrt{\sum_{i=1}^{n} (x_i - y_i)^2} \tag{3}$$

Thus, R^n is called an n dimension Euclidean space.

3   THE DEFINITION OF SPHERE DIGITAL SPACE BASED ON MANIFOLD

Digital space is the digitization of space. In the digitization of space, discrete points are used to describe the whole universe. Digital space is composed of equal meshes. Before giving the definition of sphere digital space, the definition and characteristics of planar digital space will first be presented.

3.1   *Planar digital space*

Planar digital space is the $N \times N (N = \{0, ..., n-1\})$ regular mesh based on discrete space (Gong, 1997). There are two main types of digital space in images: square mesh space (Q^2) and hexagon mesh space (H^2) (shown in Figure 1). Square mesh space is also called raster space in planar. In Euclidean space, we use $e_1 = (0, 1)$, $e_2 = (1, 0)$ as the basic coordinate system. Set $u_1(k) = (0, k)$, $u_2(k) = (k, 0)$; $v_1(k) = (0, k)$, $v_2(k) = (k/2, \sqrt{3}/2k)$ $k \in R$.

We then obtain the definitions of Q^2 and H^2:

$$Q^2 = \{x_1 u_1(k) + x_2 u_2(k), \ x_1 \text{ and } x_2 \text{ are integer}\} \tag{4}$$

$$H^2 = \{x_1 v_1(k) + x_2 v_2(k), \ x_1 \text{ and } x_2 \text{ are integer}\} \tag{5}$$

In fact, Q^2 and H^2 are all point sets with integer coordinates in different coordinate systems. Mesh space is the sample sets of digitization, and k gives rise to the resolution of image. In addition to point sets, the planar digital space can also be defined in pixels, which are the same in essence (shown in Figure 1).

3.2   The definition of sphere digital space based on manifold

Regular grid sampling structures in the plane are a common spatial framework for many applications. Constructing grids with desirable properties such as equality of area and shape is more difficult on a sphere (White *et al.*, 1998). To deal conveniently with such problems, it is necessary to construct a similar regular mesh structure as a common spatial framework for a spherical surface, just as planar. Such a mesh system is referred to as sphere digital space, which is the digitization of the spherical surface. That is, in sphere digital space a sphere can be described with discrete point samples. Therefore, it is necessary to subdivide the spherical surface according to its characteristics. There are three steps to obtain the sphere digital space, as follows.

3.2.1   Initial partition of the sphere

The Platonic solids are reasonable starting points for a spherical subdivision (shown in Figure 2). Three of the five polyhedrons have triangular faces, such as the tetrahedron (four faces), the octahedron (eight faces), and the icosahedron (20 faces). The other Platonic solids are the cube (six faces) and the pentagonal dodecahedron (12 faces). The icosahedron has the greatest number of initial

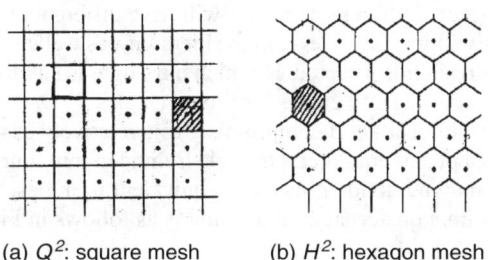

(a) Q^2: square mesh   (b) H^2: hexagon mesh

Figure 1.   Planar digital space (Gong, 1997).

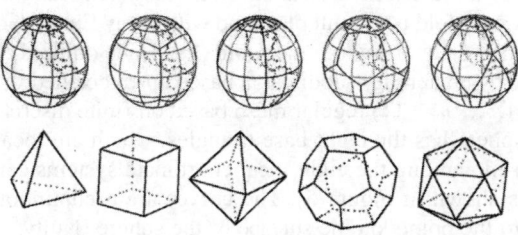

Figure 2.   Platonic solids and their spherical subdivision (White *et al.*, 1992).

293

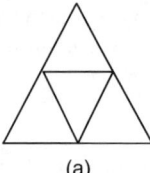 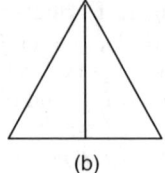

(a)   (b)

Figure 3.   (a) Quaternary subdivision and (b) binary subdivision.

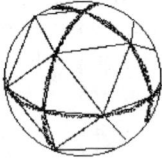 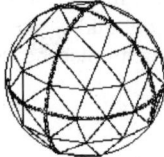 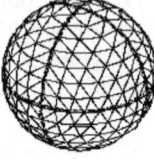

Figure 4.   The result of subdivision based on an octahedron (Dutton, 1996).

faces, and would therefore show the least distortion in the subdivision. However, the larger number of faces makes it somewhat harder to deal with the problems through the borders of the initial faces. In a word, the sphere is more easily covered by triangles, and the triangles of the initial partition need not be equilateral. Distortion could be decreased considerably by dividing each equilateral triangular side of an initial Platonic figure into equivalent scalene triangles (White *et al.*, 1998).

The octahedron has more distortion, but it has the advantage that its faces and vertices map to the important global features: meridians, the equator, and the poles (Goodchild and Shiren, 1992). Therefore, in this paper the octahedron is selected as common initial partition in which eight base triangles are produced.

3.2.2   *Subdivision of triangular cells*

There are several ways to hierarchically subdivide an equilateral triangle such as quaternary subdivision and binary subdivision (shown in Figure 3). All of these are subject to distortion when transferred to the spherical surface. Different decisions will have different effects on the uniformity of the shape and size of cells within a given level of the hierarchy, as well as on the ease of calculation. Here, the quaternary subdivision is selected, in which a triangle is subdivided by joining the midpoints of each side with a new edge, to create four sub-triangles.

The quaternary subdivision is a good compromise. It is relatively easy to work with, and nondistorting on the plane, as a planar equilateral triangle is divided into four equilateral triangles. But a spherical base triangle may be divided into four equivalent triangles. The result of subdivision based on an octahedron with a quaternary subdivision is as follows in Figure 4 (Dutton, 1996).

3.2.3   *The definition of sphere digital space based on manifold*

Manifold is the extension of Euclidean just because every point in manifold has a homeomorphism of an open set in Euclidean. So a local coordinates system can be set up for every point in manifold. It seems that manifold is a result plastered with many Euclidean spaces. It can be proved that a sphere is a two-dimensional smooth manifold (Evidence omitted).

If a sphere is divided by quaternary subdivision based on an octahedron, the sphere digital space is an 8×4^N ($N = \{0, 1, ..., n - 1\}$) regular mesh based on finite discrete space, expressed as T^2. In the first level, the sphere has the eight base triangles, which are local coordinates systems of manifold. The relationship among the eight local coordinate systems can be described by sphere spacefilling curves (as shown in Figure 5). The curves are a continuous mapping from a one-dimensional interval, to the points on the surface of the sphere (Nulty, 1993). Continuous ordering based on space-filling curves have proven to be useful in heuristics related to a number of

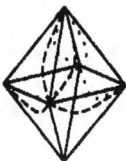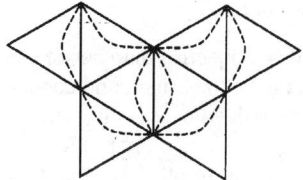

Figure 5.   Sphere spacefilling curves based on an octahedron with quaternary subdivision.

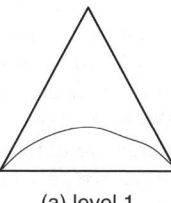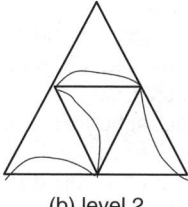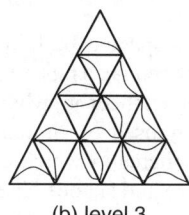

(a) level 1   (b) level 2   (b) level 3

Figure 6.   The quaternary sphere spacefilling curve.

spatial, combinatorial, and logistical problems (Bartholdi and Platzman, 1982, 1988; Platzman and Bartholdi, 2001; Nulty, 1993).

In every base triangle, quaternary sphere spacefilling curves still can be used to express the relationship between every sub-triangle. In quaternary subdivision, the relationship between sub-triangles can be depicted with quaternary sphere spacefilling curves (as shown in Figure 6). In a given resolution, sphere digital space can be continuously indexed by quaternary sphere spacefilling curves (Bartholdi, 2001a). Compared with the other model (e.g. Dutton, 1989), sphere digital space has the advantage of continuous ordering. It makes us index the sphere digital space continuously to allow quick and efficient multi-scale searches. At the same time, sphere digital space has an intrinsic disadvantage in that the triangles are equivalent to, but not equal with, each other.

4   THE PROPERTIES OF SPHERE DIGITAL SPACE

Planar digital space is a simple Euclidean space, but sphere digital space is a more complex manifold. Thus, sphere digital space is not a simple copy of planar digital space. The differences between them will be discussed in detail as follows, as this is essential to build a basic topology model in sphere digital space and to form a base for topological relationships between the entities in the sphere.

4.1   *Non-Euclidean space*

Sphere digital space is only an anisotropic manifold, not an isotropic Euclidean space (Hu, 2001). There is a Descartes coordinate in Euclidean space just for its linear structure. Euclidean space is the simplest space, while sphere digital space is a more complex space. That is to say, no single coordinates system can be set up to express every point in a spherical surface. Although a longitude and latitude coordinates system can be constructed to describe the points in a sphere, the South and North Poles are exceptions. The latitude of the North Pole is 90 degrees north, but its longitude is uncertain; the same as applies to the South Pole.

The foundation of sphere geometry is Riemannian, in which every two parallel lines have intersections. Riemannian is a perfect model for a sphere. In our daily life, however, the Euclidean model applies, while the Riemannian is more useful in global studies on such subjects as global navigation, seafaring, the atmosphere, the environment, and so on.

4.2 No homomorphous to planar

Riemannian mapping builds a bijective map between a sphere and a planar except a point (shown as Figure 7): Let the South Pole be the origin, and connect to any point Z except for the North Pole in a spherical surface with the North Pole. This line will intersect with a planar at a unique point $S(z)$. The Riemannian mapping sphere has one more point than a planar sphere, so sphere digital space is not homomorphous to planar space. In other words, the Euler's characteristic of a sphere is 2, but the Euler's characteristic of a planar is 1. Therefore the sphere is completely different from a planar.

4.3 Cells are approximately equivalent

The planar digital space is built in equal squares (or hexagons), and the distance between any two grids is absolutely equal. Thus, this distance can be denoted with only two basic vectors. For division based on an octahedron, whenever quaternary or binary subdivision is employed, sphere digital space is composed of similarly equal spherical triangles. We have shown how to create a subdivision of spheres in which cells at a given level of the hierarchy tend to be approximately equivalent (in size and shape). Therefore, most of the distances between pixels are not completely equal and there are no base vectors, which can express all triangles just as planar digital space (as shown in Figure 8).

4.4 Multi-scale and continuous ordering

Sphere digital space has the advantages of hierarchy and continuous ordering. Hierarchical models have the inherent ability to reduce the size of data, and to vary the resolution at different locations. This is useful since the density of features on the surface of the earth is not uniform; such models conserve space by increasing resolution in dense areas, and decreasing it in sparse areas. An example of such a problem is the need to increase or decrease the level of spatial details displayed in a map, when zooming in or out, respectively. This problem is sometimes handled in GIS by storing multiple versions of maps at varied levels of detail.

In sphere digital space, each cell in the subdivision can be ordered by spherical spacefilling curves in a spatially continuous way. This is useful for operations that depend on clustering and

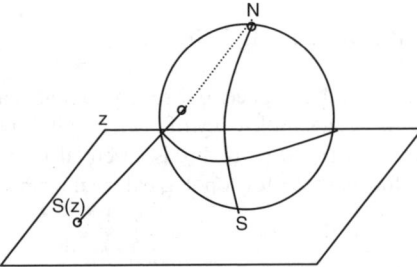

Figure 7.   Riemannian mapping.

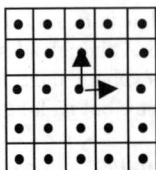

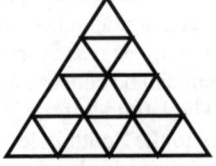

(a) Equal pixels in planar digital space   (b) Equivalent pixels in sphere digital space

Figure 8.   Equal and approximately equivalent.

spatial adjacency in general. It is precisely because sphere digital space has the property of multi-scale and continuous ordering that we can index the sphere digital space continuously to allow quick and efficient multi-scale searches.

5   POTENTIAL APPLICATIONS

Sphere digital space provides an appropriate discrete frame to express every point in a spherical surface. Therefore, it can be conveniently applied to deal with problems on earth. Before the concept of sphere digital space was presented, several authors attempted to use QTM overcome global problems such as global data generalization, global environment, sphere navigation, global continue indexing, and so on. Here, an attempt is made to discuss more basic applications to build the functional architecture of a framework for sphere digital topology. Digital topology provides a sound mathematical basis for various image-processing applications including surface detection, border tracking and thinning.

We consider a 2-D triangle mesh to represent a 2-D spherical surface digital object. In this paper, points refer to 2-D triangle mesh points unless stated otherwise. A 2-D spherical surface digital object P is defined as a triplet $(T^2; R; H)$, where H is a finite subset of T^2 and R represents the adjacency relation in the whole lattice in a specific way. There are two kinds of adjacency relationships in sphere digital space; i.e., 3 and 12. The main purpose of sphere digital topology is to study topological properties of discrete objects in sphere digital space. Sphere digital topology plays a very important role in global visualization, global indexing and sphere data models. In this representation, an object in a spherical surface is described by an array of bits. In this way, a sphere digital object can be defined as an array augmented by a neighbourhood structure. Points of T^2 associated with triangles that have a value of 1 are called black points, and those associated with triangles with a value of 0 are called white points. The set of black points normally corresponds to an object in the digital image.

As for topology relationships in discrete space, the basic topological concept in vector space, connectedness, is implicitly inherited. Computers are digital, and most image acquisition and communication efforts are currently directed toward digital approaches. Study of entities in raster (discrete) space could be more efficient than in vector space as the expression of spatial entities in discrete space is more explicit than that in connected space (Li et al., 2000). With sphere digital space the continuous indexing of the global data can be set up easily. In sphere digital space, adjacency, connectivity, regions, and boundaries can simply be defined as planar. Therefore, it is convenient to analyse the global data and make decisions with sphere digital space.

6   DISCUSSIONS AND FUTURE WORKS

The basic definition and some properties of sphere digital space were given in detail in this paper. Sphere digital space is the foundation of SGDM, which can be used to model global multi-resolution spatial data. Computers are digital, and most image acquisition and communication efforts currently focus on digital approaches. Sphere digital space provides a digital frame to conveniently analyse the global data and make decisions.

Sphere digital space, the digitization of the spherical surface, is a similar regular mesh system in which every point has a unique label and is only an anisotropic manifold but not an isotropic Euclidean space. It is not the homomorphism of the planar. Although the sphere digital space is made up of approximately equivalent cells, it still has many advantages, such as multi-scale and continuous ordering, which are useful in representing and managing the global data efficiently. Finally, the potential applications of sphere digital space were given. The basic topological model can be set up in sphere digital space so that it is convenient to analyse the global data and make decisions.

To set up the basic topological model based on sphere digital space, a great deal of in-depth research needs to be conducted on such areas as, the topological structure of sphere digital space, the basic topological components of a spatial entity in T^2, topological paradoxes associated with the definition of adjacency in T^2, and so on. This paper is simply a preliminary study on the characterization of the 2-digital sphere manifold and the Jordan–Brower separation theorem, which are all round mathematic basis of spherical spatial computing and reasoning.

ACKNOWLEDGEMENTS

The work presented in this paper was substantially supported by an outstanding youth award from the natural science foundation of China (under grant No. 40025101).

REFERENCES

Abel, D. J., and Mark, D. M. 1990. A comparative analysis of some two-dimensional orderings. *International Journal of Geographical Information System*, 4(1): 21–31.

Bartholdi, III, J. J., and Platzman, L. K. 1982. An O(nlogn) planar travelling salesman heuristic based on spacefilling curves. *Operations Research Letters*, 1: 121–125.

Bartholdi, III, J. J., and Platzman, L. K. 1988. Heuristics based on spacefilling curves for combinatorial problems in Euclidean space. Management Science, 34: 291–305.

Bartholdi, III, J. J., and Goldsman, P. 2001a. Continuous indexing of hierarchical subdivisions of the globe. *International Journal for Geographical Information Systems*, 15(6): 489–522.

Bartholdi, III, J. J., and Goldsman, P. 2001b. Vertex-labeling algorithms for the Hilbert spacefilling curve. *Software -Practice and Experience*, 31: 395–408.

Bishop, R. L., and Goldberg, S. I. 1980. Tensor Analysis on Manifolds. *Dover Publications, Inc.,* New York, 90pp.

Boothby, W. H. 2000. An Introduction to Differentiable Manifolds and Riemannian Geometry. *Springer-Verlag*, New York, 200pp.

Dutton, G. 1989. Planetary modeling via hierarchical tessellation. *Proceedings Auto Carto*, E. Anderson (Ed.), (Baltimore: ACSM), 462–471.

Dutton, G. 1996b. Encoding and Handling Geospatial Data with Hierarchical Triangular Meshes, In: Kraak, M. J., and Molenaar, M. (Eds), *Proceeding of 7th International Symposium on Spatial Data Handling*, Netherlands, 34–43.

Dutton G. 1997. Digital map generalization using a hierarchical coordinate system, *Proc. Auto Carto 13. (Seattle, WA) Bethesda, MD: ACSM/ASPRS*, 367–376.

Goodchild, M. F., and Yang, S. R. 1992. A Hierarchical Data Structure for Global Geographic Information Systems. *Computer Vision and Geographic Image Processing*, 54(1): 31–44.

Gong, W., Shi, Q. Y., and Cheng, M. D. 1992. A morphological filter without distorting image. *Proc, 11th Int. Conf. On pattern Recognition*, 3: 684–687.

Hu, P. 2001. Bottle Necks in GIS Development – Theories of GIS and the Practice of Million Image GIS. *Journal of Wuhan University*, 14(3): 77–85.

Lee, M., and Samet, H. 2000. Navigation through triangle meshes implemented as linear quadtrees. *ACM Transactions on Graphics*, 19(2): 79–121.

Li, Z. L., Li, Y. L., and Chen, Y. Q. 2000. Basic Topological Models for Spatial Entities in 3-Dimensional Space. *GeoInformatica*, 4(4): 419–433.

Nulty, W. G. 1993. Geometric Searching with Spacefilling Curves, PhD thesis. *Georgia Institute of Technology*, Atlanta, GA.

Rigaux, P., and Scholl, M. 2002. Spatial Databases: With Application to GIS. *Springer-Verlag*, 230pp.

White, D., Kimmerling J., and Overton, W. S. 1992. Cartographic and Geometric Components of a Global Sampling Design For Environment Monitoring, *Cartography & Geographical Information Systems*, 19(1): 5–22.

White, D., Kimberling, A. J., and Song L. 1998. Comparing area and shape distortion on polyhedral-based recursive partitions of the sphere. *International Journal of Geographical Information Systems*, 12(8): 805–827.

Zhao, X. S., Chen, J., and Li, Z. L. 2002. A QTM-based algorithm for generation of the voronoi diagram on a sphere. Advances in Spatial Data Handling, Published by Springer, Berlin, 269–285.

Zhao, X. S. Spherical Voronoi data model based on QTM. PhD thesis, China University of Mining technology (Beijing), Beijing, 2002, 105pp.

Advances in Spatial Analysis and Decision Making, Li, Zhou & Kainz (eds)
© 2004 Swets & Zeitlinger, Lisse, ISBN 90 5809 652 1

Design parameters for an interactive web mapping system

Lilian S.C. Pun-Cheng & Geoffrey Y.K. Shea
Department of Land Surveying and Geo-Informatics, the Hong Kong Polytechnic University, HK

ABSTRACT: Web mapping has in recent decades become a useful way of presenting and extracting information for multiple users. Because of its confined display environment, web map information has to be carefully designed to fit into different pages so that the graphics and information to be displayed will not be too dense. In addition, when incorporating with spatial analysis functionality, a selection of client-side or server-side Internet Mapping Systems has to be made to balance performance and functionality. All these issues will be examined in an application of a web map system. The development of a multi-modal public transport query system illustrates how seamless maps of varying scales are designed and synchronized with each other, how dense information in a bilingual base map and transport data are arranged into suitable pages or hierarchies, and how efficiency in retrieving relevant map pages on the web might be achieved.

1   INTRODUCTION

Web mapping has become a more and more popular way of providing information to a large group of users, country-wide or even world-wide. The map on the Internet functions as an interface or index to additional information. By pointing to a certain position on the map, photographs, drawings, texts, sounds or other maps can be linked. In general, web maps might be divided into two types: view only and interactive. The former is not greatly different from traditional paper maps. Original maps are simply scanned and displayed on the web with no navigation capabilities. Interactive maps, on the other hand, will allow a user to 'mouse over' or 'click', leading to further information or data. That is, interactive maps enable a user to surf the web and zap from page to page. In any case, the design of a web map somewhat deviates from the traditional rules of cartographic symbolization. Due to the display environment, each map page should not be too large and graphic and information density should be low. The contents of a web map hence have to be defined carefully into stages or hierarchies in displaying pages. Each is then designed with a required level of perception according to requirements of scale and accuracy. The complexity of a design will increase with an increasing amount of content and interactivity, especially if geo-spatial analysis (or simply GIS) functions are involved.

   This paper presents an example of putting a multi-modal public transport query system on the web. The maps involved not only have to clearly display the relevant transport and base map information, but also act as an interface for a user to select an origin and a destination for a back-end spatial query and analysis. A review of Internet Mapping Systems is first given as a background to achieving such functionality. This is then followed by a detailed description of the application requirements, the web map design criteria and procedures for implementation.

2   REVIEW OF INTERNET MAPPING

An Internet Mapping System (sometime referred to as a Web-enabled GIS or an Online GIS application) is a Web application fully charged with GIS functionality. Usually, the applications emerge when traditional GIS packages are extended to support the Web technology. Basically, an Internet Mapping System follows the Client-Server Model with Web browsers as the clients and the Web site serving the application as the server. There are two variations to the basic Internet mapping application: (a) client-side; and (b) server-side applications (Gifford 1999). This section will briefly discuss issues of visualization design for online GIS, taking into consideration its interactive, computational-intensive characteristics and requirements on a high-resolution graphical user interface.

3 CLIENT-SIDE INTERNET MAPPING SYSTEM

The architecture of a client-side Internet Mapping System is a highly client-dependent platform configuration that requires the client machine to take up all of the responsibilities of processing GIS operations. Since all core GIS operations such as spatial query and buffering are heavily loaded in the client machine, the hardware configuration in the client machine is critical to the success of this approach. Moreover, HTML is a document-formatting language that cannot be used to carry out even simple calculations such as additions or to display the data in the browser window. To perform the GIS operations in the client machine, the Web browser must be assisted with client-side scripting language, Web-enabled programming languages such as Java, and Web browser plug-ins whenever necessary. Another feature of client-side GIS applications, which makes it suitable for Web-based GIS applications, is data awareness. Each data-aware page contains not only the HTML information, but also proprietary vector data that support GIS operations in the client computer. Developers can create complicated GIS operations in the client computer to manipulate the proprietary vector data that usually offer more efficient and flexible data processing than the generic Web-recognized data format.

Despite the positive features mentioned above, there are several disadvantages to the client-side approach that need to be remedied in order to lead to wider acceptance by the GIS community. First, security issues related to the use of vector data in the client computer must be addressed so that the ownership and copyright of the vector data will not be abused. Second, the vector data to be transmitted between the server and client must follow some kind of standard such as the Scalable Vector Graphics (SVG) proposed by the World Wide Web Consortium (W3C) ensuring the transmitted vector data can be manipulated across multiple platforms. Third, the frequency of the transmission of vector data between the server and the client must be optimized in order to reduce the high usage of bandwidth in the Internet.

4 SERVER-SIDE INTERNET MAPPING SYSTEM

The server-side GIS application is by nature a three-tiered client-server environment and thus requires a sophisticated hardware configuration on the server platform to handle highly process-demanding requests from clients. The Web browser on the client-side plays a passive role in this approach – it only generates HTTP requests, waits for the results from the server and displays them immediately on the client's computer without taking any responsibility for processing GIS operations. The Web server, however, plays an active role in handling HTTP requests. The dedicated servers on the server-side work together as a group to extract 'live' data from the database, produce the required GIS data, format the data as an HTML document, and furnish them to the client with the help of the Web server. In this manner, the client views up-to-date, accurate information.

The server-side approach can eliminate problems relating to the incompatibility, inconsistency and unreliability of data. This is because all of the GIS operations are channelled by the Web server and processed by dedicated data servers. All clients use the same graphical user interface to perform the same set of GIS functions on the same set of GIS data stored on the database(s) provided by the servers. The server-side approach reveals, however, a major drawback that needs to be tackled immediately. Since servers must transmit GIS data to clients for every GIS operation over the Internet (e.g., users reset a display window by panning or zooming, turn a layer on or off, make a

300

spatial analysis), a heavy load on the network is unavoidable. The situation is even worse if many clients are connecting to the server with slow modems.

5   BALANCE BETWEEN CLIENT-SIDE AND SERVER-SIDE SOLUTIONS

In general, the client-side solution provides a better working environment with powerful functions for analysis. This approach is favourable for a smaller group of sophisticated users who are looking for complicated GIS functions. To achieve the best performance for this approach, additional add-on components (or *plug-ins*) for the Web browser need to be developed and extra effort to maintain software and plug-ins is also required. On the other hand, the server-side solution offers a standardized and economical GIS solution to a wider group of infrequent users who do not require a highly responsive GIS server. As all of the processing is done by the server and all of the processed information is returned to the client in the form of HTML documents, the server-side solution demands a better and more powerful hardware configuration for the server.

Employing either one of these approaches in an online GIS is a matter of choice between performance and functionality. With the advancement of Internet technologies, the classification of server-side and client-side approaches is no longer as easily distinguishable as before. Currently, mixing the two approaches to provide an optimized online GIS application is the mainstream solution. Achieving a highly responsive dedicated online GIS system that can provide complicated GIS query and analysis functions for the users to manipulate remotely is one of the main challenges facing developers.

Concerning the map format, many Web-enabled GIS applications are currently using either native Web-recognized bitmap formats such as GIF or proprietary vector formats to transfer data between the client and server, and then rendering the results on the client side with HTML. There are several drawbacks to using these bitmap data formats and the HTML presentation format in developing online GIS applications. First, neither GIF nor JPEG provides intelligent information about the data being transferred. This means that the major characteristic of GIS data – providing intelligence between graphic elements and textual information – will not be preserved in the process of data communication over the Web. If the GIS data intelligence is lost during the process of transfer, many of the interactive and sophisticated functions cannot be performed. Second, it is difficult to get high-quality results when scaling or transforming a bitmap. The Web community is still awaiting a standard vector format for exchanging vector data. In the transitional stage, a diverse proprietary vector format has been adopted by online GIS developers. This means that users have to download/install many plug-ins to display the data appropriately. The users also have to switch between different plug-ins when manipulating different GIS applications. This heterogeneity makes it difficult for users to combine data from multiple sources. Obviously this is a cumbersome design and undesirable from the point of view of the user. Third, it should be kept in mind that HTML is simply a data formatting language designed as a means of presenting static information for purposes of display only. HTML is now heavily burdened with a great many expectations about its functions that it had not originally been designed for, such as that it will present dynamic information with interactivity, serve as a means of storing specific types of data and be used as a database interface. All these involve changing its original nature. Even more significant than this is that HTML does not provide the extensibility, structure and data validation required for the large-scale development of data-centric Web applications (Bosak 1997). All of these data-centric applications are now seeking a development infrastructure that can provide a flexible and extensible data model on the one hand, and a standard and scalable graphic format to handle all kinds of graphic data including raster, vector and text on the other.

6   DESCRIBING VECTOR GRAPHICS WITH SVG FOR AN ONLINE GIS

Being XML-based, SVG will seamlessly integrate with all other emerging XML-based standards such as DOM, CSS, XLink, and so on. This means that SVG is not merely a graphics engine but is a fully scriptable, stylable, linkable, interchangeable and Unicode-ready document standard with all of the power of XML. Among the advantages of SVG that are applicable to all Web applications, several are particularly beneficial to online GIS applications:

(a) Dynamic control and greater interactivity. SVG conforms to the DOM, which means that every element inside an SVG document is reachable from JavaScript and is changeable at run-time in response to mouse events or to any function exposed by DOM. Being text-based, the generation of content is also possible, so that server-side scripts such as CGI can create graphics tailored to the user and the situation. All of these have far-reaching implications for online GIS applications. For example a support engineer is trying to update the information of a bus stop that has just been replaced. The update is being carried out on-site with a notebook computer connected to the Web via a mobile phone. An initial index map pops up when the engineer logs successfully on to his department's intranet. The engineer would then be able to delve into the map and find the workplace. When zooming into the area of interest, the engineer turns on to a display of the facility layers belonging to the department and the basic mapping layers provided by another department. The bus stop symbol pops up when the mouse is placed over it. A popup window asking for action will be displayed when the bus stop symbol is double-clicked. Finally, the information is updated.
(b) Versatile elements grouping/layering. Because SVG uses XML syntax to describe information in terms of its data type and SVG files are not proprietary binary data files, developers can easily create scripts that help all stages of the data process: data capturing, data editing, data analysis and data interchange. SVG provides greater flexibility for organizing data, such that graphics elements can be displayed in a variety of ways such as group by one layer at a time or all together.
(c) Flexible graphics display. Because vector graphics are drawn on demand, at runtime these can be displayed in any size without a loss of quality. This quality also means that it is relatively easy to substitute colours, modify line widths, change symbols, etc. on-the-fly. SVG also includes the ability to place text along the shape of a path element. It could be a curving path or any path defined by the designer. This type of flexibility goes far beyond what a bitmap image can offer. All these make SVG a very flexible and powerful graphics information standard.
(d) Better element searching. SVG keeps text in the graphics as text, rather than as a 'picture of the text'. The text remains searchable by search engines even though it is a stroked and filled in component of a piece of vector art (e.g., it is possible to search for a street name on a map, or a district name of 'Wan Chai' on a district boundary layer).

7 PUBLIC TRANSPORT QUERY SYSTEM – A WEB MAP APPLICATION

The development of a multi-modal public transport query and guiding system on the web is information-intensive, involving as it does switching from page to page and from maps of varying scales. A great deal of interactive communication is required between the users and the system; and in particular between maps of varying scales and contents, and between maps and texts of addresses or charts of route data. In addition, the clicking of users on a map might also trigger a series of GIS functions, thereby returning solutions as linked to and displayed on other maps. To carefully design this highly complex web system, it is first necessary to understand its purposes and requirements.

The system serves to answer the following several questions frequently asked by public transport commuters: (a) whether there is a direct route given the origin and destination as depicted on a map or by textual input; (b) whether one or more transfers are needed and what the options are; (c) how to get to the pick-up and drop-off locations (the stops/stations/piers) from the user's chosen origin and destination; and (d) how much the trip will cost, when and how frequent the routes are, and so on. There are not many similar developments of such a comprehensive system. A notable example is the transport for London (http://journeyplanner.tfl.gov.uk/). This paper focuses only on the pages involving a web map, while leaving the non-map textual pages alone for the time being.

In this application, the web map serves two purposes – determining origin-destination and presenting public transport locations. The maps so designed have to satisfy the following requirements for effective interactive communication, each of which will be discussed in greater depth in subsequent sections:

(a) Be seamless and scalable to cover the whole of Hong Kong's territory in a macro, meso and micro environment;
(b) Be bilingual for both Chinese and English users;
(c) Contain a high density of textual labels with the names of all streets, sites, buildings and landmarks;
(d) Dynamically display travel information such as the mode and numbers at a certain location; and
(e) Respond quickly to any selection or switching of locations.

8 SEAMLESSNESS AND SCALABILITY

In principle, there are three approaches to presenting variable scale maps on a single display: user switching between scales, having large- and small-scale maps occupy various frames of the window, and having the same map with large-scale data at the centre and small-scale data on the fringe (Harrie et al. 2002). As a responsive, clear and informative map engine is required in this public transport query system, the second option with enhanced facility is preferred so that users will not need to engage in excessive switching and will more easily identify their locations in varying scales at a glance.

The map interface is shown in Figure 1. The window consists of the main map in the middle and three small insets on the right. The user's selection of the location in which he/she is interested will determine the area shown in the main map at a scale of 1:3000 or 1:7500. Both are large-scale maps

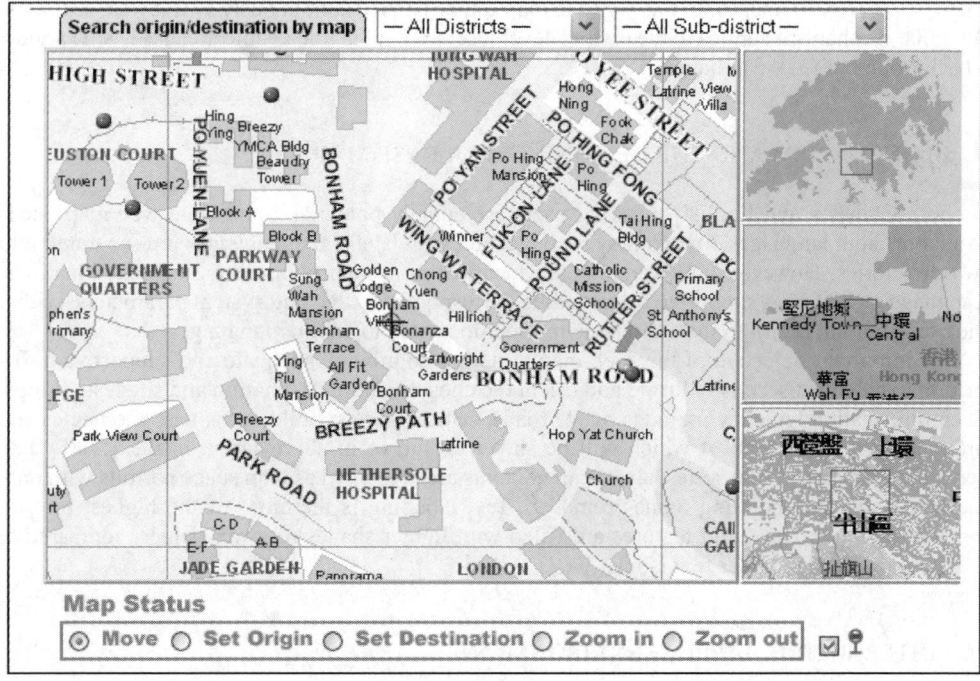

Figure 1.   The map interface with the main map and insets on the right, all synchronizing in position with each other.

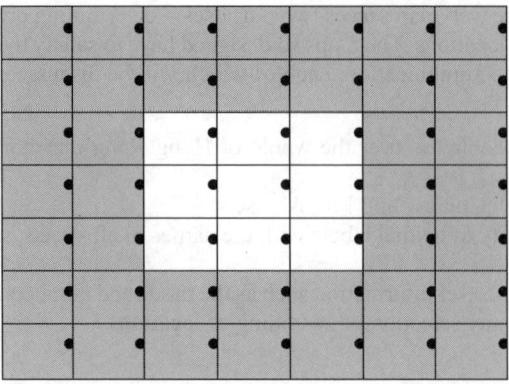

Figure 2.   The visible images (white) compared to the background images (grey).

in which detailed information on land structure can be observed, although with different degrees of labelling for areas of very high congestion. A location may be selected textually in the check box or in a smaller-scale map of both the main page and the insets. With an enhanced synchronization facility, the three insets will show the selected location in relation to the whole of Hong Kong (1:1,000,000), its region (1:250,000) and district (1:50,000) from top to bottom, respectively.

All maps are prepared and pre-processed in a Portable Network Graphics (PNG) format that occupies a much smaller file size than the GIF or JPEG format. To enable seamless panning, the main map is generated by a client-side JavaScript program. For every instant of location selection, a visible map area of 9 (3 × 3) images is displayed, while the user may pan or drag to the surrounding 49 (7 × 7) images, each of them 150 × 120 pixels in size (Figure 2). Once the user stops dragging, the central position of the visible area will be recorded. The layer will refresh itself and load with a new set of (7 × 7) images. In this way, the map can be panned continuously as if it were seamless. A similar mechanism applies to the district-level inset except that the visible area is (1 × 1) amidst a background of (3 × 3) images.

9   BILINGUALISM AND THE DESIGNING OF A BASE MAP

As both Chinese- and English-speaking users are equally potential clients, the web map needs to support both languages. To avoid over-congestion with labels, the languages are separated into two directories. However, both versions are identical in system parameters and functionalities. Language-switching is supported by providing corresponding URL and system parameters, so that the user need not return to the main page in order to switch to another language.

One important objective of this web map system is to provide adequate and supportive information to help the user easily locate and orient the chosen origin, destination and suggested stops. Such information typically includes major roads, critical streets, building names, site names and prominent landmarks, all of which will be shown in full or in abbreviation in the 1:3000 map series. Even in the 1:7500 series, almost all contents are preserved as map space permits with only the less significant building names removed. Text labelling is therefore of the highest priority when designing maps. Land features are filled with lighter shades of lower values, compared to the more prominent darker text labels.

10   INTERACTIVE MAPPING ENVIRONMENT

An interactive system-user interface has to satisfy three requirements: selected locations are panned to become visible while being synchronized with all other small map insets; origins and

Table 1. Symbology for the public transport web map query system.

Layer	Default visibility	Index (0 – bottom, 7 – top)
Base map images	Active	0
Origin symbol	Hidden	2
Destination symbol	Hidden	3
Flashing bus icon	Hidden	4
Flashing mini-bus icon	Hidden	5
Bus stop icons (red/green dots)	Hidden	6
Text box for bus stop details	Hidden	7
Some other layers		
Confirm arrow		
Cross at the centre of the main map		
Focus icon		

destinations are highlighted with prominent symbols; and the origins and destinations are highlighted with information on stops and routes. The first requirement has already been discussed in the previous section. Unlike base map information that is more static in nature, origin-destination selections are more dynamic and user-dependent. To create an interactive environment, symbols are first generated for all user selection variables. These symbols will be stored in different layers of the HTML with controlled visibility and overlapping orders (Index from 0 to 7), as shown in Table 1.

When clicking on a map position to select it as an origin or destination, the geo-referenced code will be captured and the layer of the symbol will be provoked to become visible at that exact position. In the same way, an ASP program retrieving a travel solution from the route engine will store the geo-referenced code of the public transport stop (e.g., bus, mini-bus, tram) in HTML format. At the client side, a program will transform this code to invoke a flashing icon to be displayed under the browser pixel system. Concerning the stop information, as a large number of details are involved, these will not be retrieved all at once but after a limited time upon the loading of each map page. For instance, when users click on the bus stop check box, the program will turn on the layer using geo-referenced codes from the 2-D text array inside the HTML pages. The separation of stop information from the base maps is mainly because such information often changes and needs constant updating. To conclude, the retrieval and computation of locations or the searching of routes are performed on the server's side, whereas the displaying mechanism is on the client's side.

11   RESPONSE TIME

A rapid response time is one of the critical factors affecting the popularity of a web site. To speed up the retrieval of relevant map areas, different map scale series can be synchronized in the same window. However, all maps are pre-processed as PNG images and put on the web server. Hence, every single action is simply the requesting of files from the server, and images can be cached in the client's browser. This also means the client only needs to download the same image once, thus reducing the usage of bandwidth and loading of the server.

12   CONCLUSION

While addressing and evaluating the various options in implementing an Internet mapping system, this paper has proposed a mechanism for using web maps to aid queries on public transport routes.

With the requirements carefully identified, the system employs an Internet engine balancing both the client side and the server side to perform both map query functions and GIS analyses, respectively. The developed system is enhanced with the following capabilities for a potentially very large number of users: seamless panning, synchronization of positions across maps of varying scales, hierarchical displays of relevant base maps and transport data and a rapid response time. It is yet to be evaluated with particular regard to its efficiency. In the long run, the development of eXtensible Markup Language (XML) and Scalable Vector Graphics (SVG) from the World Wide Web Consortium (W3C) may provide additional solutions for the successful development of online GIS applications.

ACKNOWLEDGEMENT

The work described in this paper was substantially supported by a grant from the Hong Kong Polytechnic University (Projects No. G-T29B and A-PC77).

REFERENCES

Blunt, E. 2001. Sydney's information highway – easy access to local council information, *GIS User*, no. 44: 26–27.

Bosak, J. 1997. XML, Java, and the future of the Web, *Sun Microsystems Technical Paper*.

Boye, J. 1999. SVG brings fast vector graphic to Web, *http://www.irt.org/articles/js176*.

Carter, J.R. 1992. Perspectives on sharing data in Geographic Information Systems, *Photogrammetric Engineering & Remote Sensing*, vol. 58, no. 11.

ESRI. 1998. ArcView Internet Map Server Extension: functional overview, an *ESRI White Paper*, Environmental Systems Research Institute, Inc.

Gifford, F. 1999. Internet GIS architectures – which side is right for you? *GIS World*, vol. 12, no. 5.

Harrie, L., Sarjakoski, L.T. and Lehto, L. 2002. A mapping function for variable-scale maps in small-display cartography, *Journal of Geospatial Engineering*, vol. 4, no. 2: 111–124.

Jing, J., Helal, A. and Elmagarmid A. 1999. Client-Server computing in mobile environments, *ACM Computing Survey*, vol. 31, no. 2: 117–157.

Margaritidis, M. and Polyzos, G.C. 2001. Adaptation techniques for ubiquitous Internet multimedia, *Wireless Communications and Mobile Computing*, vol. 1, no. 2: 141–164.

Plewe, B. 1997. GIS online: information retrieval, mapping and the Internet, Word Press, Santa Fe, USA.

Shneiderman, B. 1998. Designing the user interface: strategies for effective human-computer interaction, 3rd ed., Addison Wesley Longman, Massachusetts, USA.

Zhuang, V. 1997. Spatial engines drive Web-based GIS, in *GIS World*, vol. 10, no. 10.

Advances in Spatial Analysis and Decision Making, Li, Zhou & Kainz (eds)
© 2004 Swets & Zeitlinger, Lisse, ISBN 90 5809 652 1

ASHMR-based maintenance of inter-connectivity in multi-representations

Yanhui Wang
China University of Mining and Technology (Beijing Campus), China

Jun Chen & Jie Jiang
National Geomatics Center of China, China

ABSTRACT: As the issues of conceptual and representational differences will arise among multi-representations, maintaining inter-connectivity among multi-representations in building a multi-scale data model exists as a foundational task. The existing methods to ensure that the states of representations can be linked by bi-directional inter-level connectivity are still not satisfactory in implementation. This paper presents aggregation-based semantic hierarchical matching rules (ASHMR) as the basis of tackling inter-connectivity among multi-representations. It takes the multiple representations from road intersections in road networks as an example. A speciality is given for describing in detail the strategies for maintaining inter-connectivity among representations. This is considered a feasible approach in dealing with inter-connectivity.

1   INTRODUCTION

Since the first research project on 'multi-representation' was implemented by NGCIA in 1989, an increasing amount of attention is being paid to multi-scale (multi-representation) databases in GIS fields, for use in complex spatial analyses, visualizations, and the indispensable complementarity of cartographic generalizations. Maintaining multiple representations is one of the key problems in GIS, which means that several representations at different levels of scales representing the same feature in the real world are stored in the database (Brugger et al., 1989). In those multi-representation data model is a base and also one of necessaries, inter-connectivity means ensuring that the states of representations should be linked by bi-directional, inter-level connectivity in the modelling (Philippe and Michel, 1996; Kilpeläinen, 2000). The absence of inter-connectivity among multiple representations will lead to contradictory information, and result in changes not being promulgated from one scale level to another. When the changes are passed to the decision level, the wrong interpretations may result. As shown in Figure 1, when multiple representations from

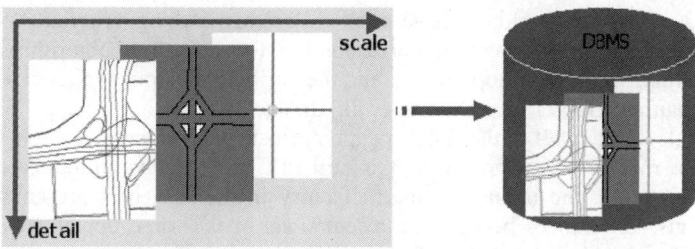

Figure 1.   When multi-representations from the same road feature are stored in a database, the inter-connectivities among them should be maintained.

the same road feature are organized in a database, the inter-connectivities among them should be maintained. Although related developments have been reported in many papers, how to maintain the hierarchical inter-connectivity among these representations in implementation remains in dispute, the various viewpoints of which are discussed in this paper. After a review in section 2 of previous work, section 3 analyses the key questions and strategies for maintaining inter-connectivity among multiple representations. An example in a multi-scale road network database is then given to illustrate the possibility.

2   RELATED WORK

There are two main approaches to acquiring multiple representations. In one approach, the data of different scales are acquired separately and the links between the scales are established afterwards. The other possibility is to derive the series of representations from a single, most detailed representation through the use of cartographic generalization, which is based on the development of automated generalization (Li, 1997, 1998). Jones (1996), Timpf (1998) and Kilpelainen (2000) gave various reasons for storing multiple representations of the same objects in the database, one of which is the relatively limited capabilities of automatic generalization.

In view of the inter-connectivity among multiple representations, Timpf and Frank (1995) applied the concept of the 'directed acyclic graph' for zooming in on geographic data at various levels of abstraction. Timpf (1998) presented a map cube model for a series of maps, connecting several of the objects of the models with the help of a tree structure. Devogele et al. (1996) chose to connect geographic data from mono-scale representations to build a multi-scale database with scale-transition relationships. These scale-transition relationships connect two sets of elements (classes, types or objects) representing the same phenomenon in the real world and carry the sequence of multi-scale operations to navigate from one representation to another. Vangenot and Spaccapietra (2002) use the stamping technique to differentiate among multiple representations of a given phenomenon and to access a particular representation.

In the context of a hierarchical data structure, multi-scale or multi-level data structures already exist on a low level; e.g., quadtrees (Samet, 1989) or hierarchical triangulated networks (Dutton, 1997), and topological structures (Bruegger and Frank, 1989). Those structures show aspects of hierarchically organized data but only for one type of hierarchy; namely, for aggregation hierarchies. Oosterom (1993) proposed using a Reactive-tree coming from an R-tree to store less detailed objects. For this structure, he proposed an 'importance' characteristic, but this characteristic could be used only in cases where objects ordered by importance are represented in a strict data structure. On the other hand, most of these models still are at the conceptual level and, except for those of Molenaar (1996) and Harrie (1998), few can provide functions of multi-representation functions.

In the context of data integration, many algorithms have been developed in the last ten years to solve different problems of practical matching. Where it is supposed that linking can basically be seen as a matching problem, objects in different representations can be assumed to share some metric, topological or attribute information, which is the principle of the relational matching technique. These related algorithms have used different matching criteria. According to the predominant criterion used in matching correspondent features, these algorithms can be classified into three kinds: geometric, topological and the attribute method (Walter et al., 1997; Sester et al., 1998; Yuan and Tao, 1999; Dueker et al., 2000). This method of matching spatial data of a similar scale works well if the data is captured using the same data model or criteria by which to define road features; however, it is still difficult to solve complicated problems in practice. What is more, the method is unsatisfactory in the case of representations at different scales because differences between representations of the same objects at different scales will lead to differences in class level, object level, geometric level, even to differences in the level of attributes and attribute values. Therefore, those features at different scales cannot be matched directly.

3 MAINTAINING INTER-CONNECTIVITY AMONG MULTI-REPRESENTATIONS

The inter-connectivity of multiple representations can only be achieved if the model is capable of explicitly inter-relating the representations and dealing with their differences at the levels of both the feature class and the representation. Thus, it is necessary to explore the available knowledge and reasonable strategies for combining corresponding representations.

Multiple characteristics among multiple representations exist, such as multiple geometries, multiple details, and so on (Chen, 2002). Although it seems that multiple representations represent inconsistent characteristics to some extent, they share intrinsic similarities, which are decided by the nature of multi-representations. This is because multiple representations are derived based on generalization operators from one scale hierarchy to another. Thus, an essentially hierarchical relationship exists among them, of which aggregation plays an important role. On the other hand, those corresponding representations also possess related semantic information, as each of them comes from the same or related entities in the real world.

In this context, semantic relations and hierarchical aggregation relations are explored. With the assistance of data-matching information, the aim is to properly educe the essential matching strategies from which aggregation-based semantic hierarchical matching rules (ASHMR) will be proposed in this paper as strategies for tackling inter-connectivity.

3.1 *Semantic hierarchical information*

When dealing with multiple interacting representations of robots in the field of artificial intelligence, Kuipers (2000) considered one of the challenges to be that of finding the natural joints for dissecting the complex natural phenomena of spatial knowledge, which is the focus of the spatial semantic hierarchy model. In view of multi-scale representations of geographical features, it appears that semantic hierarchical information among the representations and the natures should also be explored, as the corresponding representations share some semantic and hierarchy relations. In this context, semantic relations in multiple representations indicate whether some representations refer to the same or to related entities in the real world. We can distinguish three semantic relations among multiple representations: semantically equal, semantically related and semantically irrelevant. A semantically equal relation shows that some representations from different scales are derived from the same entities or phenomena in the real world. A semantically related relation predicates that some representations from different scales are not fully derived from the same entities in the real world, but that the corresponding entity classes have some superclass or subclass relation. Of course, a semantically irrelevant relation also could be defined similarly, which means that corresponding entities in the real world have no relation at all.

Since multiple representations are derived from different scales, they bring different detailed level, which will be clearly reflected in hierarchical relations. Similarly to semantic relations, hierarchical relations among multiple representations carries the information that representations at different scale levels may wholly coincide with each other (which corresponds to semantically equal relations), or interrelate implicitly but not explicitly (which corresponds to semantically related relations), and that it may be feasible to build ordered class hierarchies of representations. As hierarchies of corresponding entities in the geographic world result in hierarchies of multiple representations within a spatial database, the most important relation that of the aggregation hierarchy. As the basis for tackling the inter-connectivity of multiple representations, aggregation is a special form of association between objects, where a composite object at the smaller scale level is considered to have been assembled from those at a larger scale level. An aggregation hierarchy shows how composite objects can be built from elementary objects and how these composite objects can be put together to build more complex objects, and so on (Molenaar, 1996). An aggregation relation has close relationship with a semantic relation. For example, in the literature on semantic modelling (Brodie, 1984; Molenaar, 1996), the upward relationships of an aggregation hierarchy are called 'part of' links. These links relate a particular set of objects to a specific composite object, and on to a specific more complex object, and so on.

3.2 Data-matching information

Data-matching information refers to the knowledge available to realize links of multiple representations. In the context of multi-scale, this paper extends topological relations as follows:

Geometrical relation: When mapping corresponding representations from one scale to another, multiple representations from the same features should share a similar geometrical location; e.g., centroid, convex hull, Voronoi Diagram, etc., and all geometry operators for searching matching pairs can be derived computationally from them.

Topological relation: Topological relations among representations at one scale are often consistent with those at another scale, especially for those with connective and adjacent relations. Combined with the semantic relation in multiple representations, strong (weak) adjacency and strong (weak) connectivity can be defined as follows.

If two representation objects subject to the same complex feature possess the connectivity relation, we named the connectivity relation a strong connectivity; if they are not subject to the same feature, we name the connectivity relation a weak connectivity.

If two representation objects subject to the same class possess the first-order adjacency relation, we named the adjacency relation a strong adjacency; otherwise, we named the adjacency relation a weak adjacency. With these concepts, an auxiliary measure is to search for matching candidates at the first step and to reduce search areas, as well as to speed up computational efficiency.

3.3 Aggregation-based semantic hierarchical matching rules

From the above, semantic relations among multiple representations may be seen as the basis for maintaining inter-connectivity among them. Because of hierarchy, those representations may identify each other or make up an aggregation relation. Because of geometry, they share the same or a similar location (e.g., their geometric information is similar to each other). Even from the point of view of spatial relations, semantic relations also help to define in detail the semantically strong (weak) connectivity and semantically strong (weak) adjacency needed to confirm the search range for maintaining inter-connectivity.

On the other hand, hierarchical information combined with semantic information is used to decide how those representations may correspond, and data matching information is used to define how the correspondence can be realized. In integrating all of this information, the term 'aggregation-based semantic hierarchical matching rules (ASHMR)' may be induced to refer to rules for maintaining inter-connectivity among multiple representations. All of those rules may come from the semantic, hierarchical, geometric and spatial information among multiple representations. For example, for composite spatial representations the aggregation links might be based on the following two types of rules involving the semantic and the geometric aspects of the elementary representations:

- Rules specifying the classes of elementary representations at the smaller level building composite representations at the larger level and;
- Rules specifying the geometric characteristics (such as point, line, area) and topological relationships of these elementary representations (i.e., adjacency, connectivity, etc.).
- With the support of ASHMR, the maintenance of inter-connectivity among multiple representations could be basically achieved using the following steps:
  - Specifying the semantic relation between the source representation class and the possible object representation class;
  - Specifying the hierarchical relation among representations based on the educed semantic relation;
  - Specifying candidacy representations by data-matching information;
  - Establishing explicit rules of inter-connectivity;
  - Describing and implementing rules with specific aggregation operators, geometrical operators, topological operators, and so on.

Taking multiple representations from road intersections in a road-network database as an example, and based on predefined rules, the following demonstrates an approach to showing the correspondence of road intersections between a medium-scale and a large-scale representation.

4 AN EXAMPLE OF INTER-CONNECTIVITY IN MULTIPLE REPRESENTATIONS FOR ROAD INTERSECTIONS IN A NAVIGABLE DATABASE

The explicit maintenance of the inter-connectivity of multiple representations from the same road intersection feature in a multi-scale navigable road-network database is necessary for improving the efficiency of queries and reducing redundancies in data during the planning of routes and guiding of the navigation process. As one of the most complex and important components of road networks, an intersection represented by a point feature at scale 1 will correspond to several element features enclosed by the circle at scale 2 (see Figure 2). These representations at different scales from road intersections indicate the possible changes when moving from one representation to the next.

Taking an object class in the format of GDF (Geographic Data File Standard) as an example, the road intersection object at the more detailed level is called an intersection. An intersection could be owned by more than one road segment, and the corresponding object at the less detailed level is called a junction, being the beginning- or the end-extremity of a road element. A road element is a piece of a road segment, homogeneous in value with respect to the set of attributes and relations that describes the basic components of the road network.

In this context, the semantic relation of the corresponding intersection features of the road network follows that an intersection at the smaller scale (s1) must correspond to one or more junctions and, at the same time, may correspond to one or more road elements at the larger scale (s2). Based on the above semantic and aggregation relations between corresponding features, the reference rules are as follows:

– The geometrical position of the candidate junction at the larger scale locate the buffer the intersection at the smaller scale create at the corresponding larger scale based on the resolution transformation relation;
– If both junctions of a road element belong to the buffer, then it is considered one of the corresponding candidate features.

Thus, the formulation clauses can be split into two parts. The first clause gives a general description of the corresponding relation between representation classes in the dataset (S1) at the smaller scale (s1) and those in the dataset (S2) at the larger scale (s2), and the following two clauses describe the first clause in detail. In order to be concise, we deal with those with the help of the term 'SET' derived from algebra, which is associated by the elements subject to the same conditions.

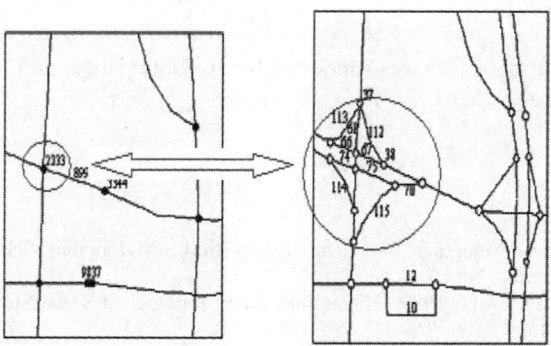

Figure 2.   Different representations for the same road intersection at different scale.

$$S1.Inter\,sec\,tion \subseteq S2.SET(\quad[1:n]\,Junction,[0:n]RoadElement)$$

(1)

$$SET([1:n)]Junction) = \{w\,/\,w \in Junction \wedge$$

$$w.geometry \quad INSIDE \quad BUFFER(Inter\,sec\,tion,resolutionS2)\}$$

(2)

$$SET([0:n)]Roadelement) = \{s\,/\,s \in Roadelement \wedge$$

$$s.FJ \in SET([1:n)]Junction) \wedge s.TJ \in SET([1:n)]Junction)\}$$

(3)

Concretely, the first clause specifies that every intersection representation in dataset S1 with a smaller scale s1 corresponds to a multi-sorted set of S2 representations. The second clause specifies that for each intersection representation, one or more Junction representations should be considered: (1) which is one junction of a road element, and (2) whose geometry lies within a given buffer surface enclosing the intersection geometry. The second predicate restricts road element representations to those whose junctions both belong to the candidate junctions.

As a result of the above formulation, new sets are created, depicted as a matrix M, and M is denoted as M = $\{J1, J2, \ldots J_{m1}, RE1, RE2 \ldots J_{m2},\}$, where m_1 denotes the number of the corresponding junctions and m_2 denotes the number of the corresponding road elements.

In terms of the above implementation, the matching results for corresponding intersections from different scales can be observed.

5 SUMMARY

The ASHMR method is basically a feasible approach to maintaining inter-connectivity among multiple representations. The available semantic hierarchical matching knowledge includes knowledge about semantic relations, aggregation relations, geometrical relations, and topological relations. In this way, representations with different spatial scales and geometric precisions can be linked and used in an integrated way. The mutual benefits lie in the fact that, on the one hand, the method opens the way to combine already existing data sets to form multi-scale databases and, on the other hand – which is of particular interest in the context of database generalization – the links can also carry information about the transitions from one dataset to another to ensure inter-connectivity among multiple representations, allowing the direct generation of a new and generalized data set.

ACKNOWLEDGEMENT

The work described in this paper was supported by the National Natural Science Foundation of China under grant No. 40025101.

REFERENCES

Brodie, M.L. 1984. On the development of data models. In Brodie, Mylopoulus, Schmidt eds, On Conceptual Modeling. Springer Verlag, New York.

Bruegger, B.P. and Frank, A.U. 1989. Hierarchies over Topological Data Structures. ASPRS-ACSM. Baltimore, pp.137–145.

Chen, June 2002. Developing Dynamic and Multi-dimensional Geo-Spatial Data Framework, Geo-Information Science, Vol. 4, No. 1, pp. 7–13. (in Chinese).

Christelle Vangenot, C.P., Stefano Spaccapietra 2002. Modelling And Manipulating Multiple Representations Of Spatial Data. In the proceedings of Symposium on Geospatial Theory, Processing and Applications, Ottawa.

Devogele, T., Trevisan, J. and Raynal, L. 1996. Building A Multi-Scale Database With Scale-Transition Relationships. In the proceedings of 7th International Symposium on Spatial Data Handling, Delft, Netherlands, pp. 337–351.

Dueker, K.J. and Butler, J.A. 2000. A Framework For Gis-T Data Sharing. www.upa.pdx.edu/cus/.

Dutton, Geoffrey 1997. Digital Map Generalization Using a Hierarchical Coordinate System. ACSM/ASPRS, Seattle, pp. 367–376.

Harrie, L.E., Incremental Generalization: A Feasibility Study, GISRUK Edinburgh, UK, 31st March –2nd April, 1998.

Jones, C.B. and Kidner, D.B. 1996. Database Design For A Multi-Scale Spatial Information System. INT. J. Geographic Information System. Vol. 10, No. 8, pp. 901–920.

Kilpeläinen, Tiina 2000. Maintenance of Multiple Representation Databases for Topographic Data. The Cartographic Journal. Vol. 37, pp. 101–107.

Kuipers, Benjamin 2000. The Spatial Semantic Hierarchy, Artificial Intelligence, Vol. 119, pp. 191–233.

Li, Zhilin 1997. Philosophical, Conceptual And Algorithmic Issues In Automated Map Generalization. In the proceedings of 2nd Workshop on Automated Map Generalization. 19–21 June.

Li, Zhilin 1998. Multi-scale Representation Of Spatial Data. In the Proceedings of Mapping Sciences '98. pp. 409–416.

Molenaar, M. 1996. Multi-scale Approaches For Geo-data. International Archives of Photogrammetry and Remote Sensing. Vol. XXXI/B3, pp. 542–554.

Oosterom Van. 1993. Reactive Data Structures for Geographic Information Systems. Oxford University Press.

Rigaux, P. and Scholl, M. 1996. Multiple Representation Modeling And Querying. In the proceedings of Conference on Object Orientation and Navigable Databases, NCGIA.

Samet, H. 1989. The Design and Analysis of Spatial Data Structures. Reading, MA: Addison-Wesley Publishing Company.

Sester, M., Anders, K. and Walter, V. 1998. Linking Objects of Different Spatial Data Sets by Integration and Aggregation. Geoinformatic, Vol. 12, No. 4, pp. 335–358.

Timpf, S. 1998, Hierarchical Structures in Map Series. PhD. Vienna Technical University, Austria.

Timpf, S. and Frank, A.U. Multi-scale DAG For Cartographic Objects, In the Proceedings of Auto-Carto 12, in Charlotte, NC, February 27–March 1, 1995, Published by ACM/ASPRS, Vol. 4, pp. 157–163.

Walter, V. and Fritsch, D. 1997, Matching Strategies For Integration Of Spatial Data From Different Sources, In the proceedings of 'International Workshop on Dynamic and Multi-Dimensional GIS, 25–26. August, Hong Kong', pp. 215–228.

Yuan, Shuxin and Chuang, Tao 1999. Development Of Conflation Components. In the Proceedings of Geoinformatics '99 Conference, pp. 1–13.

313

Advances in Spatial Analysis and Decision Making, Li, Zhou & Kainz (eds)
© 2004 Swets & Zeitlinger, Lisse, ISBN 90 5809 652 1

Surviving by specializing: a web service prospect of interactive web map for public use

Zhu Xu & Lilian S.C. Pun-Cheng
Department of Land Surveying and Geo-Informatics, The Hong Kong Polytechnic University, Hong Kong, China

Y.C. Lee
Department of Geodesy and Geomatics Engineering, University of New Brunswick Fredericton, Canada

ABSTRACT: As Internet is becoming a primary disseminator of information, many web sites have been providing maps online. Some of these sites are large scale ones devoted to interactive web mapping. Although the benefits of and the demand for web maps are obvious, there has been much criticism about their quality, doubt about their usefulness and worry about their long-term survival. This article offers a perspective on interactive web maps from the viewpoint of web service, and indicates a direction for the evolution of interactive web maps. It considers interactive web mapping as the interface and presentation layer in the overall framework of web-based spatial applications. Although interactive web maps can technically be independent web applications, we argue that it is almost necessary that they become part of other web-based spatial applications to survive in the long term. By making interactive web mapping a fundamental web spatial service, web maps serve not only the end users but also a wide range of web applications that require maps as part of their interface or/and presentation media. The necessary resources can then be acquired to turn the making and provision of high quality web maps into a specialized business. This paper also discusses the technical feasibility of turning web map applications into web map services.

1 INTRODUCTION

As the Web has been widely accepted as a prime means of disseminating information, many web sites are providing maps in their pages. Some of these maps are simply individual images showing the location of a company. Some are image maps containing hyperlinks and are used as indexes to direct users to other pages, which may show other maps but more often present other types of information, such as an introduction to a scenic site as is found in some tourism web sites. Some libraries have scanned their map collections and made them available online. As these map collections are large, some kind of online catalogue is always provided at the same time. Moreover, there have been quite a few websites devoted to providing interactive web maps for public use. These maps provide much more interactivity than the previously mentioned simple image maps. This last category of web maps is the major concern of this article. Up to the time of writing, Odden's Bookmarks [URL1], which seems to be the largest bookmark collection of map and mapping sites, listed more than 15,000 web sites providing maps, atlases or map collection catalogues online. Among them, more than 9,000 are electronic atlases and about 4,000 are scanned map collections and/or catalogues of map collections. Of course, this is not a complete list and new bookmarks are found and added everyday.

The popularity of putting maps online has been pulled by the need of web users for spatial information. The medium of the web has many advantages for map publishing, although it has limitations as well [Kraak and Borwn, 2001]. Unfortunately, at the early stage of such a practice, many web maps have been of relatively poor cartographic quality, making people doubt their usefulness. A survey of the usage of web maps provided on over 100 tourism web sites conducted by [Richmond, 2002] revealed that the majority of users found the majority of the maps not informative or not appealing or not convenient to use. Cartographic researchers and professionals have also expressed their concerns about the quality of web maps and their usefulness [Kraak and Borwn, 2001]. Godfrey [2001] complained that 'some current digital mapping is so over-simplified as to be virtually worthless' and that 'the 1:10,000 scale displays for mountain and moorland, a few lines across the screen like minimalist abstract sketch, are sad maps indee.'

The authors believe current technology has been sharp enough to make the quality of web maps comparable to that of their paper counterparts, even though the medium of the web has some inherent limitations. The poor quality of current web maps is of course the result of a lack of recognition of the importance of cartographic quality and of a lack of dual expertise in both cartography and web computing and publishing. The ultimate reason, we believe, is that no proper profit-making mode has been found or accepted in the business of providing or publishing web maps. Thus, the endeavour is short of the necessary resources to develop the required specialization. To help find the way out of this dilemma, this article offers another perspective on interactive web maps. In the next section, we examine the two basic functions of the map as a visual model of the real world and as a repository of information, and present the view of the web map as the interface and presentation layer of web GIS applications. Section 3 first discusses the need to specialize in the making of web maps if the endeavour is to survive and the maps to achieve better quality. It then suggests a web service computing model for providing web map services, and discusses how this model can help web maps gain the desired specialization and survival in the long run. Finally, we draw conclusions in section 4.

2   WEB MAPS AND WEB GIS APPLICATIONS

2.1   *Map: the visualized spatial information*

Traditionally, the paper map has been both a store of spatial information and a visual model of the real world. A map provides a picture of the world to help us understand spatial patterns, relationships, and the complexity of the environment [Robinson et al., 1995]. With a map, people can find the answers to questions on location, such as where a specific street or building is. Traditionally, a map that contains many place names has an index of places to help in the search for a location. The index is a table that can be looked up, listing sorted pairs of place names and place positions in the map. These names and positions are represented by map grid codes and page numbers in the case of a map book. People may also be able to find out how to get to a place by foot, by bus or by driving, depending on how much supporting information is contained in the map. Professional users, such as civil engineers and geologists, etc., carry out precise measurements and complicated analyses with specialized maps. Except for the first purpose, i.e. finding out what an area is like, people often find it tedious and time-consuming to search for answers from a map in response to questions as simple as where road X is. It is more painful to find public transport guidance or driving guidance from a map, not to mention more sophisticated spatial questions.

As we now know, this is because the paper map is not convenient for the retrieval and processing of information. GIS comes into play in these cases. The alphabetical index of a map mentioned above is part of what we now call a spatial database. It is the part that is suitable for presentation to users in a form other than the graphic form. The remaining part of a spatial database can be best presented to users in the graphic part of the map. As we have seen, a map is separated into two parts: the spatial database and the map image; i.e., the visualization of spatial information. The map has given up the traditional role of being a store of spatial information. The gap in conceptions of map use between users and mapmakers is now being filled by information retrieval and processing tools in GIS.

2.2 *Web map: the interface and presentation layer of a web GIS*

Although paper maps cannot match GIS in doing spatial analysis, GIS cannot itself be without maps; namely, screen maps. GIS operators interact with GIS tools and spatial databases through maps. Further, a map is often the best means of presenting the results of spatial analysis to users. This indicates that most web GIS applications require web maps. The first category of such applications may include mass customers who make heavy use of web maps, such as web Yellow Pages, location service providers, web routing applications, etc. A second category may include travel agencies and transport operators and so forth, to whom a map may not be a necessity but is highly desirable. A third category may include almost any web site because spatial information is basic to each person.

As analysed above, a web map will not be a great innovation over a paper map if it is simply a replication of the latter in a web medium. Users need better ways of using spatial information than merely browsing map images and manually searching for answers. This indicates that it is desirable that a web map be part of some web GIS application for spatial information to be better served to users. On the other hand, most web GIS applications need web maps in their interface and for the presentation of spatial information.

3 SPECIALIZING IN MAKING WEB MAPS BY MEANS OF WEB SERVICES

As many web GIS applications need web maps, it seems to the authors that the making of web maps can be specialized. In fact, the traditional map-making and publishing industry has been in charge of making maps for public use. The specialized map-making industry produces high-quality paper maps for the public. Should and can web map-making also adopt the specialized production mode of conventional paper maps?

3.1 *The need to specialize*

Currently, most web GIS applications or web sites that need maps create their own web maps. In many cases, these maps are duplicates of low quality. Such a practice is not economical. All these are potential map customers who may buy mapping services from some specialized web mapmakers instead of making their own. This is because making a map of good quality needs expertise in both cartography and web computing and publishing, and can cost a great deal even if the spatial data are free for use, as is the case in some countries. Making a map for each application is obviously a duplication of effort.

Second, the conflict between cost and quality may result in bad maps. As cartographers have realized, GIS has been enabling map users be mapmakers. Unfortunately, GIS can also be a powerful tool for the making of bad maps [Goodchild, 2003]. MacEachren [1994] wrote an introductory book to GIS users on basic cartography principals when it was noticed that GIS users could easily make cartographic mistakes. The title of the book is *Some Truth with Maps*, indicating that some maps may have lain without mapmakers' awareness. Indeed, many web maps are believed to be the result of the improper use of GIS to visualize spatial data by people who lack cartographic expertise. Van Elzakker [2001] criticized the quality of current web maps and warned that providing web maps for free might cost effectiveness or even actuality of maps. We have found from our own experience that poor web maps can greatly decrease the efficiency of web computing and interactions by dramatically increasing client-server round trip numbers, network traffic, server loads, transfer delays and the impatience of users.

On the basis of the above analysis, we argue that since the quality of web maps and the cost of making web maps are critical to the survival of web maps, it is highly desirable for the making of web maps to become a speciality. Technically, current web-computing technology is able to support the specialized production and publication of web maps.

3.2 *Web services: the enabling technology*

If web maps are produced and published by specialized providers, how can web GIS applications make use of them seamlessly? The answer lies in web computing technology. Current web computing technology has in general been able to support the integration of web applications. Although web map providers can provide web map services to web GIS applications through proprietary protocols and programming interfaces, a better approach would be by some standard means. Providing web map services to web GIS applications by means of web services is what we believe to be the proper choice. Web services are becoming a standard web-computing model for developing the so-called business-to-business (B2B) e-commerce, as opposite to the so-called business-to-customer (B2C). B2C can be considered the e-commerce mode of current web map applications that provide web maps directly to web users. Web services are considered to be the next revolution in the World Wide Web [Gardner, 2001].

Web services are different from web applications in terms of web computing. Web applications are targeted to end-users, while web services are to be applied by web applications rather than directly used by web users. Web services can be considered to consist of some packaged data and an API (application programming interface) that is ready to be integrated into web applications. Like components in component-based software developments, web services can greatly reduce the time and cost of developing web applications. We will not go into the technical details of developing web services here. Instead, we will simply present the general idea and briefly discuss the current status of developments in web services.

The general idea of web services can be illustrated with a route finding example. Imagine a route-finding web application that helps people find a way to get somewhere by public transport or by driving. The focus of the application is to find optimal routes. The application needs maps to let users input the origin and destination. It also needs maps to show the optimal routes to users. Instead of making its own maps, the application may use a web map service whenever it needs a map. To do so, the application first needs to find a web map service from the web, then bind to the service and finally send requests and receive replies. To users of the route-finding application, the maps seem to be an integral part of the route-finding application. The user will not need to go to web map sites to retrieve the map he or she needs. Further, supposing it to be a driving guide application, users may also want to know where the petrol stations are. The route-finding application does not have to maintain its own database of petrol stations. Instead, it can make use of available web services on petrol station information maintained by the proper authority, such as some petrol companies. After retrieving the petrol stations along a specific route, the route-finding application may ask the web mapping service to add the stations to their route maps. Again, the route-finding application can also publish itself as a web service to be used by other web applications, for instance the web sites of some travel agencies. The architecture of web services is illustrated in Figure 1 [Gardner, 2001].

The W3C (World Wide Web Consortium) has a specialized workgroup for the standardization of web services [URL2]. The main activities include the Web Services Description Language (WSDL), the Simple Object Access Protocol (SOAP) and Universal Discovery, Description and Integration (UDDI). WSDL is used to describe web services to facilitate the registering, publishing and finding of such services. SOAP is a standard for an XML-based information exchange between distributed applications. It is used to send requests and receive replies. UDDI is a specification for

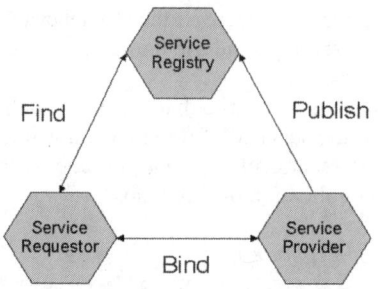

Figure 1.   Architecture of web services (cited from [Gardner, 2001]).

distributed registries of web services. A UDDI web services registry is itself a web service that can be accessed via SOAP from an application that wishes to discover web services. UDDI specifies interfaces for applications to publish web services (as WSDL documents) and to discover web services (via their WSDL documents) [Gardner, 2001]. Almost all vendors of web development tools have now enabled their products to develop web services. Also, quite a few web services are available. In the geographic information community, ArcWeb Services from ESRI [URL3] is among the pioneer works.

By turning web map application into web map services, web maps can serve both end users and web GIS applications. Thus, a wider customer base more likely to be available to businesses that make and publish web maps. Such a mode for producing and publishing web maps is compatible with the contemporary mode of using spatial information to support spatial decisions. In this mode, web GIS plays a key role while web maps are often an essential part of a web GIS in interacting with and presenting results to end users.

4   CONCLUSIONS

As pointed out by van Elzakker [2001], the Internet is undergoing an economical evolution and web map are participating in this development. This article suggests developing specialization in the making and publishing of web maps. The goal is to improve the quality of web maps and to help web maps survive in the long run. In our view, this can be achieved by means of web services, which is the web computing technology supporting the economical B2B Internet concept. We think web services will be a direction taken in the evolution of web map technology, and that the specialization of the web map business can be a successful case of B2B. Our discussion is considered to be applicable to web maps for public use; i.e., maps presenting the known to the general public. Web maps for proprietary or private use, especially those for geospatial data exploration or for stimulating visual thinking, are not within our range of discussion.

ACKNOWLEDGEMENT

The work described in this paper was partly supported by a grant from the Hong Kong Polytechnic University (Project No. G-T29B).

REFERENCES

van Elzakker, Corné P. J. M., 2001, *Users of maps on the map*, in *Web Cartography: developments and prospects* (eds. Kraak, Menno-Jan and Brown, Allan), 2001, Taylor & Francis, New York.

Godfrey, Alan, 2001, *A map user's perspective*, in *the Map library in the new millennium* (eds. Parry, R. B. and Perkins, C. R.), Library Association Publishing, 2001, London.

Goodchild, M. 2003, *Augmenting geographic reality*, open lecture at The Hong Kong Polytechnic University, Mar. 18, 2003.

Kraak, M. J. Brown, A. (eds.) *Web Cartography: developments and prospects*, 2001, Taylor & Francis, New York.

MacEachren, A. M. 1994, *Some truth with maps: a primer on symbolization and design*, Association of American Geographers.

Richmond, E., 2002, *Maps and Tourism on the Web: The Research*, Master's thesis, University of Victoria, Canada, http://office.geog.uvic.ca/mapsandtourism, last visited May 1, 2003.

Robinson, A. H. Morrison, J. L. Muehrcke, P. C. Kimerling, A. Jon, and Guptill, Stephen C., 1995, *Elements of Cartography*, 1995, John Wiley & Sons, Inc.

Tracy, G. 2001, *An Introduction to Web Services*, http://www.ariadne.ac.uk/issue29/gardner/, last visited May 2, 2003.

URL1, Odden's Bookmarks, http://oddens.geog.uu.nl, last visited May 1, 2003.

URL2, W3C Web Service Activities, http://www.w3.org/2002/ws/, last visited May 3, 2003.

URL3, ArcWeb Services, http://www.esri.com/software/arcwebservices/, last visited May 3, 2003.

Author index